Value Added Products From Food Waste

Elsa Cherian • Baskar Gurunathan
Editors

Value Added Products From Food Waste

Springer

Editors
Elsa Cherian
Department of Food Technology
Saintgits College of Engineering
Kottayam, Kerala, India

Baskar Gurunathan
Department of Biotechnology
St. Joseph's College of Engineering
Chennai, Tamil Nadu, India

School of Engineering
Lebanese American University
Byblos, Lebanon

ISBN 978-3-031-48142-0 ISBN 978-3-031-48143-7 (eBook)
https://doi.org/10.1007/978-3-031-48143-7

This Springer imprint is published by the registered company Springer Nature Switzerland AG
The registered company address is: Gewerbestrasse 11, 6330 Cham, Switzerland

Paper in this product is recyclable.

Preface

In a world dealing with the consequences of overconsumption and environmental degradation, the need to reduce waste and maximize resources has never been more pressing. Food waste, in particular, stands out as a glaring issue, accounting for a significant portion of the waste generated worldwide. The paradox of hunger juxtaposed with vast quantities of discarded food raises a moral and ethical challenge that society must confront. This book, *Value Added Products from Food Waste*, is a testament to our commitment to addressing this critical problem. It is a comprehensive exploration of innovative solutions that not only minimize food waste but also transform it into valuable resources. It is a testament to human ingenuity and the potential for sustainable practices to reshape our relationship with food. The journey from surplus and discarded food to value-added products is not merely about mitigating waste; it is about creating a more resilient, sustainable, and equitable food system. As we delve into the pages of this book, we embark on a journey that celebrates creativity, innovation, and the power of collective action. In these chapters, you will find a rich tapestry of ideas, technologies, and case studies that demonstrate the diverse ways in which food waste can be transformed into new, marketable products. From biodegradable packaging materials derived from fruit peels to nutrient-rich supplements extracted from vegetable scraps, the possibilities are boundless. The stories shared here illustrate how individuals, communities, and businesses are finding novel ways to add value to what was once considered waste.

Part I of this book mainly deals with the need of converting food waste to security of food and nutrition. It is stressed that wise use of technology can create a huge impact in the balance of the society. Part II covers value addition and sustainable management of dairy industry by-products. Unique nutritional values of various dairy by-products are explained. Dairy by-products encompass a wide range of products derived from milk processing, such as cheese, yogurt, butter, and whey. Each of these products contains a unique set of nutrients, including various vitamins, minerals, proteins, and fats, which can be beneficial for human health. Part III explains the waste utilization from cereals. In this part, the focus shifts to the effective utilization of waste generated during the processing or consumption of cereals. Cereals are a staple food for a large portion of the global population and encompass grains like wheat, rice, corn, and oats. The production and processing of cereals often lead to the generation of by-products or waste materials. This could include parts like bran, husks, and germ that are not used in the primary food production. This part likely outlines methods and strategies for making productive use of these by-products, which might involve repurposing them for other purposes like animal feed, biofuel production, or even as ingredients in various food products.

Part IV explains waste utilization from fruits and vegetables. This part likely delves into the various ways in which waste generated during the production and processing of fruits and vegetables can be put to productive use. This waste may include parts like peels, seeds, stems, and other non-edible or less-desirable portions of fruits and vegetables. Part V provides valuable insights into how waste generated from meat, poultry, and fish processing can be efficiently and sustainably utilized. This not only helps to minimize environmental impact but also has economic and agricultural benefits. It encourages practices that align with the principles of a circular economy, contributing to a more sustainable and responsible food industry. The last part covers the conversion of food waste into biofuel and electricity. The conversion of food waste into biofuel and electricity represents a significant step towards a more sustainable and circular approach to waste management and energy production. It addresses environmental concerns while also contributing to the development of renewable energy resources.

Our aim is not only to inspire but also to empower. By showcasing successful examples and offering practical insights, we hope to encourage readers from all walks of life to join the ranks of innovators who are turning food waste into a valuable resource. Whether you are a researcher, entrepreneur, policymaker, or simply someone passionate about sustainable living, this book offers something for everyone. We believe that every grain, peel, or stem that can be rescued from the bin and transformed into something valuable brings us one step closer to a more sustainable and equitable future.

Kottayam, Kerala, India — Elsa Cherian
Chennai, Tamil Nadu, India — Baskar Gurunathan

Contents

About the Editors

Elsa Cherian is a distinguished academician and researcher who presently holds the position of Associate Professor and Head of the Food Technology Department at Saintgits College of Engineering, located in Kottayam, Kerala, India. With 15 years of dedicated service in the field of education and research, she has established herself as a prominent figure in the domain of food technology. In addition to her teaching responsibilities, Dr. Elsa Cherian is a prolific researcher. She has authored a substantial number of research papers and book chapters that have added to the body of knowledge in food technology. Her contributions to scholarly literature have helped in disseminating the latest advancements, techniques, and findings in the field. These publications serve as valuable resources for students, researchers, and professionals in the food technology sector.

Baskar Gurunathan is currently a Professor in the Department of Biotechnology at St. Joseph's College of Engineering in Chennai, Tamil Nadu, India. He has 22 years of rich teaching and research experience in different fields of Biotechnology. He has published 190 research and review articles in reputed national and international journals, authored 37 book chapters and edited 6 books. He has visited Swiss Federal Institute of Technology (EFPL), Switzerland as visiting researcher in November and December, 2018. His major research areas include biofuels, bioenergy, technoeconomic analysis, environmental impact analysis, nanocatalysis, nanomedicine and food safety. He is the Fellow of the Institution of Engineers (India) and International Society for Energy Environment and Sustainability and is an active life member of various national and international professional bodies. He was listed in top 2% Scientist in the world consecutively in 2020 and 2021 by Elsevier BV and Stanford University, USA. He was bestowed upon ISTE-Periyar Best Engineering College Teacher award 2020 from Indian Society for Technical Education-Tamil Nadu Section, Prof. S. B. Chincholkar Memorial Award 2019 from Biotech Research Society, India for the outstanding work in the area of Biofuels and Food Biotechnology, ISTE-Syed Sajid Ali National Award 2016 for his outstanding research on renewable energy from Indian Society for Technical Education, New Delhi, and Young Scientist Award 2015 from International Bioprocessing Association, France. He has been actively researching for creating sustainable bioenergy solutions while contributing towards a greener and circular bioeconomy environment.

Part I
Introduction

Chapter 1
Food Waste to Food and Nutrition Security—Need of the Hour

R. Arivuchudar

Abstract The entire world is marching together toward realizing sustainable development goals (SDG) and renewing the globe by 2030. Of the 17 SDGs framed, the four SDGs, SDG1, SDG2, SDG3, and SDG12, focus on building a healthy realm with goals of No Poverty, Zero hunger, Good health and well-being, and Responsible consumption and production, respectively. The increasing population, COVID pandemic, unemployment, and increasing inflation have created food shortages and thereby food insecurity across many parts of the world. Also, in recent years, because of globalization, urbanization, and increased availability of a variety of foods, there has been a significant rise in the excessive purchase of food, leading to wastage on a large scale. This trend poses a significant threat and contributes significantly to food waste. Additionally, it raises concerns about food security, particularly in developing countries. To a great extent, this can be overcome by minimizing wastage while transporting food from farmstead to table, reducing food waste, wise reuse of the food wastage produced from various food trades, and mobilization of derivatives generated from food processing industries. As per FAO reports, 25% of the food is currently lost or wasted in the world; if redeemed, around 870 million famished population in the world could be fed, of which about 194.6 million, the highest number, will be benefited from India. The time has come to intuit this gargantuan volume of food waste and act swiftly for the benefit of people and nature. Thus, this chapter has been reviewed to provide a clear insight into the role of food waste in food and nutrition security and why it is the need of the hour. The wise use of technology, the acumen for nutritional perspectives of all parts of food, and interest in formulating novel food products from edible wastes can reduce food waste and assure food and nutrition security.

Keywords Food Security · Food Waste · Nutrition Security · Waste Management · Waste Disposal

---

R. Arivuchudar (✉)
Department of Nutrition and Dietetics, Periyar University, Salem, Tamil Nadu, India

E. Cherian, B. Gurunathan (eds.), *Value Added Products From Food Waste*,
https://doi.org/10.1007/978-3-031-48143-7_1

1.1 Introduction

Food waste and food loss are the terms of concern in the current scenario and are probable topics of research for the next few years until our sustainable development goals are met. According to FAO (2011), food loss means a reduction in either the volume or in the nutritional quality of the food actually envisioned for feeding human population. The damages are mostly because of a lack of efficiency in the food supply chain, including natural disasters, weak logistics, infrastructure, and technology, lack of skills in the understanding and managing ability of participants in the supply chain, and lack of market access. Food waste refers to discarded food that is suitable for human consumption, either it has been stored past its shelf life or left to rot. This is usually due to spoiled food but can also be due to market oversupply and consumer buying or eating habits. Food wastage includes food loss and/or food waste (FAO 2011). However, there is no uniform definition of food loss and waste for all the stages of the food chain, and there is a need to establish a universal definition (Nicholes et al. 2019; Chaboud and Daviron 2017; Lebersorger and Schneider 2011).

Food loss and food wastage have direct implications on the nation's economy, food security, and nutrition security. Wasting of food makes it difficult to maintain a sanitary and hygienic environs due to hitches in food disposal, leading to serious health risks and the outbreak of several infectious diseases (Arivuchudar 2018). As spoiled foods cause serious health risks, guidance on wise disposal of spoiled foods should be given (US Dept. of Health and Human Services 2019). The burden of hunger, malnutrition, diseases, and food insecurity, when discussed and taken to the public at large, will serve as a means to moderate food waste and loss (Rutten 2013). It is likewise studied that the lessening of food waste at every stage aids to lessen the problem of food insecurity (Saputro et al. 2021). As of Global Hunger Index 2022, India stands at 107 among 121 countries, and the grade of hunger and undernutrition is seriously high. It should be understood that the real problem arises from food loss and wastage (Saini and Khatri 2022).

Considering the urgency and importance of stoppage of food wastage as the need of the hour, the Food and Agriculture Organisation and United Nations Environment Programme have joined hands to observe the International Day of Awareness of Food Loss and Waste on September 29 since 2019. The theme for 2022 was "Stop Food Loss and Waste! For People and Planet." The importance of commemorating this day owes to the rising food and nutrition insecurity across the globe, with a goal to create mindfulness of the significance of the problems related to food loss and food waste, to find possible solutions, and stimulate inclusive efforts and action to accomplish the SDGs as mentioned earlier. SDG 12.3 explicitly aims to cut down individual food waste at the market and consumer level by 50% by 2030 and to reduce food losses along the production and supply chains (Mirage News 2021). Similarly, the problems associated with food wastage add to the environmental damage through the emission of greenhouse gases (Everett 2021). Well-planned food material purchases and adopting proper storage methods can enormously support food waste reduction.

1.2 Food Wastage: The Contemporary Portrayal

It is appraised that, globally, 14% of the food vanishes between harvest and sale, which equates to an amount 400 billion dollars in a year, whereas 17% is predicted to be wasted at retail and at the consumption stage (UNEP 2021).

In the universal food system, the lost and wasted food records 38% of total energy consumption, which in turn accounts for 8–10% of the Green House Gases globally, causing rickety climatic conditions, decreased crop yield, reduced nutritive value of crops, interferences in supply chain, and threat to food and nutrition security.

In the present day, food waste management is a primary global challenge because of the incompatible effects it poses on the economy, ecology, food security, and also nutrition security. A study on 165 nations using Ordinary Least-Squares regression (OLS) and General Linear Model (GLM) techniques has found that food wastage and poverty have a serious impact on GDP and economic growth (Conrad and Blackstone 2021), while the reduction in food loss percent and increasing food production have raised the GDP. It is also noteworthy that people prefer to raise GDP by reducing food loss rather than increasing food production (Mobaseri et al. 2021). The concern is that food wastage will not only influence the economy but will also equally influence society, and hence global strategies should be devised to find immediate solutions (Chrobog 2014).

1.3 Food Waste, Food and Nutrition Security

The expedition of food from harvest to our plates is a long mile, and food waste and loss are obvious at each stage. Food is lost at stages of production, harvest, post-harvest, processing, distribution, and consumption (Schuster and Torero 2016). The wasting of food resources in the agri-food structures influences sustainability on perspectives like depletion of natural resources, environmental pollution, and economy. The sways of food loss and waste are significant from an economic perspective (consumption and retail stages) and, from the societal perspective (reduced access and availability to food), may undermine food security (Corrado et al. 2019; Kuiper and Cui 2021).

The possible causes of food wastage at each stage and the strategies to overcome are elucidated in Table 1.1.

In India, the average food loss comprising all the food groups ranges between 4.6% and 15.8%. Global estimates show that an individual contributes to 65 kg of food waste (25% of vegetables, 24% of cereals, 12% of fruits, and 4% of other food groups) in a year, on average, menacing food security. This food waste, as a vicious cycle, markedly contributes to the wastage of nutrients like vitamin B2, vitamin B12, calcium, zinc, and choline, predominantly from cereal, fruit, and vegetable waste, followed by milk and meat product waste. Literature shows that the average

Table 1.1 Sources and prevention of food wastage

Stages	Reasons for food wastage	Prevention strategies	References
Production, handling, storage	Poor infrastructure	Government initiative	Gustavsson et al. (2011)
	Over-production	Improve market access	Beretta et al. (2013), Kaipia et al. (2013), and Garrone et al. (2014)
	Time and method of harvest	Improve harvesting technique	Kumar and Kalita (2017), Grover and Singh (2013), and Kannan et al. (2013)
	Pesticides/fertilizers	Educate farmers	Thompson (2008)
	Degradation and spillage	Improve workers' ability to adopt safe food handling practices	Miller and Welch (2013)
	Poor storage infrastructure	Improve storage facilities like warehouses, cold storage.	Willersinn et al. (2015)
Processing, packaging	Unprecedented food waste	Improve the capacity of process line	Papargyropoulou et al. (2014)
	Packaging system	Improvise packaging systems to maintain food fresh for a long period of time, Systematize labeling to support consumers	Ishangulyyev et al. (2019)
Distribution, marketing	Contamination during transportation, poor road and vehicle	Improve transportation vehicles	Martínez et al. (2014)
	Packaging management	Improve inventory systems	Plazzotta et al. (2017)
	Commercial conditions	Create online market platforms to expedite trade/ donate perishable foods	Godfray et al. (2010)
	Consumer reference	Collaborate with researchers to forecast vicissitudes in consumer demands	Lipinski et al. (2013)
Consumption	Household size, composition, income	Enable donation of unused or unsold foods from different establishments, reduce portion sizes	Parizeau et al. (2015), Jörissen et al. (2015), and Thyberg and Tonjes (2016)
	Household demographics	Avoid bulk purchases	Neff et al. (2015)
	Household culture	Promote awareness	Abiad and Meho (2018)
	Individual attitude	Educate on the importance of food and avoidance of food waste	Schanes et al. (2018), Melbye et al. (2017), and Stefan et al. (2016)
	Process of cooking	Effective use of leftovers	Silvennoinen et al. (2015)

Table 1.2 Food grains procurement and food waste in India

Year	Quantity of grains procured by the government	Quantity of food waste
2015–16	62.3 million tons	3116 tons (0.005% of total purchase)
2016–17	61 million tons	8540 tons (0.014% of total purchase)
2017–18	68.9 million tons	2664 tons (0.0038% of total purchase)
2018–19	80.2 million tons	4812 tons (0.006% of total purchase)
2019–20	85.95 million tons	1930 tons (0.002% of total purchase)
2020–21	99.24 million tons	1850 tons (0.0019% of total purchase)
2021–22	101.51 million tons	1693 tons (0.0017% of total purchase)

Source: FCI (2023), PiPa News (2022), and National Herald (2020)

amount of food wasted by an individual in a year, globally, is sufficient to meet the Recommended Dietary Allowance of 25 nutrients for a person for about 18 days. The amounts of micronutrients like vitamins C and K, zinc, copper, manganese, and selenium that are wasted every day are even higher, signifying 25–50% of their RDA value (Chen et al. 2020).

Table 1.2 implies that the procurement of grains by the Government has been raised in consecutive years with the notion of providing food to each and every individual. Similarly, the loss of food is also declining, but still, it is to be understood that if this loss is prevented, it can feed many million people. The cause for food loss can be:

- At the storage level, there is a shortage of firm management structures for ensuring food security.
- Faulty procurement.
- Diversion in crop production.
- Improper accounting.
- Climate change is plummeting agronomic yield, airing a stern danger to food security.
- During the Covid-19 pandemic, between April 2020 and September 2020, around 1571 tons of food grains were ruined and damaged. The news of concern is that, parallelly, many lost lives due to starvation.

A nexus between consumer-level food waste, nutrition, and environmental impacts (Muth et al. 2019; Birney et al. 2017) poses a real challenge for the nutritionist. Independent initiatives aimed at improving the quality of the diet can inadvertently lead to wastage of healthy food groups, under-consumed nutrients, agricultural resources, and unnecessary environmental impacts. Conversely, isolated measures to reduce food wastage may increase domestic accessibility of certain less healthy foods and over-consumed nutrients, contributing to the depletion of natural resources and environmental hazards. Nutritionists, in order to develop these manifold faces, will need to explore beyond the customary focus of the role of diet in diseases into more inter-dimensional arenas that connect food with food wastage and ecological sustainability, as the field of nutrition must ascertain the relevant fields of interest like food systems, food science, and agricultural studies, which are all interlocked with ecological sustainability (Finley et al. 2017).

For nutritionists, numerous strategies to reduce food waste are available for integration into counseling and public health messages. This includes emphasizing how individuals can save both money and the environment by avoiding food waste. Nutritionists can also enhance public knowledge on effective strategies for the selection, preparation, and storage of foods. Additionally, they may provide guidance on how schools or hostels can adjust food menus, inmates' food choices, and meal timings to minimize food waste. The combined and coordinated efforts of nutrition professionals and government policies, with a specific focus on women and children, can be effective in bringing about a change in food waste behavior. Campaigns conducted through press and media platforms can further amplify the impact of these efforts (Pearson-Stuttard et al. 2017; Afshin et al. 2015).

1.4 Necessity to Reduce Food Wastage

Food wastage, in addition to its insinuations on food security and nutrition security, emasculates the sustainability of the food systems. While food is being wasted, along with it, assets like water, land, energy, and labor, which are used for food production, are also wasted. Food prices also mount up owing to food shortage. The disposal of food waste in landfills causes greenhouse gas emissions, leading to climate change and adverse environmental effects.

Food wastage influences environmental sustainability. The major food groups that cause environmental impacts are cereals, meat, and sugar. Different countries show capricious nutrients and environmental footprints entrenched in their food waste, necessitating region-specific waste reduction intrusions (Ishangulyyev et al. 2019). The embedded environmental footprints in a person's daily food waste are depicted in Table 1.3.

Table 1.3 Embedded environmental footprints in a person's daily food waste

Factor utilized	Composition
Carbon dioxide equivalent	124 g
Freshwater	58 l
Cropland	0.36 m^2
Nitrogen	2.9 g
Phosphorus	0.48 g

Source: Foods, 2019

1.5 Preventive Measures

Many effective ways to reduce food loss are encouraging and should be explored further to monitor emerging trends in the food industry (Santeramo et al. 2018), which includes the following:

- The espousal of new technologies to improve the harvesting, storing, processing, transport, and selling process of foods
- To establish cooperatives for farmers
- To create awareness for retails and consumers
- To establish more communication platforms for all the partakers in food supply chains
- To promote reuse and recycling tactics from a circular and green economy standpoint (Santeramo 2021)

The appreciation of sectoral interconnections might aid in expanding cross-sectoral alliances to accomplish long-term financially viable, ecological, and societal goals (Santeramo and Searle 2019). An eye on the relationship between the use of water and energy to the food security status would be worthwhile, as the global agri-food systems utilize huge volumes of water (about 72%) (UN 2021) and energy (about 30%) (FAO 2020) for the production and supply chains of food. Also, the knowledge of the link between water, energy, and food security is vital to attain the United Nations SDGs from a perspective of society, economy, and environment, which depends on the effectual management of these resources. It is projected by 2050 that due to population increase, swift financial growth, urbanization, the changes in food habits, and the climate, the need for water may increase by 55% (OECD 2012; FAO 2014) and the demand for food may increase by 50% to feed the more than 9 billion people projected (FAO 2020). It is also predicted that energy consumption may grow by up to 50% by 2035. Bridging the gap in the link between water, energy, and food will help to face global challenges effectively.

An initiative by the Indian Government is the 6000-crore kick-start project, Pradhan Mantri Kisan Sampada Yojana. This national proposal is being implemented through public–private partnerships to develop an integrated supply cold chain for agricultural products. The objectives of the scheme are to:

- Create a contemporary set-up for the processing of foods by means of mega food parks or groups and individual units
- Create effectual linking of agriculturists, food processors, and food markets
- Create a strong supply chain infrastructure for perishables (india.gov.in 2019)

1.6 Food Loss and Waste to Value-Added Foods—Better Food and Nutrition Security

Food waste and byproducts can be potentially used as a sustainable alternative to ensure food and nutrition security in developing countries. Food wastes and byproducts contain sufficient quantities of macronutrients, micronutrients, nutraceuticals, and dietary fibers. Proper utilization of food waste and byproducts can address nutrient deficiency disorders, a life-threatening condition in most of the developing countries (Torres-León et al. 2018). Table 1.4 lists various food waste byproducts and details how they are transformed into value-added foods.

Table 1.4 clearly depicts the application of food wastages and derivatives (byproducts from the food processing industry) by incorporating food wastes to formulate novel food sources. This approach aims to reduce food insecurities, improve overall health, and have a direct positive impact on the nation's economy. It is necessary to investigate the efficient use of under-utilized food product processing and the use of byproducts from the food industry as edible food ingredients as a source of value-addition. Investigation of the nutraceutical and functional properties of bioactive compounds of byproducts is a budding area that further leads to the devising of valuable commercial food products. Still, numerous procedures are needed about food laws, feed regulations, and waste disposal rules to properly utilize food products and valorize the byproducts of the food industry from the food web with no harm to the security of foods and feeds. New technologies will also aid in the extraction of functional bioactive components from food wastage, instead of throwing them away, so as to warrant the sustainability of nutrition and diet.

1.7 Future Prospects and Challenges

Numerous processes are underway to reduce food waste, including advancements in food storage systems, cultivation of stress-tolerant foods, and the development of packaging materials that integrate sensors. These innovations aim to maintain the right temperature, ensure leak-proof packaging, preserve freshness, and optimize the pH within the package (Muller and Schmid 2019, Poritosh Roy et al. 2023). A number of online applications and social communities have also been developed to disperse leftover foods to food banks to reduce food waste and assure food security. Recently, the conversion of food waste into value-added products by black soldier fly pretreatment-assisted hydrothermal catalysis has been explored to prevent food waste and protect the environment (Li et al. 2023). Policy initiatives by the governments can also help reduce food waste through the food chain. At the same time, the challenge lies in implementing the technologies and educating the public.

Table 1.4 Food waste byproducts and how they are transformed into value-added foods

Food	Food waste/ byproduct	Nutrients	Applications/value-added foods	References
Rice/wheat/ barley/oats/ corn	Bran	Fat, protein ash, dietary fiber, nitrogen-free extract, magnesium, potassium, iron, manganese, B vitamins, choline, inositol, aluminum, calcium, iron, phosphorus, sodium, zinc, vitamin E	Rice bran oil, bakery foods, extruded food products, chocolates, bran wax	Panoth et al. (2019), and Palmer (2008)
	Broken kernels	Dietary fibers, phytochemicals, proteins, and vitamins	Breakfast foods, value added or commercial formulations for constipation and metabolic disorders	Huang et al. (2014)
	Husk	Tocopherols, tocotrienols, folates, dietary fiber, phytosterols, phenols, tricin	Antimicrobial and medical resource, biosensing, enzyme immobilization, green catalysis, production of drug delivery system in therapeutic drugs Breakfast cereals, bakery foods (bread, cakes, and cookies), Pasta	Papageorgiou and Skendi (2018), and Verma and Mogra (2013)
	Germs	Flavonoids, sterols, octacosanols, glutathione, unsaturated fatty acids, vitamin E, proteins, lipids, amino acids	Bakery foods, extruded food products, breakfast foods, biodegradable films	Harda et al. (2017)
Pulses	Brokens (6–13%), powder and germ mixture (7–12%), husk (4–14%)	Polyphenols, antioxidant minerals, dietary fiber	Snacks, bars, bakery foods, extruded food products, breakfast foods	Luzardo-Ocampo et al. (2019)
Mango	Peels (13–16%)	Dietary fiber (51.2–78.4%)	Can be a part of any food preparation	Serna-Cock et al. (2016)
	Seeds (9.5–25%)	Carbohydrates (58–80%), protein (6–13%), EAA and lipids (6–16%), oleic and stearic acids	Food preparation for adults and infants	Diarra (2014)
Pineapple	Stem and pomace (30–35%)	Dietary fiber (45.22%)	Alternative source of starch	Nakthong et al. (2017)

(continued)

Table 1.4 (continued)

Food	Food waste/ byproduct	Nutrients	Applications/value-added foods	References
Banana	Peel (30–40%)	Phytochemicals, flavonoids, carotenoids, catecholamines, calcium, potassium, magnesium, sodium, phosphorus, iron, zinc, palmitic acid, stearic acid, arachidic acid, myristicacid,	Chips, bread, sauces, jams, dried pulps, beer, wine	Venkateshwaran and Elayaperumal (2010)
		Linoleic acid, linoleic acid, insoluble dietary fiber, phytosterols		
Grape	Grape seed (5–6%)	Linoleic acid, tocopherol, proanthocyanidins	Grape seed oil, cookies	Iliodromiti et al. (2014)
Pomegranate	Peel	Dietary fiber, total polyphenols, stronger antioxidant capacity	Preservative, cookies, bread, noodles	Tito et al. (2021)
Citrus fruits	Peel	Phenolic acids, flavonoids, (polymethoxyflavones, flavanones, glycosylated flavanones), vitamin C, fiber	Dairy and bakery products, pectin, vital oils, enzymes, antioxidant, packaging film formation	Sormoli and Langrish (2016) Ahmed and Saeid (2021)

1.8 Conclusion

In India, though 190 million remain undernourished as per UN reports, it is appalling to note that about 40% of the food produced is wasted, which approximates Rs. 92,000 crores in a year. It is thus the need of the hour that we wake up and the nation is sensitized toward the importance of averting food waste, making valuable food products from byproducts of the food industry, thereby preventing food deprivation, and the right to food is enforced to each and every individual. Reduction in food waste will craft restorative nutritional and environmental outcomes. The prime challenge is to develop a defensible solution for the management of food waste that will enable to reap the biotic potential of biomaterials and achieve economic, social, and environmental benefits. The wise use of food waste and byproducts that are power packed with nutritional and functional properties for human consumption will help solve the nutritional problems that the globe faces, proving to be an influential tool in reducing hunger. Also, the branching out of the productive chains will provide job opportunities and sustainable income for many unemployed, thereby providing social benefits. Thus, it can be concluded that minimizing food waste, enhancing value addition, and augmenting food and nutrition security are the need of the hour.

References

Abiad MG, Meho LI (2018) Food loss and food waste research in the Arab world: a systematic review. Food Secur 10:311–322

Afshin A, Penalvo J, Del Gobbo L, Kashaf M, Micha R, Morrish K, Pearson-Stuttard J, Rehm C, Shangguan S, Smith JD, Mozaffarian D (2015) CVD prevention through policy: a review of mass media, food/menu labeling, taxation/subsidies, built environment, school procurement, worksite wellness, and marketing standards to improve diet. Curr Cardiol Rep 17:1–2

Ahmed M, Saeid A (2021) Citrus fruits: nutritive value and value-added products. In: Citrus – research, development and biotechnology [Internet]. Available from: https://doi.org/10.5772/intechopen.95881

Arivuchudar R (2018) Stint for a paradigm shift – food waste to food security. Int J Sci Res Rev 7(4):2276–2282

Ashlin Mathew, National Herald (2020). https://www.nationalheraldindia.com/india/1571-tonnes-of-food-grainswasted-between-april-and-september-due-to-covid-19-lockdow

Beretta C, Stoessel F, Baier U, Hellweg S (2013) Quantifying food losses and the potential for reduction in Switzerland. Waste Manag 33(3):764–773

Birney CI, Franklin KF, Davidson FT, Webber ME (2017) An assessment of individual foodprints attributed to diets and food waste in the United States. Environ Res Lett 12(10):105008

Chaboud G, Daviron B (2017) Food losses and waste: navigating the inconsistencies. Glob Food Sec 12:1–7

Chen C, Chaudhary A, Mathys A (2020) Nutritional and environmental losses embedded in global food waste. Resour Conserv Recycl 160:104912

Chrobog CK (2014) Wasted: understanding the economic and social impact of food waste. Doctoral Dissertation

Conrad Z, Blackstone NT (2021) Identifying the links between consumer food waste, nutrition, and environmental sustainability: a narrative review. Nutr Rev 79(3):301–314

Corrado S, Caldeira C, Eriksson M, Hanssen OJ, Hauser HE, van Holsteijn F, Liu G, Östergren K, Parry A, Secondi L, Stenmarck Å (2019) Food waste accounting methodologies: challenges, opportunities, and further advancements. Glob Food Sec 20:93–100

Diarra SS (2014) Potential of mango (Mangiferaindica L.) seed kernel as a feed ingredient for poultry: a review. Worlds Poult Sci J 70(2):279–288

Everett H (2021) Bonnie Wright teams up with food waste app Too Good to Go, 16 Oct 2021. MuggleNet. Retrieved 19 Oct 2021

FAO (2014) The Water-Energy-Food Nexus. A new approach in support of food security and sustainable agriculture. FAO, Rome

FAO, G (2011) Global food losses and food waste–extent, causes and prevention. SAVE FOOD: an initiative on food loss and waste reduction, p 9

FAO, IFAD, UNICEF, WFP, WHO (2020) The state of food security and nutrition in the world 2020. Transforming food systems for affordable healthy diets. FAO, Rome

Finley JW, Dimick D, Marshall E, Nelson GC, Mein JR, Gustafson DI (2017) Nutritional sustainability: aligning priorities in nutrition and public health with agricultural production. Adv Nutr 8(5):780–788

Garrone P, Melacini M, Perego A (2014) Opening the black box of food waste reduction. Food Policy 46:129–139

Godfray HC, Beddington JR, Crute IR, Haddad L, Lawrence D, Muir JF, Pretty J, Robinson S, Thomas SM, Toulmin C (2010) Food security: the challenge of feeding 9 billion people. Science 327(5967):812–818

Grover DK, Singh JM (2013) Post-harvest losses in wheat crop in Punjab: past and present §. Agric Econ Res Rev 26(2):293–297

Gustavsson J, Cederberg C, Sonesson U, Van Otterdijk R, Meybeck A (2011) Global food losses and food waste: extent, causes and prevention. FAO

Harda J, Amorim CA, Braga PL, Machando LDB, Oliveira RR, Neto AC, Macedo JRN, Silvia LGA, Rosa DS (2017) Characterization of biodegradable mulch black films incorporated withorganics fertilizers and rice husk ash. In: TMS 2016-146rd Annual meeting & exhibition, February 26–March 2017-San Diego, CA, USA

How technology can help avert food waste. Mirage News. 29 Sept 2021. Retrieved 19 Oct 2021

Huang Q, Shi CX, Su YB, Liu ZY, Li DF, Liu L, Huang CF, Piao XS, Lai CH (2014) Prediction ofthe digestible and metabolizable energy content of wheat milling by- products for growing pigsfrom chemical composition. Anim Feed Sci Technol 196:107–111

Iliodromiti S, Kelsey TW, Wu O, Anderson RA, Nelson SM (2014) The predictive accuracy of anti-Müllerian hormone for live birth after assisted conception: a systematic review and meta-analysis of the literature. Hum Reprod Update 20(4):560–570

india.gov.in (2019). https://www.india.gov.in/spotlight/pradhan-mantri-kisan-sampada-yojana

Ishangulyyev R, Kim S, Lee SH (2019) Understanding food loss and waste—why are we losing and wasting food? Foods 8(8). https://doi.org/10.3390/foods8080297

Jörissen J, Priefer C, Bräutigam KR (2015) Food waste generation at household level: results of a survey among employees of two European research centers in Italy and Germany. Sustainability 7(3):2695–2715

Kaipia R, Dukovska-Popovska I, Loikkanen L (2013) Creating sustainable fresh food supply chains through waste reduction. Int J Phys Distrib Logist Manag 43:262–276

Kannan EL, Kumar PA, Vishnu KE, Abraham HA (2013) Assessment of pre and post-harvest losses of rice and red gram in Karnataka. Crops 44(6):61–70

Kuiper M, Cui HD (2021) Using food loss reduction to reach food security and environmental objectives—a search for promising leverage points. Food Policy 98:101915

Kumar D, Kalita P (2017) Reducing postharvest losses during storage of grain crops to strengthen food security in developing countries. Foods 6(1):8

Lebersorger S, Schneider F (2011) Discussion on the methodology for determining food waste in household waste composition studies. Waste Manag 31(9–10):1924–1933

Li O, Liang J, Chen Y, Tang S, Li Z (2023) Exploration of converting food waste into value-added products via insect pretreatment-assisted hydrothermal catalysis. ACS Omega. Available online https://doi.org/10.1021/acsomega.3c00762

Lipinski B, Hanson C, Lomax J, Kitinoja L, Waite R, Searchinger T (2013) "Reducing food loss and waste." Working paper, installment 2 of creating a sustainable food future. World Resources Institute, Washington, DC. Available online at http://www.worldresourcesreport.org

Luzardo-Ocampo I, Cuellar-Nuñez ML, Oomah BD, Loarca-Piña G (2019) Pulse by-products. In: Food wastes and by-products, pp 59–92. https://doi.org/10.1002/9781119534167.ch3

Martínez ZN, Menacho PZ, Pachón-Ariza F (2014) Food loss in a hungry world, a problem? Agron Colomb 32(2):283–293

Melbye EL, Onozaka Y, Hansen H (2017) Throwing it all away: exploring affluent consumers' attitudes toward wasting edible food. J Food Prod Mark 23(4):416–429

Miller DD, Welch RM (2013) Food system strategies for preventing micronutrient malnutrition. Food Policy 42:115–128

Mobaseri M, Mousavi SN, Haghighi MHM (2021) Environmental effects of food waste reduction using system dynamics approach. Energy Sources A: Recovery Util Environ Eff:1–19. https://doi.org/10.1080/15567036.2021.2000066

Muller P, Schmid M (2019) Intelligent packaging in the food sector: a brief overview. Foods 8(1):16

Muth MK, Birney C, Cuéllar A, Finn SM, Freeman M, Galloway JN, Gee I, Gephart J, Jones K, Low L, Meyer E (2019) A systems approach to assessing environmental and economic effects of food loss and waste interventions in the United States. Sci Total Environ 685:1240–1254

Nakthong N, Wongsagonsup R, Amornsakchai T (2017) Industrial crops and products characteristics and potential utilizations of starch from pineapple stem waste. Ind Crop Prod 105:74–82. https://doi.org/10.1016/j.indcrop.2017.04.048

Neff RA, Spiker ML, Truant PL (2015) Wasted food: US consumers' reported awareness, attitudes, and behaviors. PLoS One 10(6):e0127881

Nicholes MJ, Quested TE, Reynolds C, Gillick S, Parry AD (2019) Surely you don't eat parsnip skins? Categorising the edibility of food waste. Resour Conserv Recycl 147:179–188

OECD (2012) The Water Challenge: sharing a precious commodity. OECD, Paris

Palmer S (2008) The top fiber-rich foods list. Today's Dietit 10(7):28

Panoth A, Lavanya D, Naik MG, Venkata CN (2019) Waste to wealth: potential of rice bran wax asedible coating. Indian Food Ind Mag 38(1):38–43

Papageorgiou M, Skendi A (2018) Introduction to cereal processing and by-products. In: Sustainable recovery and reutilization of cereal processing by-products. Woodhead Publishing, Cambridge, pp 1–25

Papargyropoulou E, Lozano R, Steinberger JK, Wright N, Ujang Z b (2014) The food waste hierarchy as a framework for the management of food surplus and food waste. J Clean Prod 76:106–115

Parizeau K, Von Massow M, Martin R (2015) Household-level dynamics of food waste production and related beliefs, attitudes, and behaviours in Guelph, Ontario. Waste Manag 35:207–217

Pearson-Stuttard J, Bandosz P, Rehm CD, Penalvo J, Whitsel L, Gaziano T, Conrad Z, Wilde P, Micha R, Lloyd-Williams F, Capewell S (2017) Reducing US cardiovascular disease burden and disparities through national and targeted dietary policies: a modelling study. PLoS Med 14(6):e1002311

Plazzotta S, Manzocco L, Nicoli MC (2017) Fruit and vegetable waste management and the challenge of fresh-cut salad. Trends Food Sci Technol 63:51–59

Roy P, Mohanty AK, Dick P, Misra M (2023) A review on the challenges and choices for food waste valorization: environmental and economic impacts. ACS Environ Au 3(2):58–75

Rutten MM (2013) What economic theory tells us about the impacts of reducing food losses and/or waste: implications for research, policy and practice. Agric Food Secur 2(1):1–13

Santeramo FG (2021) Exploring the link among food loss, waste and food security: what the research should focus on? Agric Food Secur 10:26. https://doi.org/10.1186/s40066-021-00302-z

Santeramo FG, Searle S (2019) Linking soy oil demand from the US Renewable Fuel Standard to palm oil expansion through an analysis on vegetable oil price elasticities. Energy Policy 127:19–23

Santeramo FG, Carlucci D, De Devitiis B, Seccia A, Stasi A, Viscecchia R, Nardone G (2018) Emerging trends in European food, diets and food industry. Food Res Int 104:39–47

Saputro WA, Purnomo S, Salamah U (2021) Study of food waste of farmers' households in Klaten to support food security. Anjoro: Int J Agric Bus 2(2):58–64

Schanes K, Dobernig K, Gözet B (2018) Food waste matters-a systematic review of household food waste practices and their policy implications. J Clean Prod 2018(182):978–991

Schuster M, Torero M (2016) Reducing food loss and waste. International Food Policy Research Institute (IFPRI), IFPRI book chapters; 9780896295827-03

Serna-Cock L, García-Gonzales E, Torres-León C (2016) Agro-industrial potential of the mango peel based on its nutritional and functional properties. Food Rev Intl 32(4):364–376

Silvennoinen K, Heikkilä L, Katajajuuri JM, Reinikainen A (2015) Food waste volume and origin: case studies in the Finnish food service sector. Waste Manag 46:140–145

Sormoli ME, Langrish TA (2016) Spray drying bioactive orange-peel extracts produced by Soxhlet extraction: use of WPI, antioxidant activity and moisture sorption isotherms. LWT-Food Sci Technol 72:1–8

Stefan V, Van Herpen E, Tudoran AA, Lähteenmäki L (2016) Avoiding food waste by Romanian consumers: the importance of planning and shopping routines. Food Qual Prefer 28:375–381

Saini S, Khatri P (2022). https://theprint.in/opinion/indias-real-food-problem-isnt-hunger-but-loss-and-wastage/1170972/

Thompson AK (2008) Fruit and vegetables: harvesting, handling and storage. John Wiley & Sons, UK

Thyberg KL, Tonjes DJ (2016) Drivers of food waste and their implications for sustainable policy development. Resour Conserv Recycl 106:110–123

Tito A, Colantuono A, Pirone L, Pedone E, Intartaglia D, Giamundo G, Conte I, Vitaglione P, Apone F (2021) Pomegranate peel extract as an inhibitor of SARS-CoV-2 spike binding to human ACE2 receptor (in vitro): a promising source of novel antiviral drugs. Front Chem 9. https://doi.org/10.3389/fchem.2021.638187

Torres-León C, Ramírez-Guzman N, Londoño-Hernandez L, Martinez-Medina GA, Díaz- Herrera R, Navarro-Macias V, Alvarez-Pérez OB, Picazo B, Villarreal-Vázquez M, Ascacio-Valdes J, Aguilar CN (2018) Food waste and byproducts: an opportunity to minimize malnutrition and hunger in developing countries. Front Sustain Food Syst 2:52. https://doi.org/10.3389/fsufs.2018.00052

United Nations Environment Programme portal - UNEP (2021). https://www.unep.org/resources/report/unep-food-waste-index-report-2021

UN-Water (2021) Summary progress update 2021—SDG 6—water and sanitation for all. UN-Water, Geneva

US Department of Health and Human Services, Centers for Disease Control and Prevention. How to prevent food poisoning. Available at: https://www.cdc.gov/foodsafety/prevention.html. Accessed 14 Aug 2019

Venkateshwaran N, Elayaperumal A (2010) Banana fiber reinforced polymer composites-a review. J Reinf Plast Compos 29(15):2387–2396

Verma A, Mogra R (2013) Psyllium (Plantagoovata) husk: a wonder food for good health. Int J Sci Res 4(90):1581–1585

Willersinn C, Mack G, Mouron P, Keiser A, Siegrist M (2015) Quantity and quality of food losses along the Swiss potato supply chain: stepwise investigation and the influence of quality standards on losses. Waste Manag 46:120–132

Part II
Waste Utilization from Dairy Industry

Chapter 2
Value Addition and Sustainable Management of Dairy Industry Byproducts

T. Jayasree Joshi, V. Harsha, P. Nandagopal, Asha Ashok, and Sabitra Pokhrel

Abstract Global demand for dairy products is increasing with the population, changing consumption patterns, and urbanization. The byproduct generation of the industry has also risen in tandem with this, necessitating the need for its sustainable valorization. Whey, skim milk, and ghee residue constitute the major byproducts of the dairy processing industry. The dairy byproducts are rich in organic and inorganic components, with high COD and BOD indices. This makes them a potential candidate for environmental hazards if improperly disposed of. Consequently, there is a greater demand for the conversion of these into value-added forms. Whey, a liquid byproduct produced in large quantities during the cheese-making process, is an excellent source of macro and micronutrients. Through adequate processing, whey can be effectively utilized to produce protein concentrates, lactose, and a range of other substances. The unique nutritional and functional characteristics of dairy byproduct protein isolates make them an ideal ingredient in various food processing applications. Skim milk, which is often used for milk standardization, can also be used to produce casein and its derivatives. Ghee residue can be used in the manufacturing of confectionery and chocolate products. Wastewater from a dairy plant may be utilized as an effective substrate for single-cell protein synthesis. Bioconversion of byproducts to enzymes, organic acids, biofuels, biopolymers, and bioactive components is also feasible. This chapter provides insight into various byproducts of the dairy industry and technologies for their effective valorization.

T. J. Joshi (✉) · V. Harsha · S. Pokhrel
Agricultural and Food Engineering Department, Indian Institute of Technology Kharagpur, Kharagpur, West Bengal, India

P. Nandagopal
Department of Dairy Technology, College of Dairy Science & Technology, Kerala Veterinary and Animal Sciences University, Idukki, Kerala, India

A. Ashok
Department of Dairy Engineering, College of Dairy Science & Technology, Kerala Veterinary and Animal Sciences University, Idukki, Kerala, India

E. Cherian, B. Gurunathan (eds.), *Value Added Products From Food Waste*,
https://doi.org/10.1007/978-3-031-48143-7_2

Keywords Dairy waste · Byproduct valorization · Whey · Skim milk · Whey protein concentrate · Lactose · Ghee residue

2.1 Introduction

Liquid milk and other processed milk products are essential constituents of the human diet. High demand for dairy products and an increase in the number of large-scale producers have also contributed to a substantial increase in the quantity of byproducts. The major byproducts of the dairy processing industry are whey, skim milk, and ghee residue (Ahmad et al. 2019). Lactose, protein, and fat can all be derived from the byproducts. Generally, the wastes produced by the dairy industry can be divided into two categories: fat-rich dairy substrates (e.g., cream) and low-fat dairy substrates (e.g., skim milk) (Mahboubi et al. 2017). Dairy industries also generate approximately 6–10 liters of wastewater for every one liter of milk processed. This effluent water contains fat, minerals, carbohydrates, and whey protein. The composition of wastewater varies significantly based on the types of products manufactured and the processes involved (Carvalho et al. 2013). If properly utilized, the wastewater can be effectively utilized for single-cell protein synthesis and bioconversion to enzymes, organic acids, biofuels, biopolymers, and bioactive components. In addition, due to the high amount of proteins and sugars present, dairy wastes have a high biological oxygen demand (BOD) and chemical oxygen demand (COD) value. BOD levels in untreated effluents range between 0.8 and 2.5 kg t^{-1} of milk, and COD levels are typically 1.5 times BOD levels. Therefore, it is essential to manage and dispose of waste from the dairy industry properly to maintain environmental safety. This chapter presents a review of the various byproducts of the dairy industry and technologies for their effective valorization.

2.2 Byproducts in the Dairy Industry

The session discusses in detail the different byproducts generated during the processing of milk and other dairy products.

2.2.1 Byproducts from Skim Milk

Skim milk (SM) is a byproduct of separating cream from whole milk (WM) using a cream separator or a centrifugal separator. The average composition of skim milk from cow milk includes 90.6% water, 5% lactose, 3.6% protein, 0.1% fat, and 0.7%

ash (Kolhe et al. 2009). Since SM has a lower fat content and high nutrient value, like whole milk, it has been shown to affect health positively. Whole milk products contain saturated fat, which may affect cholesterol levels in the body. Compared to WM, consuming SM has been shown to have positive effects on health, most notably in preventing cardiovascular problems (Xu et al. 2022). SM is commonly used to standardize milk for use in a broad range of dairy-based products. Protein hydrolysates, casein, and co-precipitates are the byproducts of SM. Table 2.1 shows the overview of the utilization of SM in different food applications.

2.2.1.1 Casein

Casein is manufactured through acid precipitation (acid-casein (AC)) and rennet coagulation (rennet-casein (RC)). Acidification occurs when skim milk is mixed with diluted acids such as HCl, H_2SO_4, or lactic acid. As a rule of thumb, 60 liters of sulfuric or hydrochloric acid (1 N) is needed to precipitate casein from 1000 liters of SM (Carr and Golding 2016). The protein complex gets dissociated when acid is added to the milk. The precipitation starts at a pH of 5.3, and maximum precipitation occurs when pH reaches 4.6 (isoelectric point). Manufacturing casein at a constant pH of 4.6 requires careful acid addition and thorough mixing. Curd strength can be improved by cooking the acid curd, which strengthens the hydrophobic bond. A higher pH value results in a curd that is of rubbery, large, and sticky texture, and curd with a low pH has soft, small particles, which leads to losses during the de-wheying and washing process. In rennet casein, the rennet enzyme is used for the coagulation of SM. This results in the breakage of the peptide bond (Phe-Met, residue 105–106) of k-casein and forms para-κ-casein (PKC) and glycomacropeptide (GMP). Water-soluble GMP gets into whey, which causes instability of casein molecules. Moreover, PKC forms a 3D structure with calcium ions. Continuous and batch methods are usually adopted for manufacturing RC. Heat treatment of SM is done by the HTST pasteurization method, and after that, it is cooled to 31 °C in the batch process. Rennet is added and mixed with SM in a ratio of 1:4500 in the batch process. Furthermore, in a continuous process, rennet is added to SM in a ratio of 1:7000 to 1: 8175 and agitated properly (O'Sullivan et al. 2002). After the cooking

Table 2.1 Overview of the utilization of skim milk in different food processing applications

Principle manufacturing operation involved	Products
Concentration	Plain, sweetened, or low-lactose condensed skim milk
Fermentation	Acidophilus milk, cultured buttermilk
Drying	Skim milk powder
Coagulation	Cottage cheese, baker's cheese, sapsago cheese, quarg cheese, casein
Pasteurization	Flavored milk
Sterilization	Sterilized flavored milk

Table 2.2 Application of casein products in various food products

Food product	Form of casein	Function	Reference
Cheese	Acid casein, rennet casein, caseinates	Binding agent	Warncke et al. (2022)
		Texture enhancement	Soloshenko et al. (2016)
Pasta	Sodium caseinate	Nutritional enhancement	Kumar et al. (2019)
		Improvement of sensory qualities	
Ice cream	Sodium caseinate	Texture enhancement	Kumar et al. (2016)
		Stabilization	
Meat products	Sodium caseinate	Emulsification	Petridis et al. (2010)
		Texture enhancement	
Infant foods	Caseinates	Nutritional enhancement	Raikos and Dassios (2014)
Bakery products	Casein and caseinates	Stabilization	Crowley et al. (2002)
		Nutritional enhancement	

stage, whey has to be removed from curd as earliest to avoid difficulties in washing out acid, salts, and lactose. Efficient washing is necessary to achieve better recovery of casein. Curd is then pressed to minimize the water content, followed by milling to form uniform particle size. Then, it is dried using any of the drier to achieve a moisture content of less than 12%. Furthermore, it is allowed to cool before the tempering stage. In the tempering phase, casein is kept for 8–24 h to ensure that the moisture of the coagulated particles is consistent. Afterward, casein is kneaded and stored. The utilization of SM in industrial and edible forms is well established. In the initial times, casein is used in the textiles, rubber, and paint industries. The edible casein is isolated from SM to use in food and pharmaceutical applications by taking proper precautions in quality, hygiene, and storage aspects. Casein and caseinates are used in various beverages as a stabilizer (fizzy drinks, fruit, and chocolate beverages) in the beer and wine industry to minimize bitterness and color and in dessert foods like ice cream, pudding, and frozen products as an emulsifier. It is highly recommended for use in burger patties, nuggets, and sausages due to its high emulsion stability (Petridis et al. 2010). Some of its applications in various food formulations are explained in Table 2.2.

2.2.1.2 Co-precipitates

The heating and coagulation of milk results in the development of complex byproducts. Co-precipitates contain a mixture of whey and casein proteins. Because of its high protein content, it can be used in a wide range of food production processes. The co-precipitates have low lactose content and can be a good option for lactose-intolerant people and infant food formulations (Macej and Jovanovic 2002). Processing methods and the amount of calcium present highly influence the characteristics and functions of co-precipitates. Co-precipitates are categorized into three groups based on their amount of calcium (Ca). They are low, medium, and high Ca co-precipitates, in which Ca content is 0.5–0.8%, 1.5%, and 2.5–3%, respectively (Gawande et al. 2022). Co-precipitates are prepared by heating skim milk to a degree to achieve denaturation of whey protein, followed by the formation of complexes of whey and casein. The complex protein matrix is precipitated by either acidification (4.6 pH) or a combination of acidification and the addition of calcium chloride. The precipitation is followed by whey separation, washing, pressing, and drying. The co-precipitates contain a higher quantity of sulfur-containing amino acid and cysteine. The color (whiteness) value of co-precipitate is dependent on the amount of calcium and pH. Co-precipitates can be used as a fat replacer (up to 2%) in meat products like pork sausages. Also, it has been reported to enhance the sensory attributes of the product, like flavor, juiciness, etc. (Eswarapragada et al. 2010). It can also be used in the fortification of breakfast cereals and can also be blended with other protein sources. Hence, an improved nutritional profile and functional properties can be achieved.

2.2.1.3 Protein Hydrolysates (PH)

Protein hydrolysates are another important byproduct of skim milk. Generally, two methods are adopted in the manufacturing of PH: enzyme and acid hydrolysis. During acid hydrolysis, casein or caseinates are treated with an acid (HCl or H_2SO_4) for 4–18 h at 80–100 °C. The pH of the product obtained was made up to 6–7 by using an alkali. The product is cooled to 15–25 °C, followed by centrifugation and filtering. The liquid can be spray-dried at 95–100 °C into a fine powder easily soluble in water. The protein hydrolysate powder thus obtained comprises 41–41.5% NaCl, 53.5–53.7% amino acids, and 2.0–2.5% other minerals. The loss of amino acids during processing is a significant drawback of the acid hydrolysis approach. The enzymatic hydrolysis method overcomes this drawback. In this method, proteolytic enzymes such as pepsin, rennin, trypsin, neutrase, pronase, ficin, papain, etc., are used for the production of hydrolysates. The protein hydrolysates thus produced are utilized in the fields of human nutrition, functional foods, fermentation, cosmetics, and biotechnology (Abd El-Salam and El-Shibiny 2017).

2.2.2 *Byproducts from Whey*

2.2.2.1 Whey and Its Composition

Whey is the liquid fraction obtained after paneer, cheese, chhana, and casein production. In ancient times, whey was consumed as a health drink and even used to treat various gastrointestinal ailments and skin conditions. With the industrialization and modernization of the dairy industry, milk production increased drastically. This has resulted in the conversion of liquid milk into dairy products like cheese, resulting in significant increases in the amount of whey. In many European countries like Greece, where traditionally the cheese consumption is very high (greater than 20 kg of feta cheese per capita), partial utilization of whey solids has been done through the production of whey cheeses (Philippopoulos and Papadakis 2008). The amount of whey produced is increasing nearly at the same rate as milk production (Smithers 2008). If not appropriately managed, whey is a highly polluting by/co-product with a very high COD and BOD (50,000–80,000 mg/L and 40,000–60,000 mg/L, respectively) (Lappa et al. 2019; Smithers 2008).

The protein present in whey can be categorized as major and minor proteins. Major proteins include serum albumin, α-lactalbumin, and β-lactoglobulin. Minor proteins include glycomacropeptide, lactoferrin, bovine serum albumin, immunoglobulin, and phospho-lipoproteins. The coagulation techniques used will be the main determinant of the type and composition of whey. Rennet coagulation of milk produces sweet whey (SW) and is low in acidity. The enzyme breaks down the k-casein at 105–106 Phe-Met bond into para-k-casein and glycomacropeptide. Glycomacropeptide, being hydrophilic, dissolves in whey and makes up roughly 20% of the whey protein fraction of SW, whereas it is absent from acid whey (AW). The calcium ions attach to the paracasein, neutralizing the negative charges and forming salt bridges for gel formation (Walstra et al. 2005). Acid coagulation of milk is accomplished through the addition of organic or inorganic acids into milk, and casein coagulation occurs at its isoelectric pH of 4.6. Acid whey has a higher Ca, P, and lactic acid content than SW, which has a higher protein and fat content (Silviya et al. 2016). AW and SW have similar protein contents, but the amount of free amino acids can differ depending on how much casein is hydrolyzed when cheese is being manufactured. SW may have four times the amount of free amino acids as milk, and acid whey may have ten times the amount (Božanić et al. 2014). Another emerging type of acid whey is produced as a byproduct of Greek-style yogurt and traditional Indian dairy products like Shrikhand. The fermented curd is strained to remove the whey and obtain a thick, high-protein product. The whey produced so is highly acidic and contains very few solids (Elliott 2013). Characteristics of initial milk and seasonal variations can affect the volume and physicochemical properties of whey and derivatives produced (Philippopoulos and Papadakis 2008). Table 2.3 depicts the typical constituents in SW and AW.

Table 2.3 Typical constituents of acid and sweet whey

Component	Acid whey (%)	Sweet whey (%)
Water	94–95	93–94
Lactose	3.8–4.3	4.5–5
Total protein	0.8–1	0.8–1
Whey protein	0.60–0.65	0.60–0.65
Minerals	0.5–0.7	0.5–0.7
Citric acid	0.1	0.1
Lactic acid	0.8	Traces

2.2.2.2 Whey Processing

Whey is nutrient-rich and has a highly perishable nature. So, it must be quickly processed by either chilling or pasteurization to prevent undesirable changes. The first stage of whey processing involves the separation of fat and casein fines; this improves the overall economy of the process and also prevents these components from negatively affecting the final quality of whey derivatives. Separation of whey is carried out using centrifugal separators, cyclones, or rotating filters. The fat and curd fines recovered can be incorporated into the cheese or used to manufacture other special cream products. The schematic diagram showing the overview of whey processing and its end products is shown in Fig. 2.1. Advances in membrane technology have contributed significantly to making whey processing viable; various membrane technologies are used as a separation or concentration process.

Advanced membrane technologies like microfiltration (MF), ultrafiltration (UF), reverse osmosis (RO), nanofiltration (NF), and diafiltration (DF) have been employed recently for the fractionation and concentration of different whey components. The UF process can separate macromolecules weighing between 1000 and 200000 Da. UF membrane pore sizes range from 1 to 100 nm and pressures less than 1000 kPa. The permeate obtained from UF contains water, milk sugar, soluble minerals, water-soluble vitamins, protein retentates, lipids, and colloidal salts. With the use of UF, it is possible to obtain a whey protein concentrate of 45% protein content from sweet rennet whey (Bejarano et al. 2022). Nanofiltration can permeate molecules in the range of 10–100 Da and can be effectively used for the

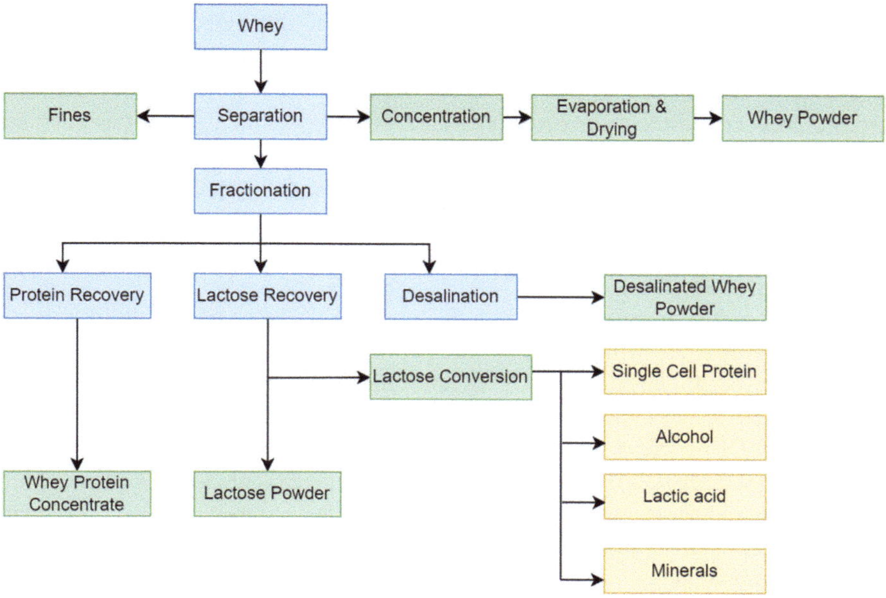

Fig. 2.1 Overview of whey processing and its end products

demineralization of whey, particularly monovalent ions. A combination of water dilution of NF retentate with diafiltration can help in an almost 50% reduction in ash content (Burling 2003).

2.2.2.3 Whey Cheese and Whey Powder

A well-established conventional method of whey utilization in various European nations is the manufacturing of whey cheeses, which recovers some fat and proteins from the whey. The process involves denaturation and coagulation of the proteins. Due to the low solid content, whey may be pre-concentrated with the addition of milk or other dairy-based products (Božanić et al. 2014; Philippopoulos and Papadakis 2008). Whey powder is another important byproduct that is made by concentrating whey to a total solid content of 40–70% and then spray/roller drying to less than 5% moisture. The concentration process is typically carried out in two or more stages in a falling-film evaporator. RO can also be used for preconcentration before evaporation. Lactose must be converted to its α-monohydrate crystalline form to reduce water absorption properties and thus prevent defects during storage (Bylund 2003).

2.2.2.4 Whey Proteins

Whey protein concentrates, isolates, and fractionates are produced by separating the proteins from the other whey components. Heat treatment can precipitate whey proteins under the optimal pH and ionization values. Lactalbumin is produced by heat precipitation of whey proteins and is a mixture of denatured β-lactoglobulin, β-lactalbumin, and other proteins. Whey is heated during the manufacturing process to denature, coagulate, and precipitate the proteins. Settling and decantation are used to recover sediment (or centrifugation). During washing, impurities such as salt and lactose are removed, and the product is recovered using centrifugation and filtering prior to drying, pulverizing, and packing. The precipitated protein formed is either insoluble or sparingly soluble, depending on the conditions at denaturation. As a result, lactalbumin can be best used in products where protein fortification is vital (Tsakali et al. 2010).

The most viable process for the manufacture of WPC is the use of ultrafiltration. Whey protein powder with different protein concentrations and low lactose and ash contents can be produced using UF by concentrating native/pre-denatured whey proteins. Lactose and mineral content can be reduced further through a subsequent diafiltration (DF) process in which deionized water is continuously added to the retentate (Tsakali et al. 2010). Whey protein isolates (WPI) contain 90% protein and 4–6% water, and the remaining portion comprises fat, lactose, and ash. The high protein content and solution clarity are the main reasons. WPIs find application in nutritional supplements, protein-based drinks, and sports drinks (Foegeding et al. 2011). WPIs are highly purified membrane retentates. Methods like ionic exchange, electrodialysis, nanofiltration, and diafiltration can be used for the purification process. After purifying, the retentate is concentrated and spray-dried.

2.2.2.5 Lactose

Lactose, also known as milk sugar, is the major component of whey solids. The main application of lactose is in the manufacture of infant milk formulas and the pharmaceutical industry as an active ingredient for the manufacturing of tablets. The lactose is recovered by crystallization from concentrated whey or deproteinized whey (Bylund 2003). NF and DF are essential for effective demineralization and for ensuring the purity of the final product. Whey is concentrated to 60–62% solids by evaporation, and lactose crystallization is achieved through seeding. Lactose crystals are then separated using centrifugal techniques, and the impurities are removed through washing. The lactose crystals obtained were dried at temperatures below 93°C to prevent the formation of β-lactose, and the dried powder was immediately packed. Lactose can be further converted into value-added derivatives like galacto-oligosaccharides, lactulose, lactitol, hydrolyzed lactose, and lactobionic acid.

2.2.2.6 Whey Beverages

Several alcoholic and non-alcoholic beverages have been produced from whey over the years. Whey beverage processing began in the 1970s, and Rivella from Switzerland is one of the oldest whey beverages. These beverages are a rich source of branched-chain amino acids such as isoleucine, leucine, and valine, giving them a high nutritional value. The major operations involved in the production of low-alcohol whey beverages include deproteinization, concentration, fermentation (typically by *Kluyveromyces fragilis* and *Saccharomyces lactis*), flavoring, sweetening, and bottling (Barukcic et al. 2008). The addition of fruit and vegetable extracts to whey enhances its sensory and nutritional qualities. Alane et al. (2017) developed a whey-mango beverage with ginger extracts with high sensory attributes. In another study, Purkiewicz and Pietrzak-Fiećko (2021) combined whey from different fruit and vegetable mixes, and the bioactive properties were assessed. They obtained the highest flavonoid content in orange whey beverages (63.06 mg/100 g).

2.2.3 *Ghee Residue*

Milk fat is processed into ghee, butter oil, and anhydrous milk fat (Illingworth et al. 2009). Ghee is heat-clarified butter fat and has excellent organoleptic properties, which makes it an excellent ingredient in many food formulations (Battula et al. 2020). Different ghee preparation methods are practiced, in which the basic principle involved is the concentration of milk fat (cream or butter), heat clarification of milk fat, reduction of moisture content to less than 0.5%, and then removal of residue. During the manufacturing process, the solid not fat (SNF) in butter or cream turns to brownish moist particles known as ghee residue (GR), which is obtained in the final stage of molten ghee filtering. The quantity of residue obtained during the process varies with the type of milk used. However, the average yield of ghee residue is 10–12% (Varma and Narender Raju 2008). GR has a smooth and soft texture and is a potential source of minerals (mainly Ca and P), protein, fat, and lactose (Selvamani et al. 2017). GR contains 11, 132, and 10 times more free fatty acids (FFA), lactones, and carbonyl compounds than ghee (Varma and Narender Raju 2008). In addition, it has good antioxidant properties and flavor potential (Wani et al. 2022). Compared to fat-soluble compounds, fat-insoluble components of GR show higher antioxidant properties. GR is one of the major byproducts used in the chocolate and confectionery industries. Thermal treatments associated with ghee production significantly influence the compositions of ghee residue. Heat treatment is generally carried out at a temperature of 110°C–120 °C for 10–20 min. Along with the raw materials, the ghee preparation method also affects the yield of ghee residue. Milk with high SNF content yields more ghee residue (Wani et al. 2022). The applications of ghee residue are presented in Fig. 2.2.

GR loses its fine, delicate texture after prolonged storage, so pre-processing is required to maintain the soft texture. The loss of moisture causes an increase in the total solid content, which is a primary cause of the hardening. Treatments commonly used are (a) soaking in boiling water (30 min), (b) soaking in sodium

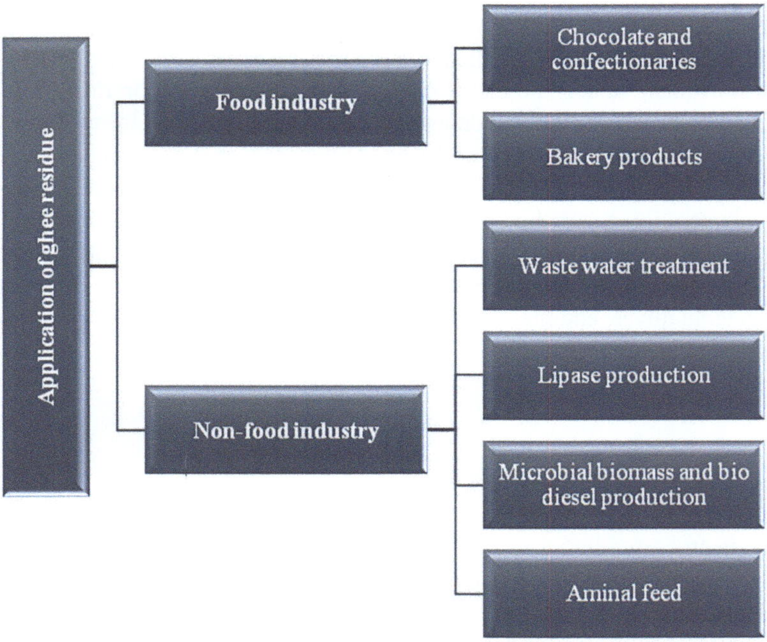

Fig. 2.2 Applications of ghee residue

bicarbonate solution (1%, 30 min), (c) washing in alcohol (50%) and then soaking in boiling water or sodium bicarbonate solution (1%, 30 min), and (d) GR from treatment "c" added with 2% vinegar and autoclave. The treatment "d" extends the shelf life of residue up to 3 months (Wani et al. 2022).

The utilization of GR in bakery products like cakes, biscuits, cookies, and muffins enhances the nutritive value of the product while maintaining market sustainability. In a study, Ranjan et al. (2020) utilized the ghee residue at four compositions from 10–40% along with other ingredients (refined wheat flour, sugar, salt, butter, milk, and egg) in the preparation of cake and muffins. Based on the sensory characteristics, the incorporation of 40% GR in 60% maida showed better acceptance in terms of flavor, color, texture, and mouthfeel. In addition, the amount of protein, fat, and calcium increased while the amount of carbohydrates and energy decreased as the percentage of GR increased. The concentrations of flavor compounds like FFA, carbonyl compounds, and lactones are high in ghee residues, which enhances the flavor profiles of bakery products. It can be used as a 50–100% replacer of butter in the preparation of cakes (Subbulakshmi et al. 1990). Ghee residue contains approximately 37% fat and can be used as a fat replacer in biscuits and cookie productions. Also, 10% ghee residue can be used as a fat substitute in cookie recipes (Sojan et al. 2019). Incorporating the ghee residue reduces production costs, making for more economical products with high consumer acceptability (Ranjan et al. 2020).

Ghee residue is a promising ingredient in the preparation of the confectionery industry, similar to bakery products, due to its flavor and brownish color. It can be used in the preparation of candy, chocolate, and milk-based sweets like burfi, khoa, etc. Hirpara et al. (2020) studied the incorporation of ghee residue in the milk-based sweet "thabdi," which is a locally available fat-rich sweet in brownish color, grainy texture, and ghee flavor (Hirpara et al. 2015). The GR is incorporated into milk and ghee for preparing thabdi. Adding 6% GR in the preparation showed better sensory qualities and enhanced storage life 28 days and 14 days at 20 °C ± 1 °C and 37 °C ± 1 °C (room temperature), respectively. Meanwhile, thabdi prepared without ghee residue had 12 days of shelf life at room temperature. Furthermore, the consistency of ghee residue aids in shortening processing time throughout production.GR is also used in other industries for biodiesel, wastewater treatment, lipase production, and animal feed production.

2.2.4 Organic Acid Production

Organic acids (OAs) are weak acid-property compounds generally produced from non-renewable sources. Due to the unavailability and depletion of these sources, renewable methods of production are being explored. One of these includes the production of organic acid from the bioconversion of agro-waste like citrus, banana, potato peels, dairy wastes, molasses, etc. OAs are generally used for food preservation, food additives, cleaning purposes, and pharmaceutical applications. The production of OA from dairy waste is described in this section. The fermentation of whey produces lactic acid, propionic acid, and acetic acid. Lactic acid has various applications in food (mainly as a preservative), chemical, and pharmaceutical industries. The major lactic acid-producing microorganism family is *Lactobacillaceae*. Cheese whey, the byproduct obtained during cheese making, is the major substrate of acetic acid and lactic acid production. Acetic acid was produced from whey cheese using the membrane-integrated fermentation method (Pal and Nayak 2016). Similar to cheese whey, skimmed whey can also be used as the substrate for the production of OA. Fermentation with *Lactobacillus helveticus* and gram-negative *Propioni bacterium freudenreichii* using skimmed whey as substrate was used for propionic acid production (Ngome et al. 2017). Cheese whey is a remarkable agro-waste that can be used to produce OA, and it is readily available in large quantities from cheese-making industries. However, more research is needed to identify suitable microbial strains and standardize the processing methodologies.

2.2.5 Enzymes

Enzymes are soluble, colloidal, and delicate organic catalysts (Bhatia 2018). The major enzymes in milk are proteinase, lipase, alkaline phosphatase, lysozyme, and lactoperoxidase (Fox et al. 2015). Dairy industry waste is a low-cost resource for the production of enzymes (Ryan and Walsh 2016). Furthermore, the production of enzymes from these substrates may offer a long-term solution to the dairy industry's pollution issues. Some bacteria, yeast, and molds can grow on lactose waste and produce enzymes such as galactosidase, amylase, protease, penicillin acylase, penicillin amidase, polygalacturonase, cutinase, inulinase, and lipase. Another option is a unique strain of *Paracoccus marcusii* that can convert lactose into β-galactosidase. The α-amylase can be produced using dairy waste along with agro-waste with the help of a semi-solid state fermentation. Aspergillus and Mucorcan grow in whey and produce protease enzymes. Manganese peroxidase, a ligninolytic enzyme, can be produced by *Bjerkandera* sp. strain BOS55 from cheese whey (Sar et al. 2021).

2.2.6 Bioactive Compounds

Microbial fermentation of dairy byproducts can be used to produce healthy products that provide functional and nutritional properties. The galacto-oligosaccharides (GOSs) are produced by the conversion of lactose using the endoenzyme β-galactosidase and can serve as a prebiotic by promoting the growth of bifidobacterial for colon health. There are still several challenges related to yield, productivity, and final product quality in the microbial production of GOSs. Microalgae have also been used in the production of GOSs from whey (Sar et al. 2021). Lactobionic acid (LBA) is a bioactive compound obtained by the oxidation of lactose using *Pseudomonas taetrolens* strain. This gluconic acid derivative has chelating, anti-aging, antioxidant, acidulant, and humectant properties and has found use in the pharmaceutical, cosmetic, and food industries. The environmentally friendly manufacture of LBA from inexpensive raw materials is a focus of research due to the high cost of chemical processes and the potential for side reactions to produce numerous hazardous byproducts. LBA can be produced from cheese whey and bovine scotta through microbial fermentation. Microbial species like *Myriococcum thermophilum*, *Acetobacter orientalis*, *Microdochium nivale*, *Zymomonas mobilis*, *Pseudomonas graveolens*, and *Sclerotium rolfsii* have also been identified for the production of LBA (Sar et al. 2021).

2.2.7 *Single-Cell Protein*

Single-cell protein (SCP) refers to the protein-rich, dehydrated cells of bacteria, yeast, algae, and fungi that are used as dietary supplements to combat the growing protein deficiency in the human population. For the production of SCP, various substrates of agro-waste have been used as starter material (Bratosin et al. 2021). According to Suman et al. (2017), dairy waste could also be used as a potential substrate for the growth of microbial culture and biomass production. The different yeasts used for the industrial production of SCP are *Kluyveromyces fragilis*, *Kluyveromyces marxianus*, *Candida pintolopepsii*, *Debaryomyces robertsiae*, and *Saccharomyces cerevisiae*, and common fungal genera mainly used are *Aspergillus*, *Rhizopus*, *Neurospora*, *Monascus*, and *Fusarium*. Different researchers investigated the potential use of different dairy wastes (such as chees, whey, milk, and cream) for the formation of fungal biomass using *Neurospora intermedia* and *Aspergillus oryzae* strains. *N. intermedia* produced 7 g/L of biomass from expired milk, while *A. oryzae* produced 11 g/L (Thunuguntla et al. 2018). According to recent studies, the biomass produced by these fungi contains 30–40% protein and can be used as animal feed. However, bioconversion of materials with high-fat content, such as cream, may be difficult.

2.2.8 *Biofuels*

Byproducts of the dairy industry have a high organic content. The byproduct can be used as a substrate for the generation of methane gas. The anaerobic digestion of dairy effluents with other wastes such as straw, cattle dung, poultry, manure, and livestock wastes is recommended. This enables maintaining the carbon/nitrogen (C/N) ratio, promotes methanogen growth, and increases biogas production. Biohydrogen is a clean energy alternative to the combustion of fossil fuels. Dairy byproducts with high organic act as a suitable substrate for the generation of biohydrogen under anaerobic conditions. Theoretically, a unit of lactose on consumption generates 8 mol of hydrogen. With the current technological developments, a unit of lactose on consumption can generate only 3 mol of hydrogen (Ryan and Walsh 2016). Microbial electro-hydro genesis cells (MECs) produce biohydrogen under controlled conditions from cheese whey. As low pH values can interfere with the function of exoelectrogens, pH must be strictly controlled in the system. Biodiesel consists of the extracted methyl ethyl esters from the plant, microalgal, or oily fungal biomass. Microalgae and fungal species in dairy waste can produce oily biomass and generate bioenergy. Microalgae species like *Chlorella*, *Scenedesmus*, *Chlamydomonas*, *Anabaena*, and *Acutodesmus* and fungal species like *Mortierella isabellina*, *Mucor* sp., *Thamnidium elegans*, and *Fusarium* species are commonly used for biogeneration (Chan et al. 2018; Cherian et al. 2022).

2.3 Sustainable Management of Dairy Industry Waste

Milk is processed into a diverse array of milk products in the industry, such as fresh milk, pasteurized milk, long-life milk or UHT milk, paneer, khoa, curd, yogurt, butter, ghee, milk powder, and cheese. Cheese production is steadily increasing compared to other dairy products as the fast food chain market expands. The world is trending toward demand-driven value-added dairy products such as cheese rather than supply-based bulk commodities such as skim milk powder. For every one kg of cheese produced, 9 liters of whey is generated (Durham and Hourigan 2007). In a typical dairy manufacturing industry, the inputs are the milk and milk products, water, compressed air, refrigerants, acids and alkali, ingredients like sugar, salt, stabilizers, emulsifiers, coloring agents, flavoring agents, and packaging materials. The output generated is milk spills, milk product spills, whey, liquid effluents from cleaning in place, air emissions like combustion gases, milk powder dust, damaged packaging material, and expired stock. The discharge of large quantities of liquid effluent is the main environmental concern in the industry (Durham and Hourigan 2007). The effluent load mainly depends on the nature of the product manufactured, the scale of operation, and the continuity of operation. Sources of wastage in the dairy industry include overfill, spillage, product purge prior to or after CIP, retained products in poorly drained pipelines, defective/returned products, milk stone, sampling for laboratory, whey, cheese/curd fines, milk powder fines, de-sludge/separators, salt whey, brine, wastewater from flushing pipelines, and deposits of heating surfaces (Hale et al. 2003). Casein, fat, whey proteins, sodium, phosphorus, potassium, high levels of organic matter, fatty acids, oil, grease, and nitrogen compounds are present in dairy sludge (Porwal et al. 2015). It also contains dissolved organic compounds like whey proteins, lactose, fat, and minerals. Dairy wastes have high BOD and chemical COD with increased concentrations of organic and inorganic solids. The characteristics of dairy waste are shown in Table 2.4.

Wetland treatment, physicochemical, biological, and biotechnological methods can be incorporated into the treatment of dairy effluents. An overview of the different treatment methods for dairy effluents is presented in Table 2.5. The physicochemical methods mainly focus on the removal of fat and protein colloids from the

Table 2.4 Characteristics of dairy wastewater and effluents

Dairy waste type	Characteristics	Range of values	Reference
Dairy wastewater	BOD	40–48,000 mg L^{-1}	Slavov (2017)
	COD	80–95,000 mg L^{-1}	
	pH	4.7–11	
	Color	White (wastewater)	
		Whey (yellowish green color)	
	Odor	Unpleasant	
	Annual temperature	17–25 °C	
	Industrial temperature (dairy effluents)	22–25 °C (winter)	
		17–18 °C (summer)	
Dairy effluents	BOD	0.8–2.5 kg per metric ton of milk	Jaganmai and Jinka (2017)
	COD	1.5 times BOD	
	Total suspended solids	100–1000 mg/l	
	Nitrogen	About 6% of BOD level	
	Phosphorus	10–100 mg L^{-1}	

effluents, while the biological treatment effectively removes organic matter from dairy waste. Aerobic and anaerobic treatments are used in combination to attain the desired effluent limit. Biotechnological processes are used to derive essential byproducts such as whey-based products, biofuels, bioplastics, organic acids, bioenergy, bioactive peptides, and enzymes (Ahmad et al. 2019). Because there is a growing interest in using dairy waste to reduce pollution, it is critical to investigate the composition and characteristics of dairy waste and to develop long-term solutions for its effective management. Most recent treatment techniques are advancing the conversion of dairy waste into valuable products. However, further research is still needed for the large-scale adoption of these techniques. However, more scientific research is required to develop standardized methodologies for the widespread adoption of these techniques.

2.4 Future Prospects

The high organic and inorganic constitutions of diary waste make them a potential candidate for environmental hazards. By proper treatment, they can be converted into value-added forms. Whey, which is produced in large quantities, is often disposed of along with the effluents. Research needs to be strengthened for the conversion of whey to protein concentrates, lactose, and a range of other substances. Isolation of macro and micronutrients from the whey can also be explored.

Table 2.5 Overview of different treatment methods for dairy effluents

Treatment method	Mode of action	Key features	Reference
Wetland treatment	Aerobic treatment	Economical, eco-friendly, and energy efficient	Slavov (2017)
	Utilizes hydrophyte plants, aggregate materials, and microorganisms	Offers considerable risk to ground and surface water due to the presence of microorganisms and dangerous, volatile substances	
Physicochemical treatment	Destruction/reduction of lipids and protein colloids are achieved through coagulation	Reduces suspended particles and colloidal materials responsible for water turbidity	Carvalho et al. (2013)
	Coagulation is achieved using lactic acid bacteria, organic and inorganic coagulants	Helps in the reduction of COD and BOD contents	Sarkar et al. (2006)
Biological treatment (aerobic/anaerobic)	Utilized microorganisms to reduce complex organic substrate and adsorb heavy metals	Most promising method for separating organic material	Yonar (2018)
	Trickling filters, activated sludge, aerated lagoons, up-flow anaerobic sludge blanket, sequencing batch reactor, biofilm reactor, anaerobic filters, membrane bioreactor, and rotating biological contactor are employed	Disposal of sludge formed during aerobic biodegradation is a major concern	Dąbrowski et al. (2017)
Biotechnological methods	Utilizes microbial fuel cell, aerobic/anaerobic microbial fermentation, anaerobic digestion	The method is used to produce biofuels, biomass, biofertilizers, enzymes, organic acids, etc.	Chokshi et al. (2016), Ozmihci and Kargi (2007) and Wan et al. (2008)

Understanding the nutritional and functional characteristics of dairy byproduct protein isolates and their applicability in various food formulations needs to be explored further. Another major underutilized byproduct is ghee residue, which can be used in the manufacturing of confectionery products. The utilization of wastewater from a dairy plant and its bioconversion to produce SCP, enzymes, organic acids, biofuels, biopolymers, and bioactive components need to be investigated.

2.5 Conclusion

The increased demand for dairy products has increased the amount of waste produced by the industry. Dairy byproducts are high in macro and micronutrients. Proper treatment methods can convert them into enzymes, organic acids, biofuels, biopolymers, and bioactive components. Whey can be used to make protein concentrates, lactose, and a variety of other products. Casein and its derivative can be produced from skim milk. Ghee residue is used in the formulation of confectionery and chocolate products. The chapter summarizes the byproducts generated in the dairy industry and their processing and utilization through value addition. Besides the nutritional attributes, high organic and inorganic matter content in dairy byproducts makes them a potential source of environmental hazards. Therefore, proper treatment is required prior to disposal. Wetland treatment, physicochemical, biological, and biotechnological methods are currently employed for the treatment of dairy effluents. Despite this, further studies are needed to standardize the methodologies and make them cost-effective for large-scale production.

References

Abd El-Salam MH, El-Shibiny S (2017) Preparation, properties, and uses of enzymatic milk protein hydrolysates. Crit Rev Food Sci Nutr 57(6):1119–1132

Ahmad T, Aadil RM, Ahmed H, Rahman U, Soares BC, Souza SL, Pimentel TC, Scudino H, Guimarães JT, Esmerino EA, Freitas MQ (2019) Treatment and utilization of dairy industrial waste: a review. Trends Food Sci Technol 88:361–372

Alane D, Raut N, Kamble DB, Bhotmange M (2017) Studies on preparation and storage stability of whey based mango herbal beverage. Int J Chem Stud 5:237–224

Barukcic I, Božanić R, Tratnik L (2008) Whey-based beverages – a new generation of diary products. Mljekarstvo 58:257–274

Battula SN, Laxmana Naik N, Sharma R, Mann B (2020) Ghee, anhydrous milk fat and butter oil. In: Dairy fat products and functionality. Springer, pp 399–430

Bejarano TE, Sepúlveda-Valencia, Rodriguez-Sandoval J (2022) Use of ultrafiltration technology to concentrate whey proteins after white cheese manufacturing. Rev Fac Nac Agron Medellin 75

Bhatia S (2018) Introduction to pharmaceutical biotechnology, Enzymes, proteins and bioinformatics, vol 2. IOP Publishing

Božanić R, Barukčić I, Lisak K (2014) Possibilities of whey utilisation. Austin J Nutr Food Sci 2(7):7

Bratosin BC, Darjan S, Vodnar DC (2021) Single cell protein: a potential substitute in human and animal nutrition. Sustainability 13(16):9284

Burling H (2003) Whey processing e utilization and products. In: Roginski H, Fuquay JW, Fox PF (eds) Encyclopedia of dairy sciences. Academic, pp 2745–2751

Bylund G (2003) Dairy processing handbook. Tetra Pak Processing Systems

Carr A, Golding M (2016) Functional milk proteins production and utilization: casein-based ingredients. Adv Dairy Chem:35–66

Carvalho F, Prazeres AR, Rivas J (2013) Cheese whey wastewater: caracterization and treatment. Sci Total Environ 445:385–396

Chan LG, Cohen JL, Ozturk G, Hennebelle M, Taha AY, de Moura Bell LN, J. M. (2018) Bioconversion of cheese whey permeate into fungal oil by *Mucor circinelloides*. J Biol Eng 12(1):1–14
Cherian E, Kalavathy G, Joshi TJ, Phoebe MGL, Gurunathan B (2022) Importance of nanocatalyst and its role in biofuel production. In: Biofuels and bioenergy. Elsevier, pp 171–182. https://doi.org/10.1016/B978-0-323-85269-2.00022-8
Chokshi K, Pancha I, Ghosh A, Mishra S (2016) Microalgal biomass generation by phycoremediation of dairy industry wastewater: an integrated approach towards sustainable biofuel production. Bioresour Technol 221:455–460
Crowley P, O'brien C, Slattery H, Chapman D, Arendt E, Stanton C (2002) Functional properties of casein hydrolysates in bakery applications. Eur Food Res Technol 215(2):131–137
Dąbrowski W, Żyłka R, Malinowski P (2017) Evaluation of energy consumption during aerobic sewage sludge treatment in dairy wastewater treatment plant. Environ Res 153:135–139
Durham RJ, Hourigan JA (2007) Waste management and co-product recovery in dairy processing Waste management and co-product recovery in dairy processing. In: Handbook of waste management and co-product recovery in food processing. Woodhead Publishing Limited, Abington, pp 332–387
Elliott J (2013) Whey too much: Greek yogurt's dark side. Modern Farmer
Eswarapragada NM, Reddy PM, Prabhakar K (2010) Quality of low fat pork sausage containing milk-co-precipitate. J Food Sci Technol 47(5):571–573
Foegeding EA, Luck PJ, Vardhanabhuti B (2011) Milk protein products. Whey Protein Products. https://doi.org/10.1016/B978-0-12-374407-4.00350-2
Fox PF, Uniacke-Lowe T, McSweeney PLH, O'Mahony JA (2015) Enzymology of milk and milk products. Dairy chemistry and biochemistry. Springer, pp 377–414
Gawande H, Arora S, Sharma V, Meena GS, Singh AK (2022) Functional characterisation of buffalo milk protein co-precipitate. Int J Dairy Technol 75(4):892–901
Hale N, Bertsch R, Barnett J, Duddleston WL (2003) Sources of wastage in the dairy industry. IDF Bull 382:7–30
Hirpara KB, Patel HG, Prajapati JP (2015) Standardization of rate of sugar addition for the manufacture of Thabdi. J Food Sci Technol 52(2):1152–1157
Hirpara P, Prajapati JP, Mehta BM, Pinto SV (2020) Development of Thabdi milk sweets of Gujarat State, India utilizing Ghee residue as an ingredient. J Appl Nat Sci 12(4):575–581
Illingworth D, Patil GR, Tamime AY (2009) Anhydrous milk fat manufacture and fractionation. Dairy Fats Relat Prod:108–166
Jaganmai G, Jinka R (2017) Production of lipases from dairy industry wastes and its applications. Int J Curr Microbiol Appl Sci 5:67–73
Kolhe AS, Ingale SR, Bhole RV (2009) Effluent of dairy technology. Shodh Samiksha aur Mulyankan 2:459–461
Kumar DD, Mann B, Pothuraju R, Sharma R, Bajaj R (2016) Formulation and characterization of nanoencapsulated curcumin using sodium caseinate and its incorporation in ice cream. Food Funct 7(1):417–424
Kumar CM, Sabikhi L, Singh AK, Raju PN, Kumar R, Sharma R (2019) Effect of incorporation of sodium caseinate, whey protein concentrate and transglutaminase on the properties of depigmented pearl millet based gluten free pasta. LWT 103:19–26
Lappa IK, Papadaki A, Kachrimanidou V, Terpou A, Koulougliotis D, Eriotou E, Kopsahelis N (2019) Cheese whey processing: integrated biorefinery concepts and emerging food applications. Foods 8(8):347
Macej OD, Jovanovic ST (2002) Co-precipitates. Mlekarstvo (Serbia and Montenegro)
Mahboubi A, Ferreira JA, Taherzadeh MJ, Lennartsson PR (2017) Value-added products from dairy waste using edible fungi. Waste Manag 59:518–525
Ngome MT, Alves JGLF, Piccoli RH, de Carmo Domingo E, Pinto SA, Bernal OLM (2017) Inoculum concentration and inoculation time for propionic acid production from whey using

mixed culture of *Lactobacillus* helveticus and *Propionibacterium freudenreichii PS-1*. Acta Sci Technol 39:543–550

O'sullivan MM, Singh H, Munro PA, Mulvihill DM (2002) The effect of cooking and washing temperature during pilot-scale rennet casein manufacture on casein hydration characteristics in disodium orthophosphate solution. Int J Dairy Technol 55(1):18–26

Ozmihci S, Kargi F (2007) Kinetics of batch ethanol fermentation of cheese-whey powder (CWP) solution as function of substrate and yeast concentrations. Bioresour Technol 98(16):2978–2984

Pal P, Nayak J (2016) Development and analysis of a sustainable technology in manufacturing acetic acid and whey protein from waste cheese whey. J Clean Prod 112:59–70

Petridis D, Vlazakis EL, Tzivanos IAK, Derlikis EM, Ritzoulis C (2010) Effects of selected ingredients and fat content on the sensory and mechanical properties of frankfurter-type sausages. J Texture Stud 41(6):880–898

Philippopoulos C, Papadakis M (2008) Current trends in whey processing and utilization in Greece. Int J Dairy Technol 54:14–19. https://doi.org/10.1046/j.1471-0307.2001.00007.x

Porwal HJ, Mane AV, Velhal SG (2015) Biodegradation of dairy effluent by using microbial isolates obtained from activated sludge. Water Resour Ind 9:1–15

Purkiewicz A, Pietrzak-Fiećko R (2021) Antioxidant properties of fruit and vegetable whey beverages and fruit and vegetable mousses. Molecules 26(11):3126

Raikos V, Dassios T (2014) Health-promoting properties of bioactive peptides derived from milk proteins in infant food: a review. Dairy Sci Technol 94(2):91–101

Ranjan R, Chauhan AK, Kumari SS, Dubey RP (2020) Nutritive value of ghee residue incorporated bakery product. Ind J Dairy Sci 73(1):21

Ryan MP, Walsh G (2016) The biotechnological potential of whey. Rev Environ Sci 15(3):479–498

Sar T, Harirchi S, Ramezani M, Bulkan G, Akbas MY, Pandey A, Taherzadeh MJ (2021) Potential utilization of dairy industries by-products and wastes through microbial processes: a critical review. Sci Total Environ 152253

Sarkar B, Chakrabarti PP, Vijaykumar A, Kale V (2006) Wastewater treatment in dairy industries – possibility of reuse. Desalination 195(1–3):141–152

Selvamani J, Radhakrishnan L, Bandeswaran C, Gopi H, Valli C (2017) Estimation of nutritive value of ghee residue procured from western districts of Tamil Nadu, India. Asian J Dairy Res 36(4):283–287

Silviya RM, Bhumika K, Dabhi Parmar SC, Aparnathi KD (2016) Whey and its utilization. Int. Int J Curr Microbiol Appl Sci 5(8):134–155

Slavov AK (2017) General characteristics and treatment possibilities of dairy wastewater – a review. Food Technol Biotechnol 55(1):14

Smithers G (2008) Whey and whey proteins – from 'gutter-to-gold'. Int Dairy J 18:695–704. https://doi.org/10.1016/j.idairyj.2008.03.008

Sojan A, Surendran A, Lukose SJ (2019) Effectiveness in utilization of ghee residue in the production of cookies and biscuit in an industrial level. Int J Sci Res 10:1342–1348

Soloshenko VA, Goncharenko GM, Petukhov VL, Grishina NB, Shishin NI, Kamaldinov EV (2016) Association of polymorphism of κ-casein gene and its relationship with productivity and qualities of a cheese production. Res J Pharm Biol Chem Sci 7(4):3214–3221

Subbulakshmi G, Periwal S, Rani PJ (1990) Studies on shelf life and utilisation of ghee residue. J Food Sci Technol 27(3):165–166

Suman G, Nupur M, Anuradha S, Pradeep B (2017) Characterization of dairy waste and its utilisation as substrate for production of single cell protein. J Biotechnol Biochem 3:73–78

Thunuguntla R, Mahboubi A, Ferreira JA, Taherzadeh MJ (2018) Integration of membrane bioreactors with edible filamentous fungi for valorization of expired milk. Sustainability 10(6):1940

Tsakali E, Petrotos K, D'Allessandro A, Goulas P (2010) A review on whey composition and the methods used for its utilization for food and pharmaceutical products. In: Proceedings of 6th international conference on simulation in food and bio industries, pp 195–201

Varma BB, Narender Raju P (2008) Ghee residue: Processing, properties and utilization. course compendium on "Technological advances in the utilization of dairy by-products". In: Centre of Advanced Studies in Dairy Technology, NDRI, Karnal, pp 176–183

Walstra P, Walstra P, Wouters JTM, Geurts TJ (2005) Dairy science and technology. CRC Press. https://doi.org/10.1201/9781420028010

Wan C, Li Y, Shahbazi A, Xiu S (2008) Succinic acid production from cheese whey using Actinobacillus succinogenes 130 Z. In: Biotechnology for fuels and chemicals: proceedings of the twenty-ninth symposium on biotechnology for fuels and chemicals. Humana Press, pp 111–119

Wani AD, Prasad W, Khamrui K, Jamb S (2022) A review on quality attributes and utilization of ghee residue, an under-utilized dairy by-product. Fut Foods 100131

Warncke M, Kieferle I, Nguyen TM, Kulozik U (2022) Impact of heat treatment, casein/whey protein ratio and protein concentration on rheological properties of milk protein concentrates used for cheese production. J Food Eng 312:110745

Xu X, Kabir A, Barr ML, Schutte AE (2022) Different types of long-term milk consumption and mortality in adults with cardiovascular disease: a population-based study in 7236 Australian adults over 8.4 years. Nutrients 14(3):704

Yonar T, Sivrioğlu Ö, Özengin N (2018) Physico-chemical treatment of dairy industry wastewaters: a review. In: Technological approaches for novel applications in dairy processing, p 179

Part III
Waste Utilization from Cereals

Chapter 3
Effective Utilization of Agricultural Cereal Grains in Value-Added Products: A Global Perspective

Meroda Tesfaye Gari, Belete Tessema Asfaw, Lata Deso Abo, Mani Jayakumar, and Gadisa Kefalew

Abstract For many years, cereal grains from agricultural sources have been the main food sources for human diets and animal feeds. All throughout the food-production chain, more byproducts are now produced in greater amounts. These byproducts are made up mostly of the germ and outer layers of wheat, rice, oat, barley, and corn that are removed through the operations of wet and dry milling, which also include other products obtained through the bread and starch-making processes. Cereal manufacturing byproducts are abundant and cheap sources of phytochemicals, such as carbohydrates, dietary fibers, lipids, proteins, and trace elements and antioxidants, polyphenols, and vitamins, which are essential chemicals for living beings. These phytochemicals also have potential nutraceutical and pharmaceutical applications. Most frequently, the byproducts derived from numerous cereal processes are utilized as cattle feed or as organic fertilizers; otherwise, they are discarded into the environment as waste materials. The world's population

M. T. Gari
College of Biological and Chemical Engineering, Addis Ababa Science and Technology University, Addis Ababa, Ethiopia

B. T. Asfaw
College of Biological and Chemical Engineering, Addis Ababa Science and Technology University, Addis Ababa, Ethiopia

Department of Chemical Engineering, Haramaya University, Haramaya Institute of Technology, Dire Dawa, Ethiopia

L. D. Abo · G. Kefalew
Department of Chemical Engineering, Haramaya University, Haramaya Institute of Technology, Dire Dawa, Ethiopia

M. Jayakumar (✉)
Department of Chemical Engineering, Haramaya University, Haramaya Institute of Technology, Dire Dawa, Ethiopia
e-mail: drjayakumarmani@haramaya.edu.et

E. Cherian, B. Gurunathan (eds.), *Value Added Products From Food Waste*,
https://doi.org/10.1007/978-3-031-48143-7_3

is increasing day by day, as is awareness of environmental sustainability. Industry's top priority is to meet the growing needs of the global market, which include devising innovative and sustainable methods for managing the byproducts derived from cereals while maintaining healthy modern lifestyles and promoting community welfare. This chapter addresses in detail the numerous cereals and their sources, value-added products from cereals, and the advantages and disadvantages of their applications and future prospects.

Keywords Cereals · Human diets · Cattle feed · Utilization · Value-added products

3.1 Introduction

Nowadays, directly through human consumption and indirectly through livestock feed (used in the production of meat), grain crops are among the most widely consumed products. The average person's diet also includes vital minerals and energy from such grain crops. In 2016, 2577.85 million tons of crops were produced, but only 1330.02 million tons of nonwheat and nonrice cereal grains were utilized largely for animal feed or brewing (Papageorgiou and Skendi 2018). Cereals include any of the nine species of plants in the *Gramineae* family, namely corn, barley, oats, rice, millet, sorghum, wheat, rye, and triticale, a rye and wheat hybrid (Skendi et al. 2020). Large-scale cereal farming, including the production of wheat, rice, and maize, adds to the global food supply (Garg et al. 2021). Better-known varieties such as buckwheat, amaranth, and quinoa but also lesser-known varieties such as chia, caihua, pit seed goosefoot, breadnut, celosia, and wattle seeds make up just a few examples of pseudocereals, which are either actual grasses or nongrasses with end uses similar to those of cereals (Skendi et al. 2020).

Cereals have long been important parts of livestock feed and critical components of human diets, and their processing heavily contributes to the value of the food supply chain. According to epidemiological research, eating whole grains is associated with a reduced risk of cardiovascular disease and chronic conditions like cancer, diabetes, and obesity. Cereal grains' high concentrations of dietary fiber, lipids, proteins, vitamins, and tocopherols have been linked to these foods' positive health effects (Galanakis 2022). Micronutrients such as vitamins are crucial for healthy growth, metabolism, reproduction, and overall wellbeing. Vitamins are divided into two groups on the basis of their solubility: fat-soluble vitamins like A, D, E, and K and water-soluble vitamins like B and C (Garg et al. 2021). Additionally, the cereal germ, hull, and bran that are removed from foods during processing contain substantial concentrations of these composites.

Moreover, even though cereals are crucial components in the food-production cycle, cereal processing is also crucial to nonfood industries. Milling, either dry milling or wet milling, is the primary process of the grain-processing sector

(Papageorgiou and Skendi 2018; Souza et al. 2022). A wide range of feedstuffs produced by the following processes are included in byproducts: cleaning, milling, or extracting cereal grains and oilseeds; brewing, distilling, or producing ethanol; and processing vegetables, fruits, and sugars (Ominski et al. 2021).

Wet milling is appropriate for maize; dry milling is typical for wheat; and pearling is appropriate for rice and barley. Biomasses of bran, cob, germ, gluten, and husks make up maize wastes (Souza et al. 2022). The grain endosperm's byproducts, the outer fibrous components and germ, separate during dry milling. Dry milling can also refer to a process that uses abrasion to polish grains like rice, oats, and barley to create byproducts that have high concentrations of bioactive compounds. Through this method, the seed coat, aleurone, and subaleurone layers and the germ are gradually removed. These grain byproducts are rich in bio-functional ingredients like phytoestrogens, phenolic compounds, minerals, lignans, vitamins, and fiber. The other wet-milling techniques are employed mostly to produce bran, germ, and starch, along with coproducts of bran and germ that contain nutrients valuable to the pharmaceutical sector (Papageorgiou and Skendi 2018). According to Véha and Balázs (2012), the bran is the top layer of the pericarp of brown rice. In addition to being used as feed, the components obtained through dry or wet milling can be utilized in a numerous food and nonfood items. Through the processes of malting and brewing, cereals can be fermented and distillated to create a plethora of highly sought-after raw materials for other industries (Papageorgiou and Skendi 2018).

The primary components of cereal grains are carbohydrates, which include starches, soluble sugars, and other carbohydrates. Humans cannot digest dietary fibers, but we can digest complex polysaccharides such as starches, lignin, pectin, hemicellulose, and cellulose (Jayakumar et al. 2023; Gundupalli et al. 2022). Starch makes up 40–70% of grain polysaccharides. Dietary fibers include contents like water-soluble pentosans and glucans and water-insoluble fibers like lignin, cellulose, and certain hemicelluloses. Fibrous materials such as hemicelluloses, proteins, polysaccharides, lipids, and phytochemicals are examples of plant-based substances that can be extracted for use in industry. Cereals are composed of 7–12% proteins and 1.7–5.7% lipids. In some cases, the extraction rates and the predominant constituents including glycol, and phospholipids, depending on the type of grain and flour (Skendi et al. 2020). However, current management systems release environmental waste from the manufacturing of grain, a process that remains unsustainable (Galanakis 2022).

This chapter analyses the uses of grain byproducts by focuses on their compounds, specifically those compounds' key traits and their applications in many fields and industries.

3.2 Methods of Processing Cereal Byproducts

Milling, which can be separated into two classes, namely dry milling and wet milling, is the first process in the cereal industry (Papageorgiou and Skendi 2018; Souza et al. 2022). Both wet-milled grains and dry-milled grains have uses in the food sector, provided that their physical and functional characteristics are properly understood. This study sought to examine how different rice flours' physical and chemical characteristics were affected by the wet- and dry-milling processes (Leewatchararongjaroen and Anuntagool 2016).

3.2.1 *Dry Milling*

One of the earliest techniques employed by the milling sector to provide milled fractions of cereal grains was dry milling. Before grinding grains, they must be cleaned and conditioned. Cleaning is crucial because bulk grains often have contaminants, at levels that vary depending on the cereal. Shriveled grains, grains with discolored germs, sprouting grains, other cereals, damaged grains, and rotted grains and various impurities like husks, seeds, extraneous matter, ergots, and dead insects are among the main grain contaminants. Cleaning removes impurities by using disk separators, sieves, magnets, and aspirators (for removing dust) and uses processes like conditioning (also known as tempering—moistening the kernels) (Galanakis 2022).

Tempering is the technique of moistening the kernels with a regulated addition of water to make the bran tougher and the inner endosperm softer. This procedure seeks to prevent the breakdown of bran, promotes progressive separation during milling, and improves the efficiency of sieving. To ensure thorough hydration, the grains are kept for 3 h in containers designed for this purpose. The type of grain, variety, and starting moisture content can affect the amount of time during which and temperature at which grain kernels need to be soaked. Conditioning, also known as tempering, is carried out in two successive processes for hard kernels such as durum. These requirements are higher for hard wheat because it needs more final moisture and conditioning time than soft wheat does. Some grains, like rye and triticale, are dried to have lower moisture contents because their endosperms are softer than those of durum wheat. Before resting in tempering chambers, corn kernels may need to be moistened up to three times to reach the required final moisture level of 18–27%. The grain's size, shape, hardness, and outer-layer adherence level to the endosperm are all crucial factors in milling. Two procedures make up dry milling: grinding and sifting (Papageorgiou and Skendi 2018). Compared to wet milling, which produces wheat with significantly reduced crystallinity and a significantly lower level of gelatinized enthalpy, dry milling results in the loss of a grain's crystalline structure (Leewatchararongjaroen and Anuntagool 2016).

3.2.2 *Wet Milling*

Wet milling, unlike dry milling, entails grinding the soaked grain and then separating the chemical constituents of the grain, such as fiber, oil, proteins, and starch. The primary goal of wet milling maize is to remove as much undamaged starch as possible. As a result, the main byproduct of wet milling is starch. The byproduct can form as waxy starch, regular starch, or higher-amylose starch, depending on how much amylose is present in the parent source. Maltodextrin and glucose syrups, other modified starches, soups, thickeners, baby meals, confectionery goods, and brewing adjuncts are the principal uses of starch in the food business. The principal purpose of the starch generated during the wet milling of corn is to convert it into sweets and ethanol. The products manufactured from starch are also exploited by the pharmaceutical industry. Additionally, starch serves a variety of nonfood purposes in the textile-processing sector, such as packing, and as a component in the manufacture of adhesives (Papageorgiou and Skendi 2018).

The flour produced via the wet-milling technique has significantly greater carbohydrate levels and reduced ash and protein contents. Additionally, flour that has been wet milled typically has less lipid and more amylose. Furthermore, compared to samples of dry-milled rice flour, samples of wet-milled rice flour on average have smaller granules. Wet-milled samples have larger swelling activity and significantly poorer solubility than flour samples dry milled at 90 °C (Leewatchararongjaroen and Anuntagool 2016).

3.3 Processing Cereal Byproducts

The principal byproducts of the wheat-milling industry, wheat germ and bran, have been found to be great sources of antioxidants, dietary fiber, proteins, phytonutrients, trace minerals, and other micronutrients (Emire 2015). The major components of dry-milling byproducts are the cereal grain's bran, endospermic tissue (aleurone layer), and germ, which are discarded during the milling process. The lignans in these grain fractions, such as phytoestrogens, phenolic compounds, fiber, minerals, vitamins, and bio-functional molecules, are abundant. The ultimate nutritional content of the byproducts significantly varies depending on the method of milling used (dry milling or wet milling). Aside from being used as feed, the grain fractions produced during dry or wet milling have a variety of uses in both food and nonfood products. Various byproducts from the processes of malting, brewing, and distilling cereals are highly sought-after raw materials for various industries (Papageorgiou and Skendi 2018).

3.3.1 Corn Byproducts

During dry milling, maize is ground between stones in a hand-driven plate mill, a hammer mill, or a powered plate mill on a larger scale. Before milling with a manual plate mill, the grain is soaked and allowed to gently ferment to increase its flavor (Ballester-Sánchez et al. 2020). The residual starch left over after the milling process and the maize hulls are byproducts of the corn wet-milling business. If the value of the corn fiber as an animal feed product is not seriously compromised, the conversion of the starch and lignocellulosic components of corn fiber into ethanol would boost the output of ethanol from a maize wet-milling process by 13% (Arvanitoyannis and Tserkezou 2008). Wet milling is a method used with maize to separate the starch from other elements, like protein, germ, fiber, and tiny foreign particles. Separation can be improved by steeping or soaking the grain in warm water that contains sulfur dioxide. Following the starch's separation from the other ingredients, several procedures are needed to extract pure starch, such as screening, washing, and ultimately drying (Liu et al. 2023).

3.3.2 Rice Byproducts

Animals are given the majority of rice byproducts, including rice husks and rice bran. Some breweries also almost entirely make their beer from rice: a mixture of broken rice, rice bran, and rice germ. Because of their higher values of vitamins, minerals, and fiber and because of their phenolic base compounds, which can help decrease cholesterol and exert antiatherogenic activity, rice byproducts have recently attracted more attention as functional foods (Mohd Esa and Ling 2016).

Before being consumed by people, rice grains must go through a variety of processing steps. Harvesting is the first step in the operations used to produce graded and polished white rice. For every kilogram of harvested paddy, the ratio of leftover product to usable product ranged from 0.41 to 3.96. The steps in the rice-milling process include cleaning, hulling, and post-hulling processing, which itself includes whitening, polishing, and grading. The type of rice and the pace of milling affect the percentages of rice byproducts: 20% husk, 8–12% bran (depending on the level of milling), and 68–72% milled rice or white rice (depending on the type) (Mohd Esa and Ling 2016). Figure 3.1 depicts the many parts of a rice grain.

3.3.3 Wheat Byproducts

When wheat is milled, significant volumes of bran and germ are created as byproducts. The endosperm is reduced to tiny particles during milling, while the bran and germ are eliminated. With recognized antioxidant capabilities concentrated mostly in the bran, wheat is a key agricultural and dietary commodity around the world. Wheat germ, a byproduct of the flour-milling sector, is one of the best and most affordable sources of critical nutrients, including calories, proteins, dietary fiber, and various beneficial microcompositions (Jiang and Niu 2011). The average concentration of waste produced by milling wheat is between 25% and 40%, and this waste is used to make bioethanol, animal feed, and succinic acid, which is blended with other ingredients to maximize the nutritional value of baked goods and to be

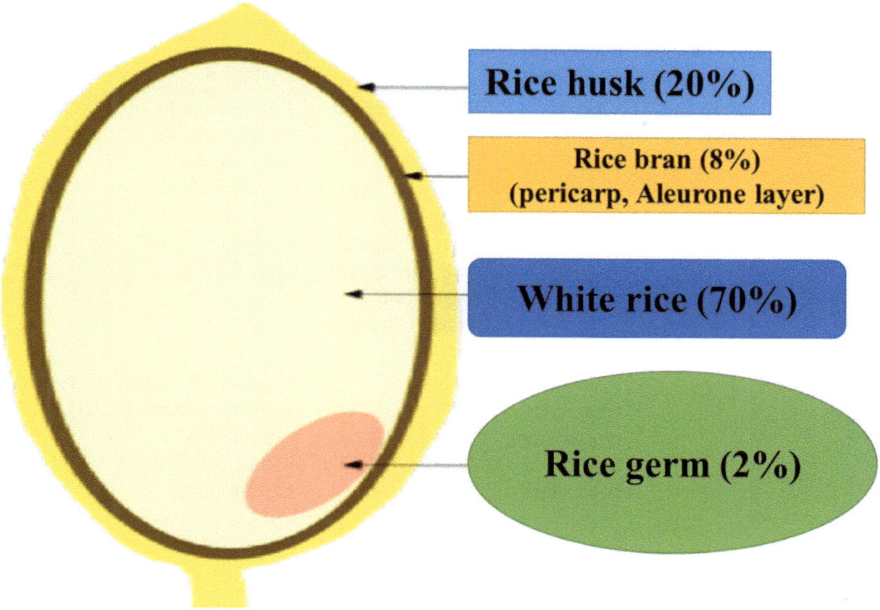

Fig. 3.1 Components of rice grain

used in pharmaceuticals, cosmetics, and meat substitutes, among many other things (Dey et al. 2021).

Different methods are used to extract the oil from wheat germ, including mechanical pressing, which extracts around 50% of the oil, and the usual method of organic solvent extraction recovers about 90% of the oil. In other words, the disadvantages can be offset by oil produced through critical extraction methods (Yılmaz Tuncel 2023). In actuality, the oils don't need to be refined, because they don't contain any solvents, and the extraction yields are comparable to those that typically need to be refined. The first two processes in extracting wheat germ oil have been shown to reduce free fatty acid and tocopherol levels (Parczewska-Plesnar et al. 2016). The quantities of natural antioxidants present in wheat grains are substantial and concentrated mostly on the outer surface. Wheat is a significant agricultural product and a staple dietary item, and it has many healthy nutritional contents. Wheat and wheat-based food ingredients rich in naturally occurring antioxidants can be optimal for the development of functional foods designed to improve the health of millions of consumers (Kosík et al. 2014).

Milling cracks wheat rather than crushes it by spacing milling stones more apart. At the same time, the rotational speed of the millstones is slowed. Gradually reducing the heat from friction and grinding and then repeatedly grinding and bolting the wheat separates the bran from the white flour. The amount of flour extracted from a given amount of wheat was similar to current extraction rates: 72–75% flour and 26–28% mill feed (Kanojia et al. 2018). Table 3.1 presents the functional components from processing cereal byproducts.

Table 3.1 Functional components from processing cereal byproducts

Cereal	Byproducts	Functional compounds	Health benefits	Examples	Reference
Rice	Bran and husk	Vitamins, proteins, dietary fiber, and oils	Proteins exhibit hypoallergenic qualities, vitamins have antioxidant characteristics, and fiber lowers the risk of cardiovascular disease.	179–389 mg/kg of tocopherols and tocotrienols (compounds made up of vitamin E) are found in rice bran.	Galanakis (2022)
Sorghumand millet	Bran	Phenolic compounds, phytosterols, and policosanols	Phenolic compounds possess policosanols, and phytosterols reduce cholesterol levels and antioxidant properties.	Policosanol content and hexane extracts significantly reduced cholesterol absorption, by up to 17%.	Carr et al. (2005)
Corn	Corncob	Oils and insoluble dietary fiber	Refining technology can be modified to maintain the natural phytosterols and derivatives in order to produce more "good" oil, considered good because of the recently acknowledged health advantages of phytosterols. The refined phytosterol-enriched corn germ oil made from the listed crude oils might be sold at a higher price to increase the incomes of corn refiners and ethanol producers.	Corn kernels and fiber contain 21.48–24.30 mg/100 g, constituting 13.36–11.90%.	Singh et al. (2001)

(continued)

Table 3.1 (continued)

Cereal	Byproducts	Functional compounds	Health benefits	Examples	Reference
Wheat	Bran and germ	Fiber and arabinoxylans	Bran fiber helps the intestinal transit move more quickly and increases fecal bulk. Arabinoxylans reduce the blood's glucose level.	The prebiotic effects of arabinoxylans on obesity and other metabolic disorders, as well as their capacity to reduce blood-cholesterol level and the post-prandial (PP) glycemic response, are credited with providing health benefits.	Onipe et al. (2015); Stevenson et al. (2012)
Oat	Bran, oat biomass	Soluble dietary fiber and β-glucans	Dietary fiber increases fecal volume, and studies have shown that glucan lowers blood cholesterol.	Oat bran has 16.0% dietary fiber level per dry matter and 5.5% glucans per dry matter.	Galanakis (2022)
Barley	Spent grains	β-Glucans and dietary fiber	Dietary fiber lowers blood sugar spikes after meals, and glucan lowers blood cholesterol.	At safe concentrations, phytosterols in barley oils can considerably reduce low-density lipoprotein cholesterol.	Moreau et al. (2007)
Rye	Bran	Condensed tannins, lignans, stilbenes, coumarins, and flavonoids	The addition of grain flour and bran, as sources of numerous beneficial bioactive components, to food products enhances the health-promoting properties of those goods.	The bran in rye gain been shown to make up between 65 and 300 mg per 100 g of the grain.	Capanoglu et al. (2022)

3.4 Applications of Cereal Byproducts

As discussed earlier, many high-value chemicals can be found in the primary and secondary byproducts of processing cereals, and they can be turned into a variety of commodities for a variety of industries. Wheat-processing byproducts are rich in nutrients and may be used in place of conventional carbon sources in medium-size formulations for microbial and industrial enzyme production (Ravindran and Jaiswal 2016). They might be used to produce, for example, single-cell proteins and microbial enzymes cheaply and at high yields (Naibaho et al. 2022).

One of cereal bran's most popular uses is to increase the nutritious fiber content of baked foods. Depending on the use, the intended amount of fiber, the appropriate glycemic index, and the required sensory qualities, different amounts of cereal bran may be used in bread. Because of the existence of phenolic compounds like pinoresinol and syringic acid, the inclusion of bran in baked goods may have negative effects on their quality and sensory characteristics, such as by altering porosity and elasticity, intensifying bitterness, and reducing nutrient absorption because of an increase in phytic acid content (Heiniö et al. 2016). Similarly, wheat bran's high oil content renders it susceptible to oxidation, which gives baked goods a disagreeable flavor. Consequently, the oil from the bran needs to be removed before including the bran.

For instance, adding wheat bran to dough at a high concentration of 15–20% leads to decreased crumb volume and textural quality, increased water immobilization, and increased water immobilization. Wheat bran has decreased stability, dough development time, extensibility, and viscosity. When wheat bran is added to dough, the gluten network is altered, becoming weaker and thinner. Amylase and wheat bran work best together to maximize the bioavailability of the minerals and reduce the detrimental effects of bran on the dough's functional characteristics (Zhuang et al. 2022).

Extruded snacks have also been created by using byproducts from the production of cereal. For instance, researchers created extruded snacks with qualities that were more desirable than those produced solely from rice flour, by combining brewers' spent grains (BSGs) with up to 30% of the dough (Nascimento et al. 2017). Contrarily, despite having comparable sensory characteristics and textures, adding wheat bran to extruded food products has problems. For instance, the hardness and density of the combination increased as the amount of bran increased, but the crispness and expansion volume decreased. Therefore, it's crucial to maximize the amount of wheat bran applied (Vandana Mishra 2013).

Using cereal bran to augment pasta had also been suggested. However, because of the higher cooking temperatures, increased cooking losses, and decreased water absorption, their integration is known to have a negative impact on the way pasta cooks (Laureati et al. 2016). However, to create gluten-free spaghetti, maize flour and oat bran rich (22%) in glucans have been used. Padalino et al. discovered that adding the mentioned hydrocolloids made the pasta more elastic and stiff, which reduced its bulkiness and adhesiveness (Padalino et al. 2011).

The manufacture of seasonings and tastes as well as the fortification of cookies have been suggested as potential uses for additional cereal manufacturing byproducts, such as rice-distilling lees (Anisha et al. 2023). To increase the nutritional value of cookies, several processing methods have also been applied, including stabilization in the oven, microwave, and autoclave. The results showed that autoclave treatment caused the greatest loss of phytic acid in the cookies, while adding 30% microwave-treated bran resulted in cookies with a lot of minerals (Galanakis 2022).

3.5 Value-Added Products from Cereals

The possibilities of using the wastes from cereals as some of the basic components of value-added products are covered in the preceding paragraphs. Waste cereals are a versatile refuse material that works well in place of other wastes like rice and other starchy grains (Lorente et al. 2023). These products have either been proven effective or are currently being used to increase finger millet intake. However, few scientific studies have been conducted on how to prepare them and meaningfully popularize them (Verma and Patel 2013).

Industrial cereal waste can provide value-added goods, including xylogalacturonans, arabinoxylans, organic acids, polysaccharides, industrially valuable enzymes, and more. Table 3.2 details the value-added products generated from cereal grains and shows the pros and cons of the extraction process.

3.6 Future Prospects and Challenges

3.6.1 Prospects

By 2050, a projected 9.7 billion people will live on the planet. To attain food security, food production must be increased and losses along the agri-food chain decreased. To lessen the effects of climate change and global warming, green materials and technologies must replace nonrenewable resources. Over time, industry experts and scientists have collaborated to investigate grains to identify possible applications for grain byproducts. Additionally, health-promoting hemicelluloses such AX and -β-glucan, which have been found in cereal brans specifically, should be incorporated into tailored cosmetics, foods, feed, and medicines. Aside from using grain byproducts as feed and biomass, no method has been formulated to use them to their full potential. This is because no new pilot- or industrial-scale fractionation technique has been more affordable or more ecofriendly.

The molar mass (MM) of AX and glucan may also be reduced via various processing techniques, as discussed in this review. The functioning of these hemicelluloses and their potential health benefits may differ depending on their molar mass.

Table 3.2 Value-added products from cereals: pros and cons

Products	Methods and conditions	Pros	Cons	Reference
Pea hulls (xylogalacturonans)	Acidic hydrolysis of HCl (0.1 mol/l, 80 °C, 24 h)	Hemicellulose and lignin reduction. Higher solubility of hemicellulose. As a result of its effectiveness, dilute acid pretreatment is widely utilized. Sugar recovery for xylose and glucose is high. Increase accessibility of cellulose for enzymatic saccharification.	Processing concentrated acid is destructive and risky. Specialized nonmetallic construction is in demand. At low pH, inhibitors can form. Salt disposal, neutralization, and sugar-content losses.	Le Goff et al. (2001)
Wheat bran (arabinoxylans)	Hydrothermal treatment (180 °C, 3 min and 5 pH)	Products produced with alkaline extractions were somewhat less brown. The enzymatic reaction produced lighter yellowish powders.	High temperatures or an alkaline pH causes compounds that degrade the development of sugar, producing brownish/ reddish tones. Tiny quantities of lignin and phenolics may also be extracted, which darken the final products.	Aguedo et al. (2014)
Rye bran (arabinoxylan)	Extraction solvent and enzymes	Because of their interactions with pentose sugars, some divalent bases, including barium hydroxide, have a stronger selectivity toward arabinoxylans AXs. Enzymes (xylanases) have high catalytic efficiency, good substrate selectivity, and the capacity to hydrolyze xylose linkages near the side-chain residues.	Arabinoxylans (AXs) may have their functional characteristics diminished via the degradation of their functional groups, caused by alkaline solvents.	Bender et al. (2017)

(continued)

Table 3.2 (continued)

Products	Methods and conditions	Pros	Cons	Reference
Wheat bran (arabinoxylan)	Twin-screw extrusion	This is an excellent technique for extracting hemicelluloses from wheat bran by using an alkaline solution. It requires less reagent and water.	Compared to batch extraction, this method provides lower extraction yields and selectivity. Complex solutions containing the biopolymers are created via twin-screw extrusion. The price and security issues associated with alcohol recycling impede the commercialization of manufacturing xylans.	Jacquemin et al. (2015)
Brewers' spent grains (total phenolic compounds)	Solid–liquid extraction (80% ethanol water, 60% ethanol water, 2 min at 1900 rpm)	Commonly used and the easiest method.	Can cause impurities. Can introduce analytical errors.	Bonifácio-Lopes et al. (2020)
Flax shives, wheat, and corn bran (ferulic acid)	Alkaline hydrolysis (0.5 M of NaOH, 4 h at 50 °C, neutralized with 6 M of HCl)	Major removal of lignin and some removal of hemicellulose. Reduces polymerization level and crystallinity.	Minimal digestibility for softwoods. A large amount of water is required for washing.	Buranov and Mazza (2009)
Brewers' spent grains (ferulic and *p*-Coumaric acid)	Alkaline hydrolysis (2% NaOH concentration, at 120 °C for 90 min)	Easily dissolves lignin. Allowing for the complete utilization of the lignocellulose with low environmental impact.	A large amount of water is needed for the washing process. High chemical recovery cost. Long pretreatment residence time.	Mussatto et al. (2007)

β-glucans, with their high molar mass, can lower cholesterol and act as glycemic-control agents, whereas lower-MM β-glucan is an anticancer agent (Choromanska et al. 2015; Schmidt 2022). Future investigations should concentrate on a variety of techniques to economically and sustainably fractionate pure AX and β-glucan while closely monitoring any changes in their respective physicochemical properties. If scaling up the separation of pure cereal byproduct hemicelluloses is not sustainable, economically viable, or fast enough for development, the use of impure fractions should be prioritized.

Cereal side streams can significantly reduce industrial expenses while enhancing the circular economy. They are made up of hemicellulose, lignin, and cellulose components. Cereal side streams, particularly in brans, are good sources of minerals, vitamins, saponins, proteins, AXs, and β-glucans (Patel 2019). Impure cereal byproducts can be employed as food additives and in bio-refineries. The uses of the impure fractions of these hemicelluloses in diverse fields should be covered in more detail in future research. Valoppi et al. (2021) demonstrated that local soluble oat bran fractions (composed of mixtures of 5% AXs, 11% proteins, and 78% glucans) and native insoluble oat bran fractions (composed of equal fractions of AXs, β-glucans, and proteins, with a trace % of cellulose and fat mixtures) can be used in combination to create stable emulsions and suspensions over time. This makes it possible to use native oat side-stream fractions for packaging, food, drinks, medicines, and cosmetics, among other uses. Additionally, scientists have found phenolic acids in oat bran in its natural state (Verma et al. 2009). These phenolic compounds, especially diferulic acid and ferulic acid, have been linked to in vivo and in vitro antihypertensive, anti-inflammatory, and antioxidant effects (Bautista-Expósito et al. 2020). Lie-Piang et al. (2021) suggested using the entire grain of wheat, meaning the husk, bran, and straw. These authors claim that in addition to maximizing resource use and promoting a circular economy, this idea has also been demonstrated to reduce the likelihood of cancer in humans, future global warming, water use, and issues with the supply of fossil fuels. The mild fractionation of impure traces of legumes, like peas and lupins, has been shown in studies to offer promising thickening and emulsifying capabilities. This shows that outstanding physicochemical properties, in addition to purity, are additional factors that determine whether a substance is sustainable (Lie-Piang et al. 2021).

3.6.2 Challenges

The development of technologies; designing efficient, ecofriendly, and sustainable extraction methods; the characteristics and compositions of cereal grains; and weather conditions are just a few of the challenges that had to be overcome in this study to effectively convert agricultural cereal grains into value-added end products. Additionally, the transformation of cereal grains into goods with value-added products has indirect or direct impacts on the human food chain. Given that humans consume the majority of cereal grains, conflict might ensue between people and businesses, particularly in developing and underdeveloped nations. Moreover, it can be difficult to turn cereal grains into goods with added value, especially at the industrial level, when steam-explosion treatment is utilized as an extraction process. However, steam-explosion treatment does not work well in isolating polymers, because it severely degrades the polymers, leading to low recovery yields. Monomeric sugar is desirable for subsequent alcoholic fermentation or xylose transformation, and it is effective at recovering monomeric sugar, which is desirable for those processes as well.

3.7 Summary

Cereal processing makes up a significant portion of the food-production chain. Malting and wet or dry milling are the two main processes used in the cereal industry. The byproducts of the cereal business are used as feed, fuel, or rubbish, despite their potential to be utilized for the extraction of bioactive components or the creation of functional foods. These days, using cereal byproducts as food additives requires that they adhere to the standards set for the safety and quality of foods intended for consumption by humans. To enable the utilization of former foodstuffs and to add value to other products by using cereal byproducts from the food chain without compromising food safety and feed safety, significant improvements need to be made in waste, food, and feed legislation. Additionally, new techniques are needed to recover food waste rather than dispose of it. Conventional techniques only partially utilize the waste generated by the cereal business, where animal feed is the primary use.

References

Aguedo M, Fougnies C, Dermience M, Richel A (2014) Extraction by three processes of arabinoxylans from wheat bran and characterization of the fractions obtained. Carbohydr Polym 105(1):317–324. https://doi.org/10.1016/j.carbpol.2014.01.096

Anisha A, Kaushik D, Kumar M, Kumar A, Esatbeyoglu T, Proestos C, Khan MR, Elobeid T, Kaur J, Oz F (2023) Volarization of brewer's spent grain for noodles preparation and its potential assessment against obesity. Int J Food Sci Technol 58:3154

Arvanitoyannis IS, Tserkezou P (2008) Original article Corn and rice waste: a comparative and critical presentation of methods and current and potential uses of treated waste. Int J Food Sci Technol 43:958–988. https://doi.org/10.1111/j.1365-2621.2007.01545.x

Ballester-Sánchez J, Fernández-Espinar MT, Haros CM (2020) Isolation of red quinoa fibre by wet and dry milling and application as a potential functional bakery ingredient. Food Hydrocoll 101:105513. https://doi.org/10.1016/j.foodhyd.2019.105513

Bautista-Expósito S, Tomé-Sánchez I, Martín-Diana AB, Frias J, Peñas E, Rico D, Casas MJG, Martínez-Villaluenga C (2020) Enzyme selection and hydrolysis under optimal conditions improved phenolic acid solubility, and antioxidant and anti-inflammatory activities of wheat bran. Antioxidants 9(10):1–22. https://doi.org/10.3390/antiox9100984

Bender D, Nemeth R, Wimmer M, Götschhofer S, Biolchi M, Török K, Tömösközi S, D'Amico S, Schoenlechner R (2017) Optimization of arabinoxylan isolation from rye bran by adapting extraction solvent and use of enzymes. J Food Sci 82(11):2562–2568. https://doi.org/10.1111/1750-3841.13920

Bonifácio-Lopes T, Vilas Boas AA, Coscueta ER, Costa EM, Silva S, Campos D, Teixeira JA, Pintado M (2020) Bioactive extracts from brewer's spent grain. Food Funct 11(10):8963–8977. https://doi.org/10.1039/d0fo01426e

Buranov AU, Mazza G (2009) Extraction and purification of ferulic acid from flax shives, wheat and corn bran by alkaline hydrolysis and pressurised solvents. Food Chem 115(4):1542–1548. https://doi.org/10.1016/j.foodchem.2009.01.059

Capanoglu E, Nemli E, Tomas-Barberan F (2022) Novel approaches in the valorization of agricultural wastes and their applications. J Agric Food Chem 70:6787. https://doi.org/10.1021/acs.jafc.1c07104

Carr TP, Weller CL, Schlegel VL, Cuppett SL, Guderian DM, Johnson KR (2005) Grain sorghum lipid extract reduces cholesterol absorption and plasma non-HDL cholesterol concentration in hamsters. J Nutr 135(9):2236–2240

Choromanska A, Kulbacka J, Rembialkowska N, Pilat J, Oledzki R, Harasym J, Saczko J (2015) Anticancer properties of low molecular weight oat beta-glucan – an in vitro study. Int J Biol Macromol 80:23–28. https://doi.org/10.1016/j.ijbiomac.2015.05.035

Dey D, Richter JK, Ek P, Gu BJ, Ganjyal GM (2021) Utilization of food processing by-products in extrusion processing: a review. Front Sustain Food Syst 4:1–18. https://doi.org/10.3389/fsufs.2020.603751

Emire SA (2015) Development of value added products from byproducts of Ethiopian wheat milling industries. J Food Process Technol 6:474. https://doi.org/10.4172/2157-7110.1000474

Galanakis CM (2022) Sustainable applications for the valorization of cereal processing by-products. Food Secur 11(2):1–15. https://doi.org/10.3390/foods11020241

Garg M, Sharma A, Vats S, Tiwari V, Kumari A, Mishra V, Krishania M (2021) Vitamins in cereals: a critical review of content, health effects, processing losses, bioaccessibility, fortification, and biofortification strategies for their improvement. Front Nutr 8:1–15. https://doi.org/10.3389/fnut.2021.586815

Gundupalli MP, Gundupalli SP, Thomas ASS, Jayakumar M, Bhattacharyya D, Gurunathan B (2022) Hydrothermal liquefaction of lignocellulosic biomass for production of biooil and by-products: current state of the art and challenges. In: Biofuels and bioenergy. Elsevier, Amsterdam, pp 61–84

Heiniö RL, Noort MWJ, Katina K, Alam SA, Sozer N, de Kock HL, Hersleth M, Poutanen K (2016) Sensory characteristics of wholegrain and bran-rich cereal foods – a review. Trends Food Sci Technol 47:25–38. https://doi.org/10.1016/j.tifs.2015.11.002

Jacquemin L, Mogni A, Zeitoun R, Guinot C, Sablayrolles C, Saulnier L, Pontalier PY (2015) Performance evaluation of a semi-industrial production process of arabinoxylans from wheat bran. Process Biochem 50(4):605–613. https://doi.org/10.1016/j.procbio.2015.01.015

Jayakumar M, Gindaba GT, Gebeyehu KB, Periyasamy S, Jabesa A, Baskar G, John BI, Pugazhendhi A (2023) Bioethanol production from agricultural residues as lignocellulosic biomass feedstock's waste valorization approach: a comprehensive review. Sci Total Environ 879:163158

Jiang ST, Niu LY (2011) Optimization and evaluation of wheat germ oil extracted by supercritical CO_2. Grasas Aceites 62(2):181–189. https://doi.org/10.3989/gya.078710

Kanojia V, Kushwaha NL, Reshi M, Rouf A (2018) Products and byproducts of wheat milling process. Int J Chem Stud 6(4):990–993

Kosík T, Lacko-Bartošová M, Kobida Ľ (2014) Free phenol content and antioxidant activity of winter wheat in sustainable farming systems. J Microbiol Biotechnol Food Sci 3:247–249

Laureati M, Conte A, Padalino L, Del Nobile MA, Pagliarini E (2016) Effect of fiber information on consumer's expectation and liking of wheat bran enriched pasta. J Sens Stud 31(4):348–359. https://doi.org/10.1111/joss.12218

Le Goff A, Renard CMGC, Bonnin E, Thibault JF (2001) Extraction, purification and chemical characterisation of xylogalacturonans from pea hulls. Carbohydr Polym 45(4):325–334. https://doi.org/10.1016/S0144-8617(00)00271-X

Leewatchararongjaroen J, Anuntagool J (2016) Effects of dry-milling and wet-milling on chemical, physical and gelatinization properties of rice flour. Rice Sci 23(5):274–281. https://doi.org/10.1016/j.rsci.2016.08.005

Lie-Piang A, Braconi N, Boom RM, van der Padt A (2021) Less refined ingredients have lower environmental impact – a life cycle assessment of protein-rich ingredients from oil- and starch-bearing crops. J Clean Prod 292:126046. https://doi.org/10.1016/j.jclepro.2021.126046

Liu Z, de Souza TSP, Holland B, Dunshea F, Barrow C, Suleria HAR (2023) Valorization of food waste to produce value-added products based on its bioactive compounds. Processes 11(3):840. https://doi.org/10.3390/pr11030840

Lorente D, Serna SD, Betoret E, Betoret N (2023) Opportunities for the valorization of waste generated by the plant-based milk substitutes industry. In: Advanced technologies in wastewater treatment. Elsevier, Amsterdam, pp 25–66

Mohd Esa N, Ling TB (2016) By-products of rice processing: an overview of health benefits and applications. J Rice Res 4(1):107. https://doi.org/10.4172/jrr.1000107

Moreau RA, Flores RA, Hicks KB (2007) Composition of functional lipids in hulled and hulless barley in fractions obtained by scarification and in barley oil. Cereal Chem 84(1):1–5. https://doi.org/10.1094/CCHEM-84-1-0001

Mussatto SI, Dragone G, Roberto IC (2007) Ferulic and p-coumaric acids extraction by alkaline hydrolysis of brewer's spent grain. Ind Crop Prod 25(2):231–237. https://doi.org/10.1016/j.indcrop.2006.11.001

Naibaho J, Butula N, Jonuzi E, Korzeniowska M, Laaksonen O, Föste M, Kütt ML, Yang B (2022) Potential of brewers' spent grain in yogurt fermentation and evaluation of its impact in rheological behaviour, consistency, microstructural properties and acidity profile during the refrigerated storage. Food Hydrocoll 125:107412. https://doi.org/10.1016/j.foodhyd.2021.107412

Nascimento TA, Calado V, Carvalho CWP (2017) Effect of brewer's spent grain and temperature on physical properties of expanded extrudates from rice. LWT Food Sci Technol 79:145–151. https://doi.org/10.1016/j.lwt.2017.01.035

Ominski K, Mcallister T, Stanford K, Mengistu G, Kebebe EG, Omonijo F, Cordeiro M, Legesse G, Wittenberg K (2021) Utilization of by-products and food waste in livestock production systems: a Canadian perspective. Anim Front 11(2):55. https://doi.org/10.1093/af/vfab004

Onipe OO, Jideani AIO, Beswa D (2015) Composition and functionality of wheat bran and its application in some cereal food products. Int J Food Sci Technol 50(12):2509–2518. https://doi.org/10.1111/ijfs.12935

Padalino L, Mastromatteo M, Sepielli G, Del Nobile MA (2011) Formulation optimization of gluten-free functional spaghetti based on maize flour and oat bran enriched in β-glucans. Materials 4(12):2119–2135. https://doi.org/10.3390/ma4122119

Papageorgiou M, Skendi A (2018) Introduction to cereal processing and by-products. In: Sustainable recovery and reutilization of cereal processing by-products. Elsevier/Woodhead Publishing, Duxford, pp 1–25. https://doi.org/10.1016/B978-0-08-102162-0.00001-0

Parczewska-Plesnar B, Brzozowski R, Gwardiak H, Bialecka-Florjanczyk E, Bujnowski Z (2016) Wheat germ oil extracted by supercritical carbon dioxide with ethanol: fatty acid composition. Grasas Aceites 67(3):e144. https://doi.org/10.3989/gya.1017153

Patel S (2019) Cereal bran: the next super food with significant antioxidant and anticancer potential. Mediterr J Nutr Metab 5(2):91–104. https://doi.org/10.3233/s12349-012-0091-1

Ravindran R, Jaiswal AK (2016) Microbial enzyme production using lignocellulosic food industry wastes as feedstock: a review. Bioengineering 3(4):30. https://doi.org/10.3390/bioengineering3040030

Schmidt M (2022) Cereal beta-glucans: an underutilized health endorsing food ingredient. Crit Rev Food Sci Nutr 62(12):3281–3300. https://doi.org/10.1080/10408398.2020.1864619

Singh V, Moreau RA, Hicks KB, Eckhoff SR (2001) Effect of alternative milling techniques on the yield and composition of corn germ oil and corn fiber oil. Cereal Chem 78(1):46–49. https://doi.org/10.1094/CCHEM.2001.78.1.46

Skendi A, Zinoviadou KG, Papageorgiou M, Rocha JM (2020) Advances on the valorisation and functionalization of by-products and wastes from cereal-based processing industry. Foods 9:1243

Souza D, Vilas-boas IT, Leite-da-silva JM, Teixeira-costa BE, Veiga-junior VF (2022) Polysaccharides in agro-industrial biomass residues. Polysaccharides 3:95–120

Stevenson L, Phillips F, O'sullivan K, Walton J (2012) Wheat bran: its composition and benefits to health, a European perspective. Int J Food Sci Nutr 63(8):1001–1013. https://doi.org/10.3109/09637486.2012.687366

Valoppi F, Wang YJ, Alt G, Peltonen LJ, Mikkonen KS (2021) Valorization of native soluble and insoluble oat side streams for stable suspensions and emulsions. Food Bioprocess Technol 14(4):751–764. https://doi.org/10.1007/s11947-021-02602-5

Vandana Mishra (2013) Extruded snacks foods, p 1–20

Véha P, Balázs A-SP (2012) Cereal processing and cereal based foods

Verma V, Patel S (2013) Value added products from nutri-cereals: finger millet (Eleusine coracana). Emir J Food Agric 25:169–176

Verma B, Hucl P, Chibbar RN (2009) Phenolic acid composition and antioxidant capacity of acid and alkali hydrolysed wheat bran fractions. Food Chem 116(4):947–954. https://doi.org/10.1016/j.foodchem.2009.03.060

Yılmaz Tuncel N (2023) Stabilization of rice bran: a review. Food Secur 12(9):1924

Zhuang K, Sun Z, Huang Y, Lyu Q, Zhang W, Chen X, Wang G, Ding W, Wang Y (2022) Influence of different pretreatments on the quality of wheat bran-germ powder, reconstituted whole wheat flour and Chinese steamed bread. LWT 161:113357. https://doi.org/10.1016/j.lwt.2022.113357

Part IV
Waste Utilization from Fruits and Vegetables

Chapter 4
Fruit Peel–Based Edible Coatings/Films

Veerapandi Loganathan, Nivetha Thangaraj, and J. Suresh Kumar

Abstract Convenience and lifestyle changes have spurred food-processing industries to produce consumable products such as fruit juices, jams, jellies, marmalades, etc. The byproducts of these fruit-processing industries are enormous, mostly the underused parts of the fruit, which are discarded as waste. Such waste includes fruit peels, seeds, rinds, pomaces, and kernels obtained during the various processing stages. These byproducts, when examined, consist of bioactive components such as dietary fibre, antioxidants, polyphenols, flavonoids, anthocyanins, etc. When these waste byproducts are properly utilised, they find diverse applications as edible coatings/films, fortified probiotics, derived metallic nanoparticles, derived carbon dots, microbiological media, biochars, derived biosorbents, etc., as discussed in the previous chapters. This chapter studies the uses of fruit peel–based edible coatings/films.

Keywords Edible coatings · Byproducts · Food processing · Waste utilisation

4.1 Introduction

A balanced diet that influences the body's overall activity and wellness has to include vegetables and fruits. Fruits and vegetables have finite postharvest lives because they are perishable. Even after the harvesting and processing, plants remain in a continuous phase of permeability and respiration, which causes their quality to

V. Loganathan (✉)
Department of Food Technology, Nehru Institute of Technology, Coimbatore, Tamil Nadu, India

N. Thangaraj
Department of Food Technology, Hindusthan College of Engineering and Technology, Coimbatore, Tamil Nadu, India

J. S. Kumar
Department of Food Technology, Saintgits College of Engineering, Kottayam, Kerala, India

E. Cherian, B. Gurunathan (eds.), *Value Added Products From Food Waste*,
https://doi.org/10.1007/978-3-031-48143-7_4

deteriorate and storage stability to decrease (Ali et al. 2019). Fruit ripening is a complex process distinguished by significant physiological and biochemical changes. This process includes the synthesis of ethylene, an increase in metabolic activity, changes in the amount of sugar, the transcription of cell wall–degrading enzymes, chlorophyll stability, and the biosynthesis of aromatic compounds (Lee and Hwang 2017; Perotti et al. 2014).

The edible films and coatings (primary packaging) are made from the edible components of plants (Tahir et al. 2019). Packaging materials can be classified into two categories, according to the working process: coatings or films. *Coating* is defined as the use of the thin layer of materials applied on food products to extend shelf life and improve the quality of food. Usually, these coating materials, formed with a thickness of 0.3 mm from natural sources, are applied directly on the surface of food products. Alternatively, various layers can be applied between food surfaces to prevent external factors such as the migration of air (oxygen, moisture) (Ferreira et al. 2016). Packaging films are prepared before products are ready for them. Many differences separate coatings from films, but in all cases, both coatings and films act as protective layers between food products and the environment, ensuring quality and a long shelf life (Pavlath and Orts 2009).

Once, citrus fruits were coated by using waxes to increase their shelf life. The packaging materials (coatings and films) are developed from the edible compounds of plants. Films are used to protect the food components after they have been formed, where coatings are applied directly on the surface of food materials (Kang et al. 2013). These packing materials plays major roles in maintaining the nutritional properties of food products that they are applied on, so more research on methods that could improve the development of coatings and films and their applications are needed. Newer methods do not aim to replace the traditional packaging process, but they can be used with other packing materials to reduce the production quantities (Pardo-Ibáñez et al. 2014). Various components are used as the edible films and coatings, such as polysaccharides, hydrocolloids, proteins, and lipids (Hassan et al. 2018). These components may be hydrophilic and may come in combinations of lipids to form complex edible films and coatings with more advanced functional properties (Corbo et al. 2015; Mehyar et al. 2014). The key function of an edible film or coating is to act as a protective barrier between food material and environmental conditions, such as moisture, oxygen, and contaminants, to preserve their flavour, aroma, and oil level while also maintaining food integrity (Menzel et al. 2020).

Because of various external factors, food products take a substantial amount of time to reach the consumer, and damage may occur during transportation, handling, and storage (Embuscado and Huber 2009). Such damage causes the product to deteriorate; dehydrate; change its colour, flavour, and appearance; and lose some of its nutritional value. Edible coating materials can protect food from microbial contaminations and physical contamination, minimise the spoilage period, reduce oxidation, and overall extend the product's shelf life. These materials are biodegradable, nontoxic, protective, affordable, and easily available (Fagundes et al. 2015).

4.2 The Mechanisms and Functions of Coatings/Films

The major functions of packaging materials include providing a barrier against external factors such as O_2, CO_2, moisture, pathogens, and the movement of water molecules. Edible coatings and films are used for food packaging, as capsules in the pharmaceutical industry for drug delivery, and in tissue engineering. In the biomedical field, edible coating materials are widely used because they have or confer numerous functional properties, including flexibility, easy biodegradability, pollution reduction, recyclability, and ecofriendliness. Also, edible films and coatings are made purely from natural sources like polysaccharides, lipids, proteins, and combinations of these components, making these edible materials safe to consume and easily digestible.

The fruit peels of apples, citrus fruits, guavas, strawberries, papayas, apricots, jackfruits, pomegranates, and raspberries are widely known for their medicinal and nutritional value. Pomegranate peels and orange peels contain high concentrations of carotenoids, enzymes, oils, vitamins, phenols, polyphenols, tannins, antidiabetic properties, antioxidants, antimicrobial properties, and other properties (Khalid et al. 2018).

4.3 Fruit Peel–Based Edible Packaging Materials

Coatings and films for foods are developed from lipids, proteins, polysaccharides, and combinations of these components. Macromolecules such as proteins and polysaccharides may be utilised to produce biodegradable packaging alternatives. Researchers have studied a variety of polysaccharides, includes starch (Vilas Dhumal et al. 2019), cellulose substitutes (Francisco et al. 2020), gum (Vilas Dhumal et al. 2019), chitosan (Vásconez et al. 2009), and alginate (Mahcene et al. 2020). Proteins also used as edible packaging alternatives include whey (Vilas Dhumal et al. 2019), gelatin (Scartazzini et al. 2019), soy proteins (dos Santos Paglione et al. 2019), and maize zein (Vahedikia et al. 2019).

The daily production of plant byproducts from the agricultural sector is significant, and all these resources might be investigated to find new materials for eco-friendly food packaging (Goncalves et al. 2017). Such resources include peels (Menzel et al. 2020), piths (Torres-León et al. 2018), kernels (González et al. 2019), rinds (Khalid et al. 2018), pomaces (Kurek et al. 2018), shells (Tóth and Halász 2019), and straws (Menzel et al. 2020). These plant-based byproducts used in biodegradable food packaging are also rich sources of beneficial nutrients like polyphenols and flavonoids. The antioxidant, antimicrobial, anti-inflammatory, and antiproliferative characteristics of the polyphenolic constituents in plants have been well documented, as has their potential to enhance the structural characteristics of foods (Lund 2021).

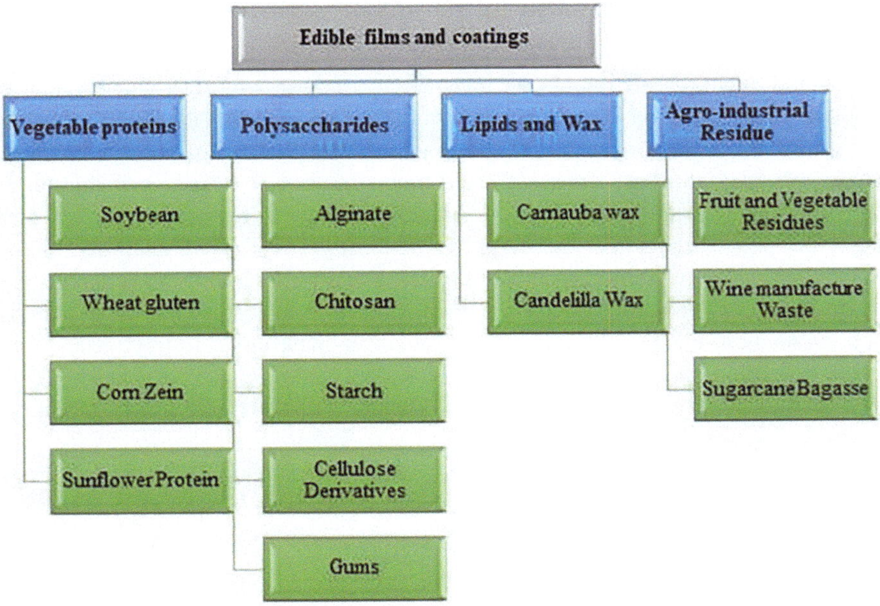

Fig. 4.1 Types of coatings used on fruits and vegetables

These materials help to protect fruits and vegetables from pathogens and insects when under preharvesting and postharvesting conditions and throughout transportation. More recently, thanks to new technologies, new edible coatings now contain various nanosize compounds such nanoparticles, nanoemulsions, and nanocomposites, which confer antioxidant and antimicrobial properties. Thanks to their renewable natures, which minimise toxicity and promote biodegradability and biocompatibility, these plant byproducts are excellent raw materials for biodegradable packaging. The use of byproducts in biodegradable packaging can reduce environmental impacts such as industry's negative contributions to climate change, which would usually be increased via the processes of decomposing and dumping plastics (Maraveas 2020). As a result of such plant-based products' ecofriendliness, recent years have witnessed an increase in research on the advantages of using plant-based byproducts in packaging materials (Fig. 4.1).

4.4 Classification of Edible Films and Coatings

Edible films and coatings can be divided into four categories on the basis of their respective compositions. Edible films and coatings include protein-based varieties, polysaccharide-based varieties, lipid-based varieties, and varieties combining lipids, proteins, and polysaccharides (Falguera et al. 2011; Miller and Krochta 1997). The main purpose of edible films and coatings is to increase the shelf life of the food products they are applied to and to act as strong, protective barriers against moisture migration, texture loss, respiration, oxidation, contamination, and pathogens (Maqbool et al. 2010; Song et al. 2011).

Furthermore, compared with conventional artificial films, they have a high selective gas permeability ratio (CO_2–O_2 ratio) and are excellent fat and oil barriers. Additionally, they can enhance the structural stability of the food by operating as carriers for food additives such as antioxidants (Song et al. 2011) and/or antibacterial agents (Maqbool et al. 2010). In some applications, the excellent mechanical and physical properties of standalone edible films might replace the use of synthetic packaging.

4.5 Protein-Based Edible Materials

Proteins are covalently bonded through peptides and in this way form the subunits of natural polymers (Hanani et al. 2014). Protein materials are made from combinations of protein components, such as gelatin/pectin, collagen, whey, gluten, starch, and keratin (Khalid et al. 2018; Moreira et al. 2011; Pardo-Ibáñez et al. 2014; Ramos et al. 2016). The main properties of protein confer physical stability to a structure. But it has weak tensile strength and puncture strength. Also, preventing the diffusion of moisture, O_2, and CO_2 is sometimes difficult. Such diffusion depends mainly on the composition of proteins and factors such as molecular weight, lipophilicity, hydrophobicity, permeability, cross-linkages, isoelectric points, and interactions between amino acid groups and carboxyl groups (Coltelli et al. 2015).

4.5.1 Gelatin

Gelatin is the only lignocellulosic material that produces a thermoreversible substance with a melting point close to room temperature. It is especially significant in the pharmaceutical, cosmetic, and food industries. Collagen, a fibrous insoluble protein that makes up the majority of the skin, bones, and connective tissues of animals, is hydrolysed under controlled conditions to produce gelatin. The amino acids in gelatin are organised in a specific manner. Gelatin is distinguished by its high concentrations of amino acids, glycine, proline, and hydroxyproline. Moreover, gelatin has a combination of single- and double-unfolded strands, which confer hydrophilicity (Ross-Murphy 1992). It is used as a gelling agent and a foaming agent in the food industry. The main sources of gelatin and pectin are apples, pineapples, citrus fruits, guavas, carrageenan, agar, guar gums, konnyaku (konjac), and animals like fish. Gelatin has good mechanical strength, protective properties, and low wettability. The lack of a sufficient vapour barrier in gelatinous films, like most protein films, restricts their use as edible coatings and composites. Using cross-linking polymer molecules to alter the nanocomposition of protein films can make these films more effective.

4.5.2 Collagen

Collagen is a major protein component of plant and animal connective tissues. Almost 80% of the protein in animal tissues is composed of several varieties of collagen. A substantial proportion of these varieties contains a triple-helix structure, though each type has a distinct composition of amino acids. Type I collagen is a prominent cellular membrane thanks to its biological features and availability for production. Many structures are produced with collagen, such as sponges, films, and membranes (Sisken et al. 1993). It acts as a strong barrier against O_2 and CO_2 thanks to its polar nature. Collagen is insoluble in water, but decreasing its pH increases its solubility. It can be used to protect citrus fruits by applying a layer on the surface of fruits, thereby reducing the transfer of gases between fruit and environment.

4.5.3 Starch

To make sustainable films that are clear or translucent, colourless, and unflavoured, starch, a natural polysaccharide, is used as the major ingredient (Skurtys et al. 2015). Starch granules are typically hydrophilic and semicrystalline. The granules absorb water molecules in their nearby environment thanks to the free hydroxyl groups in the subunits of starch, which results in swelling that lasts until a threshold level has been reached. The minimum quantity of starch has to be present for the swelled granules to fill a given volume at 95 °C, and the starch turns into a gel upon cooling (Pelissari et al. 2019). A fruit peel contains large amounts of cellulose, sugar, and starch. It is insoluble in water, but starch can become somewhat soluble under heat. The gelatinisation of starch depends on the type of starch molecule, heating temperature, and level of water content. Heating starch with a low concentration of fruit peel changes the concentrations of its contents in a way that increases its water-vapour permeability. Films and coatings made from the peels of starches have been shown to be odourless, transparent, and tasteless and to confer gas-barrier properties, structural stability, hydrophilicity, antimicrobial properties, oxidation properties, and protection against vitamin and mineral losses. Modified starch is also used to fortify rice with vitamins and minerals to increase its nutrition and market value.

4.6 Polysaccharide-Based Edible Materials

Polysaccharides are carbohydrate macromolecules containing two or more monosaccharides (Saberi et al. 2016). These monosaccharides connect via glycosidic linkages. The glycosidic linkages of monosaccharides determine the physical properties of the polysaccharides that they form. The cellulose, starch derivatives, pectin derivatives, carrageenan, agar, alginate, acacia, guar, microbial fermentation gums, pullulan, and chitosan in polysaccharides are utilised in edible films or coatings (González et al. 2011). These are available in the form of stabilisers, gelling agents, binding agents, and thickening agents. Polysaccharides have various functions: They confer low permeability to oxygen, partial permeability to CO_2, and protection against moisture loss and water transmission.

4.6.1 Pectin

In many vegetables and fruits, the central layer is typically an abundant collection of high-molecular-weight polysaccharides called pectin. Pectin can be obtained from the cell walls of fruits and vegetables and is extracted mainly from citrus peels and apple pomaces. Pectin may be isolated from plant cell walls and is a component of plant fibre (Valdés et al. 2015). The carboxyls in uronic acid are either completely (high-methoxy pectin) or substantially (low-methoxy pectin) methyl-esterified, generating a complex polysaccharide residue of 1,4-linked D-galacturonic acid (Antoniou et al. 2015; Chen et al. 2020). Pectin is soluble in water with anionic linkages and is cross-linked with polyvalent cations. It is classified into two types on the basis of its level of methylation. High-methoxy pectin exhibits more than 50% methylation, whereas low-methoxy pectin exhibits less than 50% methylation (Coffin and Fishman 1994); in other words, when more than 50% of the carboxyl groups in pectins are methylated, they are called high-methoxy pectins, whereas when less than 50% of the carboxyl groups in pectins are methylated, they are called low-methoxy pectins. Pectin-based films and their derivatives confer numerous benefits, such as mechanical strength and good barriers against water, oil, and oxygen, but they have poor moisture properties (Al-Tayyar et al. 2020). Pectin is used as a coating material on citrus fruits, raspberries, apricots, yoghurts, ice creams, and jams. In addition, pectin can act as a stabilising agent, a thickening agent, a gelling agent, and a plasticising agent (Jahromi et al. 2020).

4.6.2 Cellulose and Its Derivatives

The most abundant organic molecule in nature is cellulose, which also seems to be the most widely used polymer. In plants and their tissues, it acts as the main structural component. The naturally occurring long-chain monomer called cellulose has a positively significant impact on the human food chain. The pharmaceutical and cosmetic industries make considerable usage of cellulose's mainly semisynthetic polymers. Cellulose derivatives come in two primary classes: cellulose ethers and cellulose esters, each of which has a unique physicochemical composition and unique mechanical behaviours (Krässig 1996). More important in the pharmaceutical industry are organic cellulose esters. For example, cellulose acetate (CA), cellulose acetate phthalate (CAP), cellulose acetate butyrate (CAB), cellulose acetate trimelitate (CAT), hydroxypropyl methylcellulose phthalate (HPMCP), and others are utilised in commercial products or pharmacological studies (Heinämäki et al. 1994; Lecomte et al. 2003; Liu and Williams III 2002).

4.6.3 Chitosan

Chitosan is a consumable material that is synthesised from chitin. The shells of crabs, for example, are composed mostly of chitin. According to Shahidi et al. (1999), chitosan is the most widely utilised natural, nontoxic substance in the development of edible films. Chitosan has excellent characteristics alone—without adding any compounds or inhibitors—such as a high O_2 barrier, CO_2 permeability, and antimicrobial activity against pathogens. Outstanding mechanical properties have also been observed in chitosan. Chitosan has an exceptionally high viscosity that is comparable to that of natural adhesives. Fruits and vegetables have longer shelf lives thanks to translucent or invisible chitosan coatings. Its surface is typically smooth, sparkly, cohesive, and free of surface fractures (Ribeiro et al. 2007).

4.7 Lipid-Based Edible Materials

For decades now, edible films derived from lipids have been used to prolong the shelf life of vegetables and fruits. They give food a glossy appearance. Lipids are derived from plants such as herbs and are composed of paraffins, glycerides, resins, waxes, and fatty acids (Debeaufort et al. 1998). The functional groups in lipids include phospholipids, glycerides, monoglycerides, diglycerides, triglycerides, fatty alcohols, and phosphatides (Morillon et al. 2002). Lipids have both hydrophobic properties and hydrophilic properties, acting as excellent barriers against water and oils (Galus and Kadzińska 2015). Lipids reduce the rate of oxidation, gas transportation, and anaerobic respiration (Robertson 2009). Finally, lipid-based films improve product quality, flavour, ripening, and structural stability. Applying lipids to food surfaces coats them with a strong barrier because of their high level of adhesion when mixed with emulsions like emulsified proteins and dual-coating emulsions.

4.8 Characterizations of Edible Coatings/Films

Improving the structural properties of food can improve its overall quality. Various methods are available for analysing the structural properties of coatings and films. Fourier transform infrared spectroscopy (FTIR) (Keshani-Dokht et al. 2018), nuclear magnetic resonance (NMR) spectroscopy (Pawar and Jadhav 2015), confocal laser scanning microscopy (CSLM) (Pawar and Jadhav 2015), scanning electron microscopy (SEM) (Shiku et al. 2003), and atomic force microscopy (AFM) are used for analysing the physicochemical properties of molecules and the internal and external structures of materials.

Coatings and films have and confer properties such as mechanical strength, barriers against pathogens, thermal properties, colours, and sensory and textural properties. The rheological properties of coating materials can be assessed by using a compact modular rheometer during oscillatory tests. The moisture content of coatings has been analysed via gravimetric methods and the thickness has been measured with a Fowler micrometer (with an accuracy value of 0.0001 mm). Water activity (aw) tests have been carried out with a water-activity meter (Zhang and Han 2006). Colour is one of the main properties of any food material, determining the product's transparency and appearance. The colour of films has been measured by using a colorimeter and an ultraviolet-visible (UV-vis) spectrophotometer with parameters L* (lightness), a* (red/green), and b* (yellow/blue) (Rhim et al. 1999).

The mechanical properties of edible films include tensile strength, breakability, elasticity, and toughness. Tensile properties can be assessed via the standard method that measures the relationship between stress and strain during stretching. Tensile strength is the maximum force that a material can withstand during stretching. The film-puncture test has been calculated by using an analyser. Finally, toughness is a material's capacity to absorb energy while deforming without fracturing (Sánchez Aldana et al. 2015).

4.9 Conclusion

Coating materials come with many advantages, but also many limitations, such as weak tensile strength, weak puncture strength, fragility, weak moisture barriers, and weak oxygen barriers. The number and the degree of these drawbacks depend on the material's components. The polysaccharide based edible films and coatings are hydrophilic in nature and have very poor moisture prevention properties. The lipid based coating materials give positive influence on food products. The outstanding properties and performance levels of edible films and coatings have improved the products of the food industry. Research in this area has recently improved, but several problems remain unsolved in food packaging, particularly processing complications. The latest approach to packaging food in the packaging sector applies edible films via edible-film technology, which minimises food waste. The substances used to develop these films, whether they are organic or synthetic, are edible. Carbohydrate, protein, lipid, paraffin, and oil constitute the majority of their structural elements, and they all form effectively. Edible films are made from natural edible ingredients and are quickly replacing synthetic packaging materials because the former are safe to consume. However, several problems still need to be solved to expand the commercial value of using edible films and coatings.

References

Ali S, Nawaz A, Ejaz S, Haider ST-A, Alam MW, Javed HU (2019) Effects of hydrogen sulfide on postharvest physiology of fruits and vegetables: an overview. Sci Hortic 243:290–299

Al-Tayyar NA, Youssef AM, Al-Hindi RR (2020) Edible coatings and antimicrobial nanoemulsions for enhancing shelf life and reducing foodborne pathogens of fruits and vegetables: a review. Sustain Mater Technol 26:e00215

Antoniou J, Liu F, Majeed H, Zhong F (2015) Characterization of tara gum edible films incorporated with bulk chitosan and chitosan nanoparticles: a comparative study. Food Hydrocoll 44:309–319

Chen Y, Xu L, Wang Y, Chen Z, Zhang M, Chen H (2020) Characterization and functional properties of a pectin/tara gum based edible film with ellagitannins from the unripe fruits of Rubus chingii Hu. Food Chem 325:126964

Coffin DR, Fishman ML (1994) Physical and mechanical properties of highly plasticized pectin/starch films. J Appl Polym Sci 54(9):1311–1320

Coltelli M-B, Wild F, Bugnicourt E, Cinelli P, Lindner M, Schmid M, Weckel V, Müller K, Rodriguez P, Staebler A (2015) State of the art in the development and properties of protein-based films and coatings and their applicability to cellulose based products: an extensive review. Coatings 6(1):1

Corbo MR, Campaniello D, Speranza B, Bevilacqua A, Sinigaglia M (2015) Non-conventional tools to preserve and prolong the quality of minimally-processed fruits and vegetables. Coatings 5(4):931–961

Debeaufort F, Quezada-Gallo J-A, Voilley A (1998) Edible films and coatings: tomorrow's packagings: a review. Crit Rev Food Sci Nutr 38(4):299–313

dos Santos Paglione I, Galindo MV, de Medeiros JAS, Yamashita F, Alvim ID, Grosso CRF, Sakanaka LS, Shirai MA (2019) Comparative study of the properties of soy protein concentrate films containing free and encapsulated oregano essential oil. Food Packag Shelf Life 22:100419

Embuscado ME, Huber KC (2009) Edible films and coatings for food applications, vol 9. Springer, Dordrecht

Fagundes C, Palou L, Monteiro AR, Pérez-Gago MB (2015) Hydroxypropyl methylcellulose-beeswax edible coatings formulated with antifungal food additives to reduce alternaria black spot and maintain postharvest quality of cold-stored cherry tomatoes. Sci Hortic 193:249–257

Falguera V, Ceron JPQ, Jiménez A, Muñoz A (2011) Películas y recubrimientos comestibles: estructuras, funciones activas y tendencias en su uso. Tendencias En Ciencia y Tecnología de Los Alimentos 22(6):292–303

Ferreira ARV, Alves VD, Coelhoso IM (2016) Polysaccharide-based membranes in food packaging applications. Membranes 6(2):22

Francisco CB, Pellá MG, Silva OA, Raimundo KF, Caetano J, Linde GA, Colauto NB, Dragunski DC (2020) Shelf-life of guavas coated with biodegradable starch and cellulose-based films. Int J Biol Macromol 152:272–279

Galus S, Kadzińska J (2015) Food applications of emulsion-based edible films and coatings. Trends Food Sci Technol 45(2):273–283

Goncalves I, Nunes C, Mendes S, Martins LO, Ferreira P, Coimbra MA (2017) CotA laccase-ABTS/hydrogen peroxide system: an efficient approach to produce active and decolorized chitosan-genipin films. Carbohydr Polym 175:628–635

González A, Strumia MC, Igarzabal CIA (2011) Cross-linked soy protein as material for biodegradable films: synthesis, characterization and biodegradation. J Food Eng 106(4):331–338

González A, Barrera GN, Galimberti PI, Ribotta PD, Igarzabal CIA (2019) Development of edible films prepared by soy protein and the galactomannan fraction extracted from Gleditsia triacanthos (Fabaceae) seed. Food Hydrocoll 97:105227

Hanani ZAN, Roos YH, Kerry JP (2014) Use and application of gelatin as potential biodegradable packaging materials for food products. Int J Biol Macromol 71:94–102

Hassan B, Chatha SAS, Hussain AI, Zia KM, Akhtar N (2018) Recent advances on polysaccharides, lipids and protein based edible films and coatings: a review. Int J Biol Macromol 109:1095–1107

Heinämäki JT, Colarte AI, Nordström AJ, Yliruusi JK (1994) Comparative evaluation of ammoniated aqueous and organic-solvent-based cellulose ester enteric coating systems: a study on free films. Int J Pharm 109(1):9–16

Jahromi M, Niakousari M, Golmakani MT, Mohammadifar MA (2020) Physicochemical and structural characterization of sodium caseinate based film-forming solutions and edible films as affected by high methoxyl pectin. Int J Biol Macromol 165:1949–1959

Kang H-J, Kim S-J, You Y-S, Lacroix M, Han J (2013) Inhibitory effect of soy protein coating formulations on walnut (Juglans regia L.) kernels against lipid oxidation. LWT Food Sci Technol 51(1):393–396

Keshani-Dokht S, Emam-Djomeh Z, Yarmand M-S, Fathi M (2018) Extraction, chemical composition, rheological behavior, antioxidant activity and functional properties of Cordia myxa mucilage. Int J Biol Macromol 118:485–493

Khalid S, Yu L, Feng M, Meng L, Bai Y, Ali A, Liu H, Chen L (2018) Development and characterization of biodegradable antimicrobial packaging films based on polycaprolactone, starch and pomegranate rind hybrids. Food Packag Shelf Life 18:71–79

Krässig H (1996) Cellulose, polymer monographs, vol 11. Gordon and Breach Science Publishers, Amsterdam, pp 6–42

Kurek M, Garofulić IE, Bakić MT, Ščetar M, Uzelac VD (2018) Development and evaluation of a novel antioxidant and pH indicator film based on chitosan and food waste sources of antioxidants. Food Hydrocoll 84:238–246

Lecomte F, Siepmann J, Walther M, MacRae RJ, Bodmeier R (2003) Blends of enteric and GIT-insoluble polymers used for film coating: physicochemical characterization and drug release patterns. J Control Release 89(3):457–471

Lee Y, Hwang KT (2017) Changes in physicochemical properties of mulberry fruits (Morus alba L.) during ripening. Sci Hortic 217:189–196

Liu J, Williams RO III (2002) Long-term stability of heat–humidity cured cellulose acetate phthalate coated beads. Eur J Pharm Biopharm 53(2):167–173

Lund MN (2021) Reactions of plant polyphenols in foods: impact of molecular structure. Trends Food Sci Technol 112:241–251

Mahcene Z, Khelil A, Hasni S, Akman PK, Bozkurt F, Birech K, Goudjil MB, Tornuk F (2020) Development and characterization of sodium alginate based active edible films incorporated with essential oils of some medicinal plants. Int J Biol Macromol 145:124–132

Maqbool M, Ali A, Ramachandran S, Smith DR, Alderson PG (2010) Control of postharvest anthracnose of banana using a new edible composite coating. Crop Prot 29(10):1136–1141
Maraveas C (2020) Production of sustainable and biodegradable polymers from agricultural waste. Polymers 12(5):1127
Mehyar GF, Al-Qadiri HM, Swanson BG (2014) Edible coatings and retention of potassium sorbate on apples, tomatoes and cucumbers to improve antifungal activity during refrigerated storage. J Food Process Preserv 38(1):175–182
Menzel C, González-Martínez C, Vilaplana F, Diretto G, Chiralt A (2020) Incorporation of natural antioxidants from rice straw into renewable starch films. Int J Biol Macromol 146:976–986
Miller KS, Krochta JM (1997) Oxygen and aroma barrier properties of edible films: a review. Trends Food Sci Technol 8(7):228–237
Moreira MDR, Pereda M, Marcovich NE, Roura SI (2011) Antimicrobial effectiveness of bioactive packaging materials from edible chitosan and casein polymers: assessment on carrot, cheese, and salami. J Food Sci 76(1):M54–M63
Morillon V, Debeaufort F, Blond G, Capelle M, Voilley A (2002) Factors affecting the moisture permeability of lipid-based edible films: a review. Crit Rev Food Sci Nutr 42(1):67–89
Pardo-Ibáñez P, Lopez-Rubio A, Martínez-Sanz M, Cabedo L, Lagaron JM (2014) Keratin–polyhydroxyalkanoate melt-compounded composites with improved barrier properties of interest in food packaging applications. J Appl Polym Sci 131(4):39947
Pavlath AE, Orts W (2009) Edible films and coatings: why, what, and how? In: Edible films and coatings for food applications. Springer, Dordrecht, pp 1–23
Pawar HA, Jadhav P (2015) Isolation, characterization and investigation of Cordia dichotoma fruit polysaccharide as a herbal excipient. Int J Biol Macromol 72:1228–1236
Pelissari FM, Ferreira DC, Louzada LB, dos Santos F, Corrêa AC, Moreira FKV, Mattoso LH (2019) Starch-based edible films and coatings: an eco-friendly alternative for food packaging. In: Starches for food application. Academic Press, London, pp 359–420
Perotti VE, Moreno AS, Podestá FE (2014) Physiological aspects of fruit ripening: the mitochondrial connection. Mitochondrion 17:1–6
Ramos M, Valdés A, Beltran A, Garrigós MC (2016) Gelatin-based films and coatings for food packaging applications. Coatings 6(4):41
Rhim JW, Wu Y, Weller CL, Schnepf M (1999) Physical characteristics of a composite film of soy protein isolate and propyleneglycol alginate. J Food Sci 64(1):149–152
Ribeiro C, Vicente AA, Teixeira JA, Miranda C (2007) Optimization of edible coating composition to retard strawberry fruit senescence. Postharvest Biol Technol 44(1):63–70
Robertson GL (2009) Food packaging and shelf life: a practical guide. CRC Press, Boca Raton
Ross-Murphy SB (1992) Structure and rheology of gelatin gels: recent progress. Polymer 33(12):2622–2627
Saberi B, Thakur R, Vuong QV, Chockchaisawasdee S, Golding JB, Scarlett CJ, Stathopoulos CE (2016) Optimization of physical and optical properties of biodegradable edible films based on pea starch and guar gum. Ind Crop Prod 86:342–352
Sánchez Aldana D, Contreras-Esquivel JC, Nevárez-Moorillón GV, Aguilar CN (2015) Characterization of edible films from pectic extracts and essential oil from Mexican lime. CyTA J Food 13(1):17–25
Scartazzini L, Tosati JV, Cortez DHC, Rossi MJ, Flôres SH, Hubinger MD, di Luccio M, Monteiro AR (2019) Gelatin edible coatings with mint essential oil (Mentha arvensis): film characterization and antifungal properties. J Food Sci Technol 56:4045–4056
Shahidi F, Arachchi JKV, Jeon Y-J (1999) Food applications of chitin and chitosans. Trends Food Sci Technol 10(2):37–51
Shiku Y, Hamaguchi PY, Tanaka M (2003) Effect of pH on the preparation of edible films based on fish myofibrillar proteins. Fish Sci 69(5):1026–1032
Sisken BF, Walker J, Orgel M (1993) Prospects on clinical applications of electrical stimulation for nerve regeneration. J Cell Biochem 51(4):404–409

Skurtys O, Velásquez P, Osorio F (2015) Wetting behavior of chitosan solutions on blueberry epicarp with or without epicuticular waxes. In: Water stress in biological, chemical, pharmaceutical and food systems. Springer, New York, pp 509–518
Song Y, Liu L, Shen H, You J, Luo Y (2011) Effect of sodium alginate-based edible coating containing different anti-oxidants on quality and shelf life of refrigerated bream (Megalobrama amblycephala). Food Control 22(3–4):608–615
Tahir HE, Xiaobo Z, Mahunu GK, Arslan M, Abdalhai M, Zhihua L (2019) Recent developments in gum edible coating applications for fruits and vegetables preservation: a review. Carbohydr Polym 224:115141
Torres-León C, Ramírez-Guzman N, Londoño-Hernandez L, Martinez-Medina GA, Díaz-Herrera R, Navarro-Macias V, Alvarez-Pérez OB, Picazo B, Villarreal-Vázquez M, Ascacio-Valdes J (2018) Food waste and byproducts: an opportunity to minimize malnutrition and hunger in developing countries. Front Sustain Food Syst 2:52
Tóth A, Halász K (2019) Characterization of edible biocomposite films directly prepared from psyllium seed husk and husk flour. Food Packag Shelf Life 20:100299
Vahedikia N, Garavand F, Tajeddin B, Cacciotti I, Jafari SM, Omidi T, Zahedi Z (2019) Biodegradable zein film composites reinforced with chitosan nanoparticles and cinnamon essential oil: physical, mechanical, structural and antimicrobial attributes. Colloids Surf B: Biointerfaces 177:25–32
Valdés A, Burgos N, Jiménez A, Garrigós MC (2015) Natural pectin polysaccharides as edible coatings. Coatings 5(4):865–886
Vásconez MB, Flores SK, Campos CA, Alvarado J, Gerschenson LN (2009) Antimicrobial activity and physical properties of chitosan–tapioca starch based edible films and coatings. Food Res Int 42(7):762–769
Vilas Dhumal C, Pal K, Sarkar P (2019) Synthesis, characterization, and antimicrobial efficacy of composite films from guar gum/sago starch/whey protein isolate loaded with carvacrol, citral and carvacrol-citral mixture. J Mater Sci Mater Med 30:1–14
Zhang Y, Han JH (2006) Plasticization of pea starch films with monosaccharides and polyols. J Food Sci 71(6):E253–E261

Chapter 5
Bioenzymes from Wastes to Value-Added Products

Gamachis Korsa, Chandran Masi, Digafe Alemu, Abera Beyene, and Abate Ayele

Abstract Bioenzymes are defined as garbage enzymes or fruit enzymes that are natural, multi-purpose cleaners made from waste or vegetable/fruit peels (often citrus). Bioenzymes contribute to reducing some waste and transforming it into a material that is affordable, readily available, and has a wide range of potential uses for society at large. The location of biomass waste, with a focus on various waste items in different areas, is highlighted for consumption technologies aimed at producing enzymes for value-added products.. Innovations in the bioenzyme industry are introducing a portfolio of sustainable and eco-efficient enzyme products to compete in a market currently dominated by valuable-based products, and therefore, this has become a subject of intensive exploration. Due to its nutrient-rich and organic composition, ideally agricultural, industrial, and food waste can be consumed as a useful resource for the production of enzymes through various fermentation processes. Such conversion of each waste is potentially more profitable than its conver-

G. Korsa · D. Alemu · A. Ayele
Department of Biotechnology, Addis Ababa Sciences and Technology University, Addis Ababa, Ethiopia

Center of Excellence for Biotechnology and Bioprocess, Addis Ababa Science and Technology University, Addis Ababa, Ethiopia

C. Masi (✉)
Department of Biotechnology, Addis Ababa Sciences and Technology University, Addis Ababa, Ethiopia

Center of Excellence for Biotechnology and Bioprocess, Addis Ababa Science and Technology University, Addis Ababa, Ethiopia

Department of Food Technology, Dhanalakshmi Srinivasan Engineering college (Autonomous), Perambalur, India
e-mail: biochandran@gmail.com.in

A. Beyene
Department of Industrial Chemistry, Addis Ababa Sciences and Technology University, Addis Ababa, Ethiopia

Center of Excellence for Nanotechnology, Addis Ababa Science and Technology University, Addis Ababa, Ethiopia

E. Cherian, B. Gurunathan (eds.), *Value Added Products From Food Waste*,
https://doi.org/10.1007/978-3-031-48143-7_5

sion to plants, animal feed, soils, health, and fuel for transportation environments. As a result, multiple approaches for generating energy from wastes are being explored worldwide. The potential for advanced forms and improvements to transform complicated, natural-rich biomass waste into a variety of bioenzymes and bioproducts with an advanced circular low-cost has been demonstrated in this paper.

Keywords Bioenzyme · Enzyme · Fruit · Waste · Vegetable waste

Abbreviations

AChE	Acetylcholinesterase
CAT	Catalase
CFU	Colon forming unit
DM	Dry mass
EE	Ecoenzyme
E-Pt	Pt nanoparticles
FW	Food waste
GA	Glucoamylase
LCA	Life cycle assessment
LCC	Life cycle cost
MSW	Municipal solid waste
OFMSW	Organic fraction municipal solid waste (OFMSW)
ProK	Proteinase K enzymes
Pt	Platinum
SDG	Sustainable Development Goal
SmF	Sub-merged fermentation
SSF	Solid-state fermentation

5.1 Introduction

The majority of waste generated in the agriculture and food industries is high in proteins, carbohydrates, and lipids, which provide a breeding ground for a wide range of microorganisms (Gaur et al. 2020; O'Connor et al. 2021). Food waste will continue to gain market share in the creation of high-value items in the future. Bio-based chemicals are expected to account for 15% of the global chemical market by 2025 (Van Dorst et al. 2019; Sharma et al. 2020; Mamo et al. 2021). Household or municipal trash is typically created from a variety of sources as a result of various human activities. Fruit waste peels, pomace, and seed fractions are suitable feed stocks for recovering bioactive substances such as phenolics, pectin, lipids, dietary fibers, and so on. Only recently has the extraction of active compounds from waste moved closer to environmental certifications and LCA (Life Cycle Assessment) and

LCC (Life Cycle Cost) studies, which are likewise geared toward environmental product certifications (Shaddel et al. 2019; Lucarini et al. 2021). According to the Food and Agriculture Organization (FAO), global corn production exceeded 1 billion tons in 2019. Food Waste Explorer contains 27,069 data points from various sources, including scientific peer-reviewed papers, manufacturer data, and other data sources, covering 587 nutrients, 698 bioactives, and 49 toxicants (Suri et al. 2022).

Organic fraction municipal solid waste (OFMSW) is one of many anthropogenic biomass residues that are widely available. Through various thermo-chemical and bio-chemical conversion pathways, several alternative energies (biogas, electricity, and heat) and material products (renewable diesel, fertilizer, bio-oil) can be derived from wastes and biomass residues that are currently at various stages of economic and technical maturity (Liu and Rajagopal 2019; Kumar et al. 2020a; Salah et al. 2021).

Over 120 companies are known to manufacture industrial enzymes worldwide, with more than 80% of the companies holding up to 90% of the market in Europe and North America and none in Africa (Fernandes et al. 2022; Streimikyte et al. 2022). East Africa has a diverse microbial population that could provide new enzymes for industrial use (Arun and Sivashanmugam 2015; Singla et al. 2019). Researchers are working to increase the productivity of food items, particularly fruits, vegetables, and cereals, to fulfill the need as the world's population grows. For example, India is the world's second-largest producer of fruits and vegetables, contributing for 10% and 14% of worldwide production, respectively (panda et al. 2016; Bharathiraja et al. 2017; Di Donna et al. 2020).

Ecoenzymes are derived from fruit and vegetable fermentation, which can also be coupled with vegetables, non-chlorinated water, and molasses, and contain anti-microbial/pathogenic protease, amylase, and lipase enzymes. Ecoenzyme can also regulate both Gram-negative and Gram-positive bacteria (Agarwal and Kaur 2014; Arun and Sivashanmugam 2015; Rasit et al. 2019; Yong et al. 2021). Environmental pollution has become a serious problem in the age of globalization as the world's population has grown (Aktamovich and Mirzayevich 2022; Erdoğan et al. 2022). To meet the need, more food resources, particularly fruits, vegetables, and cereals, are produced. The global production of fruits and vegetables is expanding, yet roughly 30–40% of the overall production is discarded as waste due to a variety of factors (Pandey et al. 2020; Srivastava et al. 2021). Sustainable Development Goal (SDG) 12 advocates for sustainable consumption and production patterns. One of the goals is to reduce food waste and losses by half by 2030. According to the Food Wastes Index 2021 study (UNEP, 2021), worldwide food waste totaled 931 million tons in 2019, with the majority of the waste coming from households (61%), food service (26%), and retail (13%) (Birwal et al. 2017; Yong et al. 2021).

According to a study by Capanoglu et al. (2022), the production of solid trash has recently increased. This includes waste from agriculture, industry, households, and animals as well as waste from the home. Waste valorization using biotechnological methods is evolving into a sustainable, environmentally friendly method of handling this issue in the face of these challenges. Natural resources are under a lot of stress, as reported by Zhu et al. (2023), because of imbalances in the rise of the human population and the quickening urbanization process. Particularly, there is a large increase in trash. The United Nations has forecast that the worldwide demand for food might rise by 70%, with the majority of this growth occurring in developing nations. By 2050, the current global population of 8 billion people might reach a maximum of 9.7 billion (Yong et al. 2020; Li Y et al. 2020a).

5.2 Sources of Bioenzymes from Wastes

Agricultural, domestic, and industrial wastes as well as food preparation waste are all sources of waste, as given in Table 5.1 (Awasthi et al. 2022; Chilakamarry et al. 2022; Karić et al. 2022).

Table 5.1 Some sources of bioenzymes from different areas

S.No.	Bioenzymes from different substrates	Enzyme products	Reference
1	Oil palm empty fruit bunch wastes	Cellulase	Kim et al. (2014)
2	Agro-industrial products	Cellulase	Ortiz et al. (2015)
3	Agricultural wastes	Cellulase	Saratale et al. (2014)
4	Fruit wastes	Cellulase	Ferraz et al. (2018)
5	Waste date palm seeds	Cellulase	Swathy et al. (2020)
6	Citrus wastes	Cellulase	O'Shea et al. (2015)
7	Fruit wastes	Cellulase	Esparza et al. (2020)
8	Agriculture waste	Cellulase	Chaudhary and Padhiar (2020)
9	Fruit wastes	Cellulase	Srivastava et al. (2021)
10	Agriculture waste	Cellulase	Biswal and Mandavgane (2021)
11	Municipal solid waste (MSW)	Cellulase	Zhao et al. (2011)
12	Green tea waste	Laccase	Sermyagina et al. (2021)
13	Agro wastes	Laccase	Dave et al. (2021)
14	Agricultural wastes	Laccase	Liu et al. (2011)
15	Black gram residue	Laccase	Kumar et al. (2016)
16	Kitchen waste biomass	Laccase	Janveja et al. (2014)
17	Saw dust wastes	Laccase	Daâssi et al. (2016)
18	Corn cob	Laccase	Nuhu et al. (2020)
19	Wheat straw	Laccase	Birhanli and Yeşilada (2013)
20	Orange peel	Invertase	Nehad and Atalla (2020)
21	Molasses and sugar cane bagasse	Invertase	Veana et al. (2014)
22	Garbage and citrus wastes	Invertase	Nazim and Meera (2017)

(continued)

Table 5.1 (continued)

S.No.	Bioenzymes from different substrates	Enzyme products	Reference
23	Fruit and vegetable wastes	Invertase	Rasit and Mohammad (2018)
24	Fruit and vegetable wastes	Invertase	Galintin et al. (2021)
25	Poultry droppings	Invertase	Hauwa and Ahmed (2019)
26	Red sword bean	Invertase	Zhou et al. (2019)
27	Fruit and vegetable wastes	Invertase	Das and Mondal (2013)
28	Ragi husk	Xylanase	Manasa et al. (2020)
29	Rice straw	Xylanase	Kumar et al. (2017b)
30	Rice bran	L-Asparaginase	Prabhu and Jayadeep (2017)
31	Agro wastes	L-Asparaginase	Xie et al. (2022)
32	Cow dung	Xylanase	Vijayaraghavan et al. (2014)
33	Waste paper recycling	Cellulase and xylanase	Pathak et al. (2014)
34	Palm kernel cake	Mannanase	Chen et al. (2022)
35	Lignocellusic agro waste	Mannanase	Ja'afar and Shitu (2022)
36	Coffee grounds and tea waste	Mannanase	Pangsri and Pangsri (2017)
37	Coffee industry wastes	Mannanase	Favaro et al. (2020)
38	Açaí seed	Mannanase	Lima et al. (2021)
39	Agro waste	Pectinase	Thangaratham and Manimegalai (2014)
40	Gut waste of sardines	Alkaline protease	Ramkumar et al. (2018)
41	Agricultural wastes	Protease	De Castro et al. (2014)
42	Tannery solid waste	Alkaline protease	Ravindran et al. (2011)
43	Food waste	Protease	Escaramboni et al. (2022)
44	Grass carp scale waste	Protease	Li et al. (2022)
45	Fish waste-derive waste	Protease	Akyüz A and Ersus S (2021)
46	Wheat bran waste	Protease	Espoui et al. (2022)
47	Agroindustrial wastes	Protease	De Castro et al. (2014)
48	Soy fiber residue waste	Protease	Abraham et al. (2013)
49	Shrimp waste	Protease	Gaonkar and Furtado (2021)
50	Orange peel waste	Protease	Chimbekujwo et al. (2020)
51	Fruit and vegetable waste	Amylase	Chakraborty and Mohan (2019)
52	Vegetable, grape tomato pomace and waste	Amylase	Kumar et al. (2017a)
53	Citrus fruit peels waste	Amylase	Srimathi et al. (2020)
54	Meat industry waste	Amylase	Chen et al. (2020)
55	Marigold flower wastes	Amylase	Liu et al. (2020)
56	Tomato pomace	Amylase	Kumar et al. (2017a)
57	Soy and bread waste	Amylase	Cerda et al. (2016)
58	Sago industrial waste	Amylase	Melnichuk et al. (2020)
59	Kitchen waste or organic waste	Amylase	Mojumdar and Deka (2019)
60	Kitchen waste or organic waste	Amylase	Bhatt et al. (2020)
61	Domestic food waste	Glucoamylase	Kiran et al. (2014)

(continued)

Table 5.1 (continued)

S.No.	Bioenzymes from different substrates	Enzyme products	Reference
62	Agricultural wastes	Alpha-amylase	Afrisham et al. (2016)
63	Cassava bagasse	Alpha-amylase	Sivamani and Baskar (2018)
64	Agro-industrial waste	Amylase	Saxena and Singh (2011)
65	Bread pieces waste	Glucoamylase	Melikoglu et al. (2015)
66	Tomato skin and pomace	Pectinase	Sengar et al. (2020)
67	Citrus and bagasse waste	Pectinase	Biz et al. (2016)
68	Sawdust waste	Pectinase	Sethi et al. (2016)
69	Pomelo peel waste	Pectinase	Jalil and Ibrahim (2021)
70	Organic solid waste	Lipase	Selvakumar and Sivashanmugam (2017)
71	Corn oil	Lipase	Coradi et al. (2020)
72	Malt bagasse waste	Lipase	Eichler et al. (2020)
73	Organic solid waste	Lipase	Selvakumar, P., & Sivashanmugam (2019)
74	Vegetable oil-refining wastes	Lipase	Santis-Navarro et al. (2011)
75	Waste from castor bean waste	Lipase	Herculano et al. (2016)
76	Garbage and citrus wastes	Lipase	Nazim and Meera (2017)
77	Sweet potato leaves waste	Lipase	Mesa et al. (2020)
78	Soybean meal waste	Lipase	Mesa et al. (2020)
79	Tobacco waste	Lipase	Banoˇzi'c et al. (2019)
80	Durian leaf waste	Lipase	Kam et al. (2020)
81	Kitchen waste	Lipase	Bolzonella et al. (2018)
82	Coffee waste	Lipase	Battista et al. (2016)
83	Fruit and vegetable wastes	Lipase	Li et al. (2013)
84	Persian lemon	Lipase	Medina-Torres et al. (2019)
85	Fruit and vegetable wastes	Amylase, protease, and lipase	Neupane and Khadka (2019)
86	Organic waste	Amylase, cellulase, and lipase	Gopinath et al. (2014)
87	Animal food carcass wastes	Bio-organic fertilizer	Liu et al. (2016)
88	Tomato and orange residues	Protease, amylase, and lipase	Rasit et al. (2019)

Recycling wastes provides a number of benefits, including reducing garbage disposal, saving natural resources, particularly nonrenewable ones like petroleum, lowering the amount of energy necessary to make new things, and reducing pollution (Sharma et al. 2021).

For example, every year, 1.3 billion ton of food is thrown away globally, accounting for 28% of all agricultural land. The amount of gray water produced is determined by the amount of water available and local practices. This includes factors such as whether activities like personal and clothes washing occur at home or at the water source (Nazim and Meera 2017; Sharma et al. 2020).

5.3 Production Mechanisms of Wastes for Bioenzymes (Fig. 5.1)

Solid state fermentation (SSF) refers to any biotechnological process in which organisms grow on non-soluble materials or solid substrates in the absence or near absence of free water (Masaphy et al. 2022; Meftahi et al. 2022; Yafetto 2022). The best and most effective pretreatments for biowastes are those that do not require particle size reduction, preserve the pentose (hemicellulose) fraction, avoid the formation of possible inhibitors of hydrolytic enzymes and fermenting microorganisms, use minimal energy, have low operating costs (operating costs, capital costs, and biomass costs), use little or no chemicals, and use cheap chemicals (Castro and Sato 2013; Gunam et al. 2019). For maximum production, batch, fed-batch, and continuous fermentation modes can be selected. During fermentation processes, fed-batch and continuous modes can overcome substrate limitations. Cell immobilization can also boost productivity by increasing the biomass content in the bioreactor, resulting in a higher concentration of bioprocesses in the reactor (Fig. 5.1) (Naiket al. 2020; Galintin et al. 2021; Ding et al. 2022). By conversion of biomass to biological products and biofuels, such as agriculture and cattle waste, food waste, kitchen waste, green waste, seaweed, algae, sewage sludge, agro-industries, forestry residues, and other organic industrial waste, fermentation, as shown in Fig. 5.2, provides a sustainable way to manage degradable biowaste (Agarwal and Kaur 2014; Sodhi, P. S., & Ocean, Y. K. 2018; Jain et al. 2022; Prabhu et al. 2022).

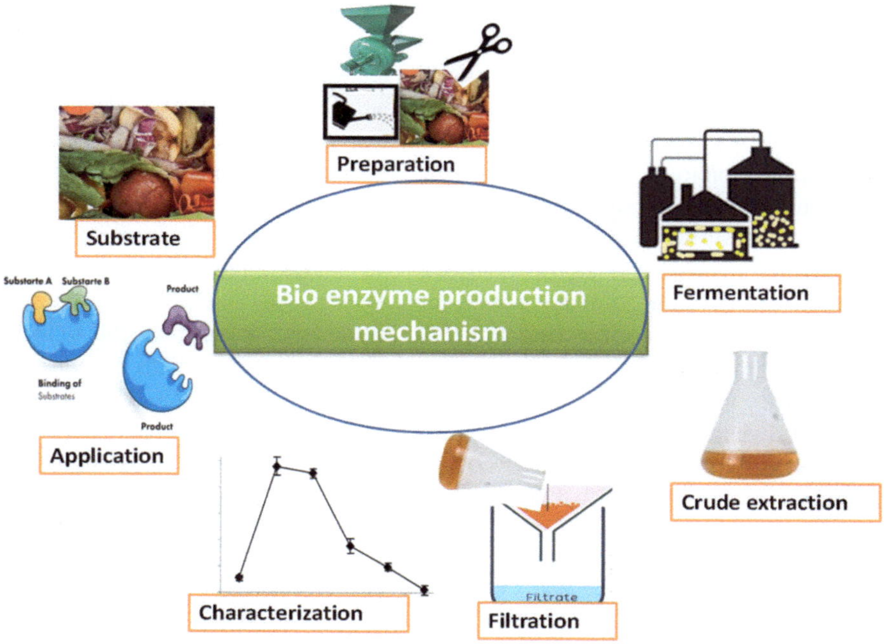

Fig. 5.1 The production mechanisms of bioenzyme (Dinget al. 2022; Karimi et al. 2022)

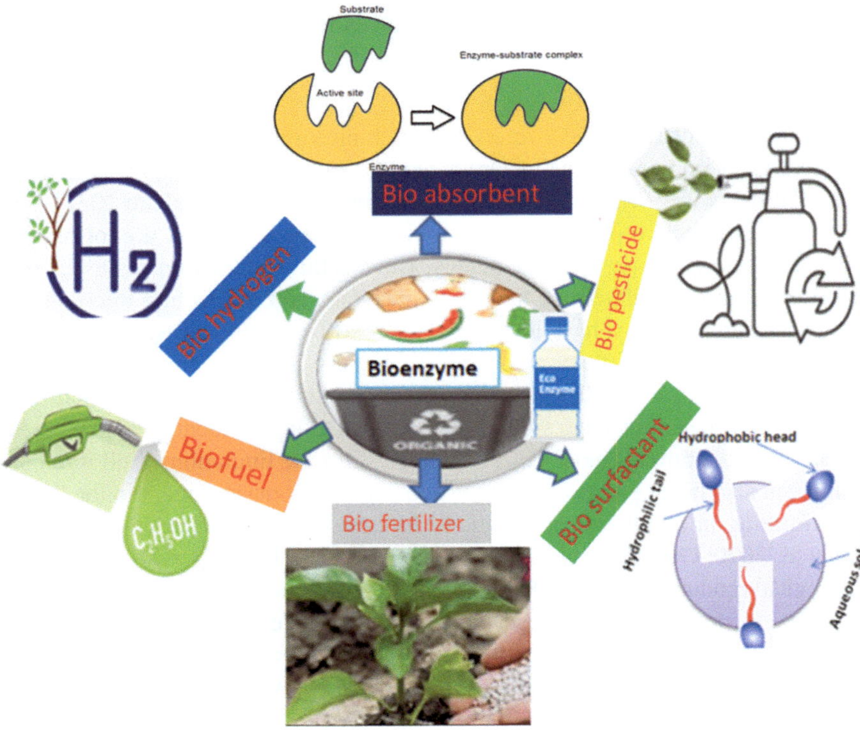

Fig. 5.2 Different applications of bioenzymes (Sivakumar et al. 2022)

Higher moisture content is required for the maturation of microscopic microorganisms, resulting in a reduced output (Bolte et al. 2022; Fan et al. 2022; Negi et al. 2022). The steps of SSF are as follows: substrate pretreatment entails increasing the mechanical or biological binding of polymeric substrates, such as polysaccharides and proteins, in order to minimize the size of the components. It is crucial to note that the composition, chemical nature, mechanical properties, particle size (including inter and intra-particle gaps), water retention capacity, and surface area of the substrates employed in SSF vary considerably (Hong et al. 2019; Shi et al. 2020). Given that enzymes boost the extraction yield of secondary metabolites due to their release from complicated substrates, they are used as part of a pretreatment in a SmF as a downstream process (Verduzco-Oliva and Gutierrez-Uribe 2020; López-Gómez and Venus 2021). The cultivation of microorganisms in liquid nutrient broth is known as submerged fermentation. This method can be used to make industrial enzymes (Gupta et al. 2022; Lübeck and Lübeck 2022; Mitri et al. 2022). This entails culturing carefully selected microorganisms in controlled vessels with a nutrient-rich broth (fermentation medium) and a high oxygen content. Microorganisms release the desired enzymes into the solution while they break down the nutrients (Castro and Sato 2014; Yankey et al. 2022). Amylases, proteases, lipases, pectinases, and other enzymes that accelerate biological reactions are widely used in both domestic and industrial applications, as illustrated in Table 5.2.

Table 5.2 The raw materials for bioenzyme production mechanisms and optimization

Raw materials	Mechanisms	Strain	Optimization					
			Temp	pH	Enzyme Conc.	Dosage	Time	Reference
Laboratory-cultured activated sludge	Precipitation and re-suspension	–	–	–	5.4 U/mL	Ratio of 1:1	2 hrs.	Jung et al. (2002)
Agro-industrial wastes	Solid-state fermentation	*Penicillium restrictor*	–	–	0.1% w/v	1200 mg/L	–	Damasceno et al. (2008)
Fruit peel waste	Fermentation		35 °C	7	–	–	60 hrs.	Arun and Sivashanmugam (2015)
Pulp and paper biosludge	Anaerobic digestion	*Bacillus licheniformis*	–	–	–	–	–	Bonilla et al. (2018)
–	–	–	55 ± 2 °C	–	30 mg/g	7.5 µg/L	360 hrs.	Zhou et al. (2018)
Organic waste (fruit peels)	AF and ASP	–	35 °C	–	56.5 U/m	–	16 hrs.	Selvakumar and Sivashanmugam (2019)
Hair waste	Fermentation, ultrafiltration, and lyophilization	–	–	–	7556–9373 U/cm^2	–	24 hrs	Abu Yazid et al. (2016)
–	–	*Bacillus licheniformis*	40 °C	–	7 mg/mL	–	1 hr	El Hadj-Ali et al. (2007)
Flower waste	–	–	–	–	10 to 15%	–	120 hrs.	Kumar et al. (2020a)
Fruit and vegetable waste	–	–	–	6–8	10%	–	240 hrs.	Chakraborty and Mohan (2019)
Citrus fruit peel	Agar well diffusion method	*E. Coli*, *Pseudomonas aeruginosa*	37 °C	3.6	15%	–	24 hrs.	Srimathi et al. (2020)
–	–	–				–		

AF Anaerobic Fermentation, *ASP* Ammonium Sulfate Precipitation, *°C* Celsius, *U/g* unit per gram, *U/mg* unit per milligram, *U/ml/min* unit per milliliter per minute, *U/cm*^{2} unit per centimeter square, *Hr* hours, *SSF* Solid State Fermentation, *SF* Submerged State Fermentation

These hydrolytic enzymes, also known as bioenzyme or garbage enzymes, can be easily prepared at home from vegetable and fruit whey (Kanwar et al. 2012; Liu and Smith 2021; Raheem et al. 2020). Bioenzyme has recently attracted considerable attention as the most cost-effective and environmentally friendly alternative, as given in Table 5.2. Bioenzyme, also known as garbage enzyme or fruit enzyme, is a multi-purpose, natural cleaner made from wastes such as citrus peels or a mixture of complex organic substances such as proteins, salts, and other materials that are naturally occurring byproducts of the bacteria/yeast that will be used to make the bioenzyme (Agarwal and Kaur 2014; Raheem et al. 2020; Aartheeswari and Kirthiga 2021).

5.4 Application of Bioenzymes for Different Purposes

One of the most important needs of food science and technology to achieve Global Sustainability Goals is the introduction of different bioprocesses, such as enzymatic and microbiological exploitation of food waste to ensure safe food production (Singhania et al. 2021; Wohlgemuth et al. 2021; Torres - Lenon et al. 2021; Melendez et al. 2022). The selection of an appropriate substrate has an impact on the success of enzyme synthesis for various applications (Lillford and Hermansson 2021; Wiltschi et al. 2020; Mamimin et al. 2021; Sharma et al. 2021).

5.4.1 Bio Adsorbent

Adsorption is a prominent and cost-effective approach for eliminating a variety of contaminants. In terms of initial cost, elasticity and simplicity of layout, ease of operation, and insensitivity to harmful contaminants, adsorption has been demonstrated to be superior to alternative strategies for water reuse, as shown in Fig. 5.2 (Sirajudheen et al. 2021; Solanki et al. 2022). Activated carbon is the most prominent and commonly exploited adsorbent material for the treatment of many pollutants, particularly organic pollutants (Jamil et al. 2020; Li Z et al. 2020b).

The synergetic effect produced from using rich bio-waste as a substrate for biological treatment is a promising protocol for environmental health and pollution reduction, as well as the development of green, environmentally friendly activated materials for dye adsorption (Abdelraof et al. 2019; Hasanin et al. 2020; Youssef et al. 2019). As a result, improving the binding and adsorption efficiency of pectin-derived or other polymer bio adsorbents through enzymatic or chemical modification of pectin structure is advantageous (Li and Buschle-Diller 2017; Luo et al. 2020; Panwar et al. 2021).

5.4.2 *Biosurfactant Production*

Biosurfactants (e.g., rhamnolipids, surfactin, sophorolipids, and peptidolipids) are produced as secondary metabolites by fungi, yeasts, and bacteria during their growth, as given in Fig. 5.2 (Nadaf et al. 2021; Banerjee et al. 2022). *Pleurotus djamor* has been known to produce biosurfactants (surfactin and rhamnolipids) on sunflower seed shells as substrates supplemented with sunflower seed oil as a carbon source underneath the SSF (Jimenez-Penalver et al. 2016; Velioglu, Z., & Urek, R. O. (2016).

Biosurfactant production has received a lot of attention in recent years because of its environmentally beneficial characteristics, such as low toxicity, high efficiency, and high biodegradability, as compared to chemical surfactant synthesis, which is mostly based on petroleum (Jimémenz-Peñalver et al. 2019; Domínguez Rivera et al. 2019). The yield of biosurfactant is related to the decomposition of hydrocarbons, as the cell retains the biosurfactant and the outer cell surface reduces the hydrophobic cell surface, enabling the alkanes to attach and be transferred to the interior of the cells (Jimoh and Lin 2019; Lopez-Prieto et al. 2019). Microbial species such as *Rhodococcus*, *Pseudomonas*, *Bacillus*, and *Candida* spp. produce a variety of biosurfactants (Litvinenko et al. 2019; Cazals et al. 2020; Markande et al. 2021).

5.4.3 *Bioethanol and Biofuel Production*

Bioconversion of organic solid wastes instead of food as a source of bioethanol using solid-state fermentation (SSF) is appropriate and desirable for sustainability and renewable energy production, as given in Fig. 5.2 (Chilakamarry et al. 2022; Molina-Peñate et al. 2022; Khalid et al. 2022). In order to attain zero waste generation, the researchers suggested a fascinating integrated bioconversion of potato peel by SSF for the production of bioethanol and manure employing yeast and fungus (*Aspergillus niger*, *Aspergillus variabilis*, and *Saccharomyces cerevisiae*) (Liu et al. 2015; Song et al. 2019; John et al. 2020).

Municipal solid wastes, which can solve domestic garbage disposal and so limit environmental concerns that may arise as a result of such problems, are a viable choice for raw materials that have the potential for bioethanol production as an alternative to agricultural cellulosic residues (Nikku et al. 2019; Alio et al. 2020; Baramee et al. 2020). Continuous depletion of fossil-derived crude oil increases environmental issues due to greenhouse gas emissions from conventional fuel burning and the disposal of food waste (FW) in landfills, illuminating the approach to biofuel generation. Food waste biofuel production is the most promising, cost-effective, and environmentally benign alternative to sustainable development and circular bioeconomy (Li and Yang 2016; Kannah et al. 2020). Thermal approaches have received a lot of attention in the previous few years for converting biomass into biofuels, including bio-oil, biogas, and biochar (Qin et al. 2021; Ren et al. 2022).

Wet biomass conversion techniques such as hydrothermal carbonization (HTC) are commonly used to convert food waste, municipal sludge, animal manures, and water-containing waste feedstock into a solid product, termed hydrochar, with improved calorific value, hydrophobicity, and homogeneity. Hydrochar appears to be a good solid-fuel alternative with enormous potential (Huang et al. 2018; Paul & Joshi 2022).

The ultimate quality of hydrochar is strongly influenced by the production parameters (Karmee et al. 2015; Wang et al. 2018; Pal et al. 2019). Corn is being used for a broad range of applications, from food to biofuels. Under optimal conditions, co-culturing *Aspergillus niger* and *Saccharomyces cerevisiae* on potato peel and mash wastes yields 6.18% v/v and 9.3% v/v of bioethanol, respectively, as a step toward sustainable potato waste management and contributing to the circular bioeconomy (Hosseinzadeh-Bandbafha et al. 2022; Narisetty et al. 2022).

After 72 h of incubation, *E. aerogenes*-mediated saccharification and fermentation of pineapple ethanol stillage yields 1.67%, or 1.42 g ethanol/g sugar, whereas co-culturing of *Enterobacter aerogenes* and *S. cerevisiae* yields around 1.762 g ethanol/g sugar from the same substrate and incubation period (Selvakumar and Sivashanmugam 2019; Kumar et al. 2020b; Qureshi et al. 2020; Sharma et al. 2021).

5.4.4 Biopesticide Production

Citrus waste-based bio-fertilizers have good antibacterial qualities and eliminate hazardous heavy metals from the soil due to their high pH and lignocellulose (Zema et al. 2018; Inam et al. 2022; Suriet al. 2022). Chemical pesticides have long been known to have serious health consequences as well as environmental consequences. In recent years, the shift to biopesticides has become more attractive. *Bacillus thuringiensis* is a typical biopesticide producer that produces a crystalline protein called -endotoxin. Several investigations examined SSF and sub-merged fermentation (SmF) synthesis of -endotoxin by *B. thuringiensis* (Jisha & Benjamin, 2014; Mejias et al. 2020; Sala et al. 2020).

The use of soy fabric residue as a substrate for the production of viable cells (3.8×10^9 colony forming unit (CFU)/g dry mass (DM)) and spores (1.3×10^8 spores/g DM) of *B. thuringiensis* under non-sterile environments facilitated the scale up to 10 L SSF reactors (9.5×10^7 CFU/g DM; 1.1×10^8 spores/g DM), reduced production costs, and simplified waste management (Ballardo et al. 2016; Mishra et al. 2016). Some nanomaterials possessing peroxidase or oxidase-like catalytic properties have been investigated and employed in pesticide analysis, as shown in Fig. 5.2. Despite the fact that the development of inorganic nanoenzymes has made pesticide detection more sustainable and affordable, bioenzymes such as acetylcholinesterase (AChE) are still required in these circumstances (Huang et al. 2019; Liu et al. 2020).

Pesticide residue identification is important since it affects food and environmental safety. Although several sensors for pesticide analysis have been produced and used, the majority of them rely on the usage of fragile bioenzymes, which decreases the sensors' advantages (Mahawar et al. 2020; Liu et al. 2021). Future studies will need to replicate the study on bigger scales and field test the finished product's pesticide activity before it can be incorporated into the expanding global market for biopesticides, which is supposed to hit USD 11,438.1 million in 2026 (Mahawar et al. 2020; Molina-Penate et al. 2022).

5.4.5 *Biofertilizer*

Rose waste compost was made from a variety of rose waste, poultry manure, and sawdust combinations (Mahapatra et al. 2022; Yadav et al. 2022). When combined with fertilizer application, the compost significantly increased organic matter, accessible potassium and phosphorus, nitrogen, and other micronutrients in the soil, resulting in increased cut rose yield and quality. By substituting this recycled fertilizer (biofertilizer) for traditional synthetic fertilizers used in hydroponic cultivation, it may be possible to achieve sustainable urban food production in proximity to retailers and customers (Liu et al. 2016; Idrovo-Novillo et al. 2019).

Liquid organic fertilizers made from organic products produced from household waste, such as vegetable and fruit waste, are known as ecoenzymes (EEs). EEs can improve physical qualities, soil chemical and biological properties, and product quality by adding organic matter (Rani et al. 2020; Novianto 2022).

Organic acid content plays an important role in correcting physical attributes, soil chemicals, and microbes for plants. It acts as a soil amendment for soil, providing nutrients, substance, and growth regulatory metabolites for plants, as well as protecting the roots from pests and diseases and stimulating the root system to grow in an optimal way (O'Connor et al. 2021). In comparison to the use of mineral fertilizers, biofertilizers added to nutritional composition in tiny amounts for plants shown are shown in Fig. 5.2. Organic fertilizer, on the other hand, must be applied at frequent intervals and must be maintained continuously. Liquid organic fertilizer created from domestic waste fermentation, such as leftover vegetables and fruits, can be converted into material for producing liquid natural fertilizers in the form of EE (Rasit et al. 2019; Novianto 2022).

5.4.6 *Nanomedicines for Enhanced Cancer Therapy*

Bioenzymes, which accelerate events inside living systems, have a lot of potential for cancer therapy, especially when combined with nanoparticles to increase tumor site accumulation (Liu et al. 2022). Nanomedicines can carry harmful bioenzyme straight into cancer cells, causing them to die. This can be used to cure cancer.

Bioenzyme-based nanomedicine improves the therapeutic efficacy of treatments by modifying the tumor microenvironment, such as pH, glucose concentration, hypoxia, redox levels, and heat shock protein expression (Qin et al. 2021; Ding et al. 2022). Because bioenzyme activity is easily affected by temperature, the photothermal effect induced by NIR light can be exploited to adjust the activity of bioenzyme nanomedicine for precision cancer treatment. Cheng's team attempted to encapsulate glucoamylase (GA), catalase (CAT), and proteinase K (ProK) enzymes by using platinum (Pt) nanoparticles. This process resulted in the formation of enzyme-embedded Pt nanoparticles (E-Pt) (Wang et al. 2017; Hu et al. 2021; Mitchell et al. 2021).

Due to their non-cytotoxicity and biodegradability, biopolymers and their composites have immense potential in the creation of biocompatible medical devices, implants, and drug delivery carriers, as shown in Fig. 5.2. They replicate a body part and can successfully replace a damaged organ or structure that is impairing normal body functionality, such as a braided vascular prosthetic blood vessel and a PLA-based simulating ligament (Rebelo et al. 2017; O'Connor et al. 2022).

To make therapeutic soap, researchers used *Citrus sinensis* seed oil that had been re-covered using the Soxhlet extraction process and saponified it with natural lye solution in various proportions. The seed oil included 36% linoleic acid and 27% oleic acid, according to the study. Citrus seed oil soap has remarkable antimicrobial, antifungal, antiparasitic, and antioxidant characteristics (Mancini et al. 2018; Atolani et al. 2020; Li et al. 2020a). Nanomedicines have been commonly used to increase the effectiveness of cancer therapy, as demonstrated in Fig. 5.2. Exemplifying enzymes in this nanoparticle improves their stability and circulation, providing for high bioavailability and efficient tumor accumulation (Ke et al. 2019; Zhang et al. 2020; Bahreyni et al. 2020; Liao et al. 2020).

Chito-oligosaccharides (COSs) are the partially hydrolyzed products of chitin, which is abundant in the shells of crustaceans, the cuticles of insects, and the cell walls of fungi. These oligosaccharides have received immense interest in the last few decades due to their highly promising bioactivities, such as their antimicrobial, antitumor, and anti-inflammatory properties (Liang & Han, 2020; Taokaew and Kriangkrai 2023).

5.4.7 *Bio Soil Stabilization*

Many bioenzymes have recently been introduced to be cost-effective soil stabilizers. Terrazyme, a bioenzyme, was employed in this investigation to see how it affected the Black Cotton soil's unconfined compressive strength, as shown in Fig. 5.2. With a longer curing period, Terrazyme-treated Black Cotton soil exhibits a significant improvement in Unconfined Compressive strength (Agarwal and Kaur 2014). Soil stabilization is a strategy for improving the qualities of soil using alternative techniques such as chemical, biological, or mechanical means to transform soil into a substance with acceptable attributes. The bearing capacity and shear strength of the stabilized soil have increased, whereas permeability and compressibility have decreased (Rahayu & Situmeang. 2021; Heidemann et al. 2020).

The current period of ground improvement began in the 1960s and 1970s, when generalized limitations of aggregates and fuel resources pushed engineers to investigate alternatives to the traditional method of replacing bad soils on construction projects with shipped-in aggregates with better engineering properties. Modern practices are characterized by recurrent endeavor to ensure adequate subgrade stability, particularly in weaker soils (Mekonnen et al. 2020; Fazal et al. 2021; Mupambwa et al. 2022).

5.4.8 *Biohydrogen*

Biohydrogen biomass is divided into three generations: first, second, and third. The first generation of biomass includes crops and starches with high sugar content, such as potato, sweet sorghum, sugar beet, pumpkin, wheat, oily plants, and seeds, as well as their wastes after treatment, that are grown for plants and animals. Acid pretreatment breaks the polymeric linkages and increases cellulose availability and biodegradability by converting hemicellulose into monomers (Yadav et al. 2020; Zhang et al. 2020; Sampath et al. 2020). Mechanical comminution and irradiation are two physical pretreatment techniques of biohydrogen production (Donkor et al. 2022; Nahak et al. 2022). Ozone does not leave out hazardous, acidic, or alkaline residues in the treatment of resources, yet it is regarded as a powerful oxidizing agent (Hitam and Jalil 2020).

The production of biohydrogen was also found to be 30% greater when only food waste was used as a substrate, as shown in Fig. 5.2. When flowers were mixed with sewage sludge, the investigations demonstrate increased microbial activity, which accounted for the enhanced bio-hydrogen generation (Yang and Wang 2018; Mohan et al. 2020; Sharma et al. 2021).

5.5 Future Prospective of Bioenzyme Products

Cost-effective solutions for the industrial synthesis of value-added compounds are provided by the manufacturing of industrially significant enzymes from the abundantly available agro-industrial food waste. The most recent advancements in biocatalytic systems aim to either improve the catalytic efficiency of existing commercial enzymes or generate new enzymes with unique characteristics (Castro et al. 2015; Sharma et al. 2022). The development of fermentation techniques in the latter decades of the previous century, specifically intended for the manufacture of enzymes, made it possible to produce enzymes as processed, well-characterized preparations even on a large scale (Zhang et al. 2017; Muthusamy et al. 2022).

Microbial transformations based on biotechnology offer effective, affordable, and sustainable methods for producing products with added value. Harnessing organically abundant leftovers brings up new opportunities for the development of enzymes, pigments, biofuels, bioactive substances, biopolymers, and other products with extensive commercial and therapeutic uses. The use of cutting-edge technologies is possible, including strain improvement, enzyme immobilization, genome editing, morphological engineering, ultrasound/supercritical fluid extraction, pulse electric field extraction, etc. (Jeno et al. 2021; Sodhi et al. 2022).

Global population growth has significantly increased the need for sources of food as well as the sectors that manufacture food, which routinely produce significant volumes of food waste. Waste from the food processing industry includes peels, seeds, straws, stalks, and other materials. Since ancient times, agro-industrial waste has drawn a lot of attention from people all over the world since it is typically burned or dumped, threatening both human health and the environment (Souza et al. 2022; Naik et al. 2023).

There are possible applications for waste materials obtained from coffee manufacturing, such as spent coffee grounds (SCGs) from post-consumption and coffee leaves and blossoms from cultivation and processing, as well as coffee pulps, husks, and silverskin. By creating an appropriate infrastructure and networks between researchers, industries, and policymakers, it is possible to fully utilize these coffee byproducts and reduce the environmental and financial burdens of processing coffee in a sustainable way (Banožić et al. 2019; Lee et al. 2023).

Agro wastes are essential substrates that can be used to make enzymes at a minimal cost as opposed to being dumped and left to naturally degrade. Additionally, it has been determined that soil is a significant resource for microbial production of significant products. Five bacteria species, including *E. coli*, *B. subtilis*, *S. aureus*, *E. faecalis*, and *P. aeroginosa*, were isolated and biochemically characterized (Hussain et al. 2023).

The production of millions of tons of leftovers by the fruit and vegetable sector might result in significant financial losses. The wastes and byproducts from fruits and vegetables are rich in bioactive compounds that have antioxidant, antibacterial, and other effects. The development of biorefinery methods is crucial for the long-term synthesis of important chemicals. The byproducts and trash from fruits and vegetables can be used to make biofuels, food bioactive chemicals, and other items (Rasit and Kuan. 2018; Biundo et al. 2023; Zhu et al. 2023).

5.6 Conclusion

Bioenzyme production may be done with a variety of resources, and it is regarded as one of the most sustainable resources on the planet. Food waste, agricultural waste (bagasse, molasses, husks, seeds, stems, leaves, straw, tails, shells, mash, stubble, strip, and roots), and some industrial waste have all played a significant role in the industry and ecology. The best and most effective pretreatments are those that do not require particle size reduction, preserve biomass and fermenting microorganisms, use the least amount of energy, have low operating costs (operating costs, capital costs, and biomass costs), use little or no chemicals, and use inexpensive chemicals. According to the findings of the study, a wide range of high-titer industrial enzymes can be produced from a variety of sources, and the enzymes produced can be used in industrial applications and value-added product development processes to create biobased/bioenzyme products that can be used as a sustainable resource for stuff like bio adsorbent, biopesticide, biofuel or bioethanol, bio soil stabilizer, biohydrogen, biofertilizer, and biosurfactant for sustainable environment life.

Acknowledgments The authors express their gratitude to Addis Ababa Science and Technology University for their support. We extend our sincere gratitude to Addis Ababa Science and Technology University for their invaluable support in the creation of this chapter. Their support extended beyond the chapter's development, encompassing the facilitation of various workshops dedicated to honing the scientific art of writing for journals and articles. We appreciate the university's commitment to fostering academic excellence and research skills among its community.

Author's Contribution All authors, Gamachis Korsa, Chandran Masi, Digafe Alemu, Abera Beyene, and Abate Ayele, have equally contributed to this chapter. Furthermore, these authors have approved the latest version of the manuscript and agree to be held accountable for the content therein.

Availability of Data and Materials The datasets used or analyzed during the preparation of the manuscript are available from the corresponding author upon reasonable request.

Declarations Ethics approval and consent to participate are not applicable.

Competing Interests The authors declare that they have no competing interests.

Consent for Publication Not applicable.

Funding No fund was received from any funding agency or organization toward making this manuscript.

References

Aartheeswari S, Kirthiga B (2021) Production of an Ecofriendly Enzyme Biocleaner from Fruit Wastage;International Journal for Research in Engineering Application & Management (IJREAM) ISSN : 2454-9150 Vol-06, Issue-12, MAR 2021.

Abdelraof M, Hasanin MS, El-Saied H (2019) Ecofriendly green conversion of potato peel wastes to high productivity bacterial cellulose. Carbohydrat Poly 211:75–83. https://doi.org/10.1016/j.carbpol.2019.01.095

Abraham J, Gea T, Sánchez A (2013) Potential of the solid-state fermentation of soy fibre residues by native microbial populations for bench-scale alkaline protease production.J. Biochem Eng 74:15–19. https://doi.org/10.1016/j.bej.2013.02.008

Afrisham S, Badoei-Dalfard A, Namaki-Shoushtari A, Karami Z (2016) Characterization of a thermostable, CaCl2-activated and raw-starch hydrolyzing alpha-amylase from *Bacillus licheniformis* AT70: production under solid state fermentation by utilizing agricultural wastes. J Mol Cataly B: Enzymatic 132:98–106. https://doi.org/10.1016/j.molcatb.2016.07.002

Agarwal P, Kaur S (2014) Effect of bio-enzyme stabilization on unconfined compressive strength of expansive soil. Int J Res Eng Technol 3(5):30–33. http://www.ijret.org

Aktamovich IR, Mirzayevich KB (2022) The role of ecological values in the private perspective in the process of globalization. Eur J Life Saf Stab 2660-9630(15):68–73. http://www.ejlss.indexedresearch.org/index.php/ejlss/article/view/492

Akyüz A, Ersus S (2021) Optimization of enzyme assisted extraction of protein from the sugar beet (*Beta vulgaris* L.) leaves for alternative plant protein concentrate production. Food Chem 335:127673. https://doi.org/10.1016/j.foodchem.2020.127673

Alio MA, Tugui OC, Rusu L, Pons A, Vial C (2020) Hydrolysis and fermentation steps of a pre-treated sawmill mixed feedstock for bioethanol production in a wood biorefinery. Bioresour Technol 310:123412. https://doi.org/10.1016/j.biortech.2020.123412

Arun C, Sivashanmugam P (2015) Investigation of biocatalytic potential of garbage enzyme and its influence on stabilization of industrial waste activated sludge. Proc Safety Environ Prot 94:471–478. https://doi.org/10.1016/j.psep.2014.10.008

Atolani O, Adamu N, Oguntoye OS, Zubair MF, Fabiyi OA, Oyegoke RA et al (2020) Chemical characterization, antioxidant, cytotoxicity, anti-toxoplasma gondii and antimicrobial potentials of the *Citrus sinensis* seed oil for sustainable cosmeceutical production. Heliyon 6(2):e03399. https://doi.org/10.1016/j.heliyon.2020.e03399

Awasthi MK, Sindhu R, Sirohi R, Kumar V, Ahluwalia V, Binod P et al (2022) Agricultural waste biorefinery development towards circular bioeconomy. Renew Sustain Energy Rev 158:112122. https://doi.org/10.1016/j.rser.2022.112122

Bahreyni A, Mohamud Y, Luo H (2020) Emerging nanomedicines for effective breast cancer immunotherapy. J Nanobiotechnol 18(1):1–14. https://doi.org/10.1186/s12951-020-00741-z

Ballardo C, Abraham J, Barrena R, Artola A, Gea T, Sánchez A (2016) Valorization of soy waste through SSF for the production of compost enriched with *Bacillus thuringiensis* with biopesticide properties. J Environ Mgt 169:126–131. https://doi.org/10.1016/j.jenvman.2015.12.029

Banerjee B, Kaur G, Priya A (2022) Naturally occurring bioactive biosurfactants. In: Green sustainable process for chemical and environmental engineering and science. Acad Press, pp 337–350. https://doi.org/10.1016/B978-0-323-85146-6.00015-2

Banožić M, Banjari I, Jakovljević M, Šubarić D, Tomas S, Babić J, Jokić S (2019) Optimization of ultrasound-assisted extraction of some bioactive compounds from tobacco waste. Mol 24(8):1611. https://doi.org/10.3390/molecules24081611

Baramee S, Siriatcharanon AK, Ketbot P, Teeravivattanakit T, Waeonukul R, Pason P et al (2020) Biological pretreatment of rice straw with cellulase-free xylanolytic enzyme-producing *Bacillus firmus* K-1: structural modification and biomass digestibility. Renewab Ener 160:555–563. https://doi.org/10.1016/j.renene.2020.06.061

Battista F, Fino D, Mancini G (2016) Optimization of biogas production from coffee production waste. Bioresour Technol 200:884–890. https://doi.org/10.1016/j.biortech.2015.11.020

Bharathiraja S, Suriya J, Krishnan M, Manivasagan P, Kim SK (2017) Production of enzymes from agricultural wastes and their potential industrial applications. Adv Food Nutr Res 80:125–148). Acad Press. https://doi.org/10.1016/bs.afnr.2016.11.003

Bhatt B, Prajapati V, Patel K, Trivedi U (2020) Kitchen waste for economical amylase production using *Bacillus amyloliquefaciens* KCP2. Biocatalys Agr Biotechnol 26:101654. https://doi.org/10.1016/j.bcab.2020.101654

Birhanli E, Yeşilada Ö (2013) The utilization of lignocellulosic wastes for Laccase production under semisolid-state and submerged fermentation conditions. Turkish J Biol 37(4):450–456. https://doi.org/10.3906/biy-1211-25

Birwal P, Deshmukh G, Priyanka SS, Saurabh SP (2017) Advanced technologies for dairy effluent treatment. J Food Nutr Popul Health 1(1):7. http://www.imedpub.com/food-nutrition-and-population-health/

Biswal D, Mandavgane SA (2021) Biomass wastes: a potential feedstock for cellulase production. In: Cur Stat Fut Scop Microbial Cel. Elsevier, pp 347–359. https://doi.org/10.1016/B978-0-12-821882-2.00017-X

Biundo A, Stamm A, Gorgoglione R, Syrén PO, Curia S, Hauer B, Capriati V, Vitale P, Perna F, Agrimi G, Pisano I (2023) Regio-and stereoselective biocatalytic hydration of fatty acids from waste cooking oils en route to hydroxy fatty acids and bio-based polyesters. Enzyme Microbial Technol 163:110164. https://doi.org/10.1016/j.enzmictec.2022.110164

Biz A, Finkler ATJ, Pitol LO, Medina BS, Krieger N, Mitchell DA (2016) Production of pectinases by solid-state fermentation of a mixture of citrus waste and sugarcane bagasse in a pilot-scale packed-bed bioreactor. J Biochem Eng 111:54–62. https://doi.org/10.1016/j.bej.2016.03.007

Bolte EE, Moorshead D, Aagaard KM (2022) Maternal and early life exposures and their potential to influence development of the microbiome. Genome Med 14(1):1–29. https://doi.org/10.1186/s13073-021-01005-7

Bolzonella D, Battista F, Cavinato C, Gottardo M, Micolucci F, Lyberatos G, Pavan P (2018) Recent developments in biohythane production from household food wastes: a review. Bioresour Technol 257:311–319. https://doi.org/10.1016/j.biortech.2018.02.092

Bonilla S, Choolaei Z, Meyer T, Edwards EA, Yakunin AF, Allen DG (2018) Evaluating the effect of enzymatic pretreatment on the anaerobic digestibility of pulp and paper biosludge. Biotechnol Rep 17:77–85. https://doi.org/10.1016/j.btre.2017.12.009

Capanoglu E, Nemli E, Tomas-Barberan F (2022) Novel approaches in the valorization of agricultural wastes and their applications. J Agr Food Chem 70(23):6787–6804. https://doi.org/10.1021/acs.jafc.1c07104

Castro RJS, Sato HH (2013) Synergistic effects of agroindustrial wastes on simultaneous production of protease and α-amylase under solid state fermentation using a simplex centroid mixture design. Indust Crop Prod 49:813–821. https://doi.org/10.1016/j.indcrop.2013.07.002

Castro RJS, Sato HH (2014) Production and biochemical characterization of protease from *Aspergillus oryzae*: an evaluation of the physical–chemical parameters using agro-industrial wastes as supports. Biocatalys Agr Biotechnol 3(3):20–25. https://doi.org/10.1016/j.bcab.2013.12.002

Castro RJS, Ohara A, Nishide TG, Bagagli MP, Dias FFG, Sato HH (2015) A versatile system based on substrate formulation using agro-industrial wastes for protease production by *Aspergillus Niger* under solid state fermentation. Biocatalys Agr Biotechnol 4(4):678–684. https://doi.org/10.1016/j.bcab.2015.08.010

Cazals F, Huguenot D, Crampon M, Colombano S, Betelu S, Galopin N et al (2020) Production of biosurfactant using the endemic bacterial community of a PAHs contaminated soil, and its potential use for PAHs remobilization. Sci Total Environ 709:136143. https://doi.org/10.1016/j.scitotenv.2019.136143

Cerda A, El-Bakry M, Gea T, Sánchez A (2016) Long term enhanced solid-state fermentation: inoculation strategies for amylase production from soy and bread wastes by *Thermomyces* sp. in a sequential batch operation. J Environ Chem Eng 4(2):2394–2401. https://doi.org/10.1016/j.jece.2016.04.022

Chakraborty D, Mohan SV (2019) Efficient resource valorization by co-digestion of food and vegetable waste using three stage integrated bioprocess. Bioresour Technol 284:373–380. https://doi.org/10.1016/j.biortech.2019.03.133

Chaudhary K, Padhiar A (2020) Cellulase production by fungi from agro wastes under solid state fermentation. Biosci Biotech Res Comm 13(3):1495–1501. https://doi.org/10.21786/bbrc/13.3/76

Chen DMC, Bodirsky BL, Krueger T, Mishra A, Popp A (2020) The world's growing municipal solid waste: trends and impacts. Environ Res Lett 15(7):074021. https://doi.org/10.6084/m9.figshare.12102510

Chen M, Wang M, Zhang Y, Zhag H, Du Q, Jin P (2022) Biosynthesis of hyaluronan in engineered *Escherichia coli* via the secretion of thermophilic exo-mannanase using palm kernel cake as the carbon source.J. Biochem Eng 177:108254. https://doi.org/10.1016/j.bej.2021.108254

Chilakamarry CR, Sakinah AM, Zularisam AW, Sirohi R, Khilji IA, Ahmad N, Pandey A (2022) Advances in solid-state fermentation for bioconversion of agricultural wastes to value-added products: opportunities and challenges. Bioresour Technol 343:126065. https://doi.org/10.1016/j.biortech.2021.126065

Chimbekujwo KI, Ja'afaru MI, Adeyemo OM (2020) Purification, characterization and optimization conditions of protease produced by *Aspergillus brasiliensis* strain BCW2. Afr Scient 8:e00398. https://doi.org/10.1016/j.sciaf.2020.e00398

Coradi PC, Lima RE, Padia CL, Alves CZ, Teodoro PE, da Silva Candido AC (2020) Soybean seed storage: Packaging technologies and conditions of storage environments. Journal of Stored Products Research, 89, 101709.

de Castro AM, López JA, dos Reis Castilho L, Freire DMG (2014) Technoeconomic analysis of a bioprocess for the production of multienzyme solutions from the cake of babassu industrial processing: evaluation of five different inoculum propagation strategies. Biomass Conversion and Biorefinery, 4, 237–247.

Daâssi D, Zouari-Mechichi H, Frikha F, Rodríguez-Couto S, Nasri M, Mechichi T (2016) Sawdust waste as a low-cost support-substrate for Laccase production and adsorbent for azo dyes decolorization. J Environ Health Sci Eng 14(1):1–12. https://doi.org/10.1186/s40201-016-0244-0

Damasceno FR, Freire DM, Cammarota MC (2008) Impact of the addition of an enzyme pool on an activated sludge system treating dairy wastewater under fat shock loads. J Chem Technol Biotechnol: Int Res Pro Environ Clean Technol 83(5):730738. https://doi.org/10.1002/jctb.1863

Das A, Mondal C (2013) Studies on the utilization of fruit and vegetable waste for generation of biogas. Int J Eng Sci 3(9):24–32. www.researchinventy.com

Dave BR, Parmar P, Sudhir A, Singal N, Subramanian RB (2021) Cellulase production under solid state fermentation using agro waste as a substrate and its application in saccharification by *Trametes hirsuta* NCIM. J Microbiol Biotechnol Food Sci 203:203–208. https://doi.org/10.15414/jmbfs.2014-15.4.3.203-208

Di Donna L, Bartella L, De Vero L, Gullo M, Giuffrè AM, Zappia C et al (2020) Vinegar production from *Citrus bergamia* by-products and preservation of bioactive compounds. European Food Res Technol 246(10):1981–1990. https://doi.org/10.1007/s00217-020-03549-1

Ding M, Zhang Y, Li J, Pu K (2022) Bioenzyme-based nanomedicines for enhanced cancer therapy. Nano Converg 9(1):1–20. https://doi.org/10.1186/s40580-022-00297-8

Domínguez Rivera Á, Martínez Urbina MÁ, Lópezy López VE (2019) Advances on research in the use of agro-industrial waste in biosurfactant production. World J Microbiol Biotechnol 35(10):1–18. https://doi.org/10.1007/s11274-019-2729-3

Donkor KO, Gottumukkala LD, Lin R, Murphy JD (2022) A perspective on the combination of alkali pre-treatment with bioaugmentation to improve biogas production from lignocellulose biomass. Bioresour Technol 1:126950. https://doi.org/10.1016/j.biortech.2022.126950

Eichler P, Bastiani DC, Santos FA, Ayub MA (2020) Lipase production by *Aspergillus brasiliensis* in solid-state cultivation of malt bagasse in different bioreactors configurations. Anais da Acad Brasileira de Ciênc 92:e20180856. https://doi.org/10.1590/0001-3765202020180856

El Hadj-Ali N, Agrebi R, Ghorbel-Frikha B, Sellami-Kamoun A, Kanoun S, Nasri M (2007) Biochemical and molecular characterization of a detergent stable alkaline serine-protease from a newly isolated *Bacillus licheniformis* NH1. Enz Microbial Technol 40(4):515–523. https://doi.org/10.1016/j.enzmictec.2006.05.007

Erdoğan S, Onifade ST, Altuntaş M, Bekun FV (2022) Synthesizing urbanization and carbon emissions in Africa: how viable is environmental sustainability amid the quest for economic growth in a globalized world? Environ Sci Pollut Res 1-14:24348. https://doi.org/10.1007/s11356-022-18829-4

Escaramboni B, Garnica BC, Abe MM, Palmieri DA, Fernández Núñez EG, de Oliva NP (2022) Food waste as a feedstock for fungal biosynthesis of amylases and proteases. Waste Biomass Valor 13(1):213–226. https://doi.org/10.1007/s12649-021-01511-0

Esparza I, Jimenez-Moreno N, Bimbela F, Ancín-Azpilicueta C, Gandía LM (2020) Fruit and vegetable waste management: conventional and emerging approaches. J Environ Mgt 265:110510. https://doi.org/10.1016/j.jenvman.2020.110510

Espoui AH, Larimi SG, Darzi GN (2022) Optimization of protease production process using bran waste using *Bacillus licheniformis*. Korean J Chem Eng 1-10:674. https://doi.org/10.1007/s11814-021-0965-3

Fan X, Yuan G, Liu W (2022) Response strategies of N-fixation by epiphytic bryophytes to water change in a subtropical Montane cloud forest. Ecol Indi 135:108527. https://doi.org/10.1016/j.ecolind.2021.108527

Favaro CP, Baraldi IJ, Casciatori FP, Farinas CS (2020) β-Mannanase production using coffee industry waste for application in soluble coffee processing. Biomol Ther 10(2):227. https://doi.org/10.3390/biom10020227

Fazal AE, Thyagaraj KJ, Aravindan S, Shafi SF, Shahzaib M (2021) Stabilization of residual soil from wastewater treatment plant by using bio-enzyme (*Terrazyme*). In: IOP Conf Series: Earth Environ Sci, vol 822. IOP Publishing, p 012032. https://doi.org/10.1088/1755-1315/822/1/012032

Fernandes CD, Oechsler BF, Sayer C, de Oliveira D, de Araújo PHH (2022) Recent advances and challenges on enzymatic synthesis of biobased polyesters via polycondensation. European J Poly 1:111132. https://doi.org/10.1016/j.eurpolymj.2022.111132

Ferraz JLDAA, Souza LO, Soares GA, Coutinho JP, de Oliveira JR, Aguiar-Oliveira E, Franco M (2018) Enzymatic saccharification of lignocellulosic residues using cellulolytic enzyme extract produced by Penicillium roqueforti ATCC 10110 cultivated on residue of yellow mombin fruit. Bioresource technology, 248, 214–220.

Galintin O, Rasit N, Hamzah S (2021) Production and characterization of eco enzyme produced from fruit and vegetable wastes and its influence on the aquaculture sludge. Biointerf Res Appl Chem 11(3):10205–10214. https://doi.org/10.33263/BRIAC113.1020510214

Gaonkar SK, Furtado IJ (2021) Valorization of low-cost agro-wastes residues for the maximum production of protease and lipase haloextremozymes by *Haloferax lucentensis* GUBF-2 MG076078. Process Biochem 101:72–88. https://doi.org/10.1016/j.procbio.2020.10.019

Gaur VK, Sharma P, Sirohi R, Awasthi MK, Dussap CG, Pandey A (2020) Assessing the impact of industrial waste on environment and mitigation strategies: a comprehensive review. J Hazard Mat 398:123019. https://doi.org/10.1016/j.jhazmat.2020.123019

Gopinath LR, Christy PM, Mahesh K, Bhuvaneswari R, Divya D (2014) Identification and evaluation of effective bacterial consortia for efficient biogas production. IOSR J Environ Sci Tox Food Technol 8(3):80–86. www.iosrjournals.org

Gunam IBW, Antara NS, Anggreni AAMD, Setiyo Y, Wiguna IPE, Wijaya IMM, Putra IWWP (2019) Chemical pretreatment of lignocellulosic wastes for cellulase production by *Aspergillus Niger* FNU 6018. AIP Conf Proceed 2155(1):020040. AIP Publishing LLC. https://doi.org/10.1063/1.5125544

Gupta K, Bardhan P, Saikia D, Rather MA, Loying S, Mandal M, Kataki R (2022) Microbial fermentation: basic fundamentals and its dynamic prospect in various industrial applications. Indust Microbiol Biotechnol 1:107–128. https://doi.org/10.1007/978-981-16-5214-1_4

Hasanin MS, Hashem AH, El-Sayed A, Essam S, El-Saied H (2020) Green ecofriendly bio-deinking of mixed office waste paper using various enzymes from *Rhizopus microsporus* AH3: efficiency and characteristics. Cel 27(8):4443–4453. https://doi.org/10.1007/s10570-020-03071-3

Hashem AH, Hasanin MS, Khalil AMA, Suleiman WB (2020) Eco-green conversion of watermelon peels to single cell oils using a unique oleaginous fungus: *Lichtheimia corymbifera* AH13. Waste Biomass Valor 11(11):5721–5732. https://doi.org/10.1007/s12649-019-00850-3

Hauwa H, Ahmed AS (2019) Production of bioenzymes with banana peels by some selected fungi isolated from poultry droppings in Sokoto, north eastern Nigeria. Int J Adv Scient Res Eng 5(6):161–170. https://doi.org/10.31695/IJASRE.2019.3323.1

Heidemann M, Bressani LA, Flores JA (2020) Residual shear strength of a residual soil of granulite. Soil Rock 43(1):31–41. https://doi.org/10.28927/SR.431031

Herculano PN, Moreira KA, Bezerra RP, Porto TS, de Souza-Motta CM, Porto ALF (2016) Potential application of waste from castor bean (*Ricinus communis* L) for production for xylanase of interest in the industry. 3 Biotechnol 6(2):1–10. https://doi.org/10.1007/s13205-016-0463-1

Hitam CNC, Jalil AA (2020) A review on biohydrogen production through photo-fermentation of lignocellulosic biomass. Biomass Conv Bioref 1-19:8465. https://doi.org/10.1007/s13399-020-01140-y

Hong J, Park S-H, Kim S, Kim S-W, Hahn J-S (2019) Efficient production o lycopene in *Saccharomyces cerevisiae* by enzyme engineering and increasing membrane flexibility and NAPDH production. Appl Microbiol Biotechnol 103(1):211–223. https://doi.org/10.1007/s00253-018-9449-8

Hosseinzadeh-Bandbafha H, Nazemi F, Khounani Z, Ghanavati H, Shafiei M, Karimi K et al (2022) Safflower-based biorefinery producing a broad spectrum of biofuels and biochemicals: a life cycle assessment perspective. Sci Total Environ 802:149842. https://doi.org/10.1016/j.scitotenv.2021.149842. https://doi.org/10.1039/D0CS00215A

Hu C, Wang J, Liu S, Cai L, Zhou Y, Liu X et al (2021) Urchin-shaped metal organic/hydrogen-bonded framework nanocomposite as a multifunctional nanoreactor for catalysis-enhanced synergetic therapy. ACS Appl Mater Interf 13(4):4825–4834. https://doi.org/10.1021/acsami.0c19584

Huang R, He L, Zhang T, Li D, Tang P, Zhao Y, Feng Y (2018) Fabrication and adsorption behavior of magnesium silicate hydrate nanoparticles towards methylene blue. Nano 8(5):271. https://doi.org/10.3390/nano8050271

Huang L, Sun DW, Pu H, Wei Q, Luo L, Wang J (2019) A colorimetric paper sensor based on the domino reaction of acetylcholinesterase and degradable γ-MnOOH nanozyme for sensitive detection of organophosphorus pesticides. Sensor Actuator B: Chem 290:573–580. https://doi.org/10.1016/j.snb.2019.04.020

Hussain H, Maikaji FS, Salihu M (2023) Effect of substrate concentration and temperature for bio-enzyme production using enterococcus faecalis isolated from waste dump soil in Sokoto Metropolis. Nigeria J Appl Sci Environ Manag 27(4):665–671. https://doi.org/10.4314/jasem.v27i4.4

Idrovo-Novillo J, Gavilanes-Terán I, Veloz-Mayorga N, Erazo-Arrieta R, Paredes C (2019) Closing the cycle for the cut rose industry by the reuse of its organic wastes: a case study in Ecuador. J Clean Prod 220:910–918. https://doi.org/10.1016/j.jclepro.2019.02.121

Inam A, Mutaf T, Deniz I (2022) Sustainable biorefineries for circular bioeconomy. Biomass biofuel Biochem. Elsevier, pp 3–28. https://doi.org/10.1016/B978-0-323-89855-3.00020-0

Ja'afar JAN, Shitu A (2022) Utilization of lignocellulosic agro-waste as an alternative carbon source for industrial enzyme production. Waste Mgt Pro Valor 2:221–233. https://doi.org/10.1007/978-981-16-7653-6_12

Jain A, Sarsaiya S, Awasthi MK, Singh R, Rajput R, Mishra UC et al (2022) Bioenergy and bioproducts from bio-waste and its associated modern circular economy: current research trends, challenges, and future outlooks. Fuel 307:121859. https://doi.org/10.1016/j.fuel.2021.121859

Jalil MTM, Ibrahim D (2021) Partial purification and characterization of pectinase produced by *Aspergillus Niger* LFP-1 grown on pomelo peels as a substrate. Tropical life Sci Res 32(1):1. https://doi.org/10.21315/tlsr2021.32.1-1

Jamil S, Loganathan P, Kandasamy J, Listowski A, McDonald JA, Khan SJ, Vigneswaran S (2020) Removal of organic matter from wastewater reverse osmosis concentrate using granular activated carbon and anion exchange resin adsorbent columns in sequence. Chemos 261:127549. https://doi.org/10.1016/j.chemosphere.2020.127549

Janveja C, Rana SS, Soni SK (2014) Optimization of valorization of biodegradable kitchen waste biomass for production of fungal cellulase system by statistical modeling. Waste and Biomass Valorization, 5, 807–821.

Jeno JGA, Viveka R, Varjani S, Nagappan S, Nakkeeran E (2021) Current trends and prospects of transforming food waste to biofuels in India. In: Waste bioref. Elsevier, pp 391–419. https://doi.org/10.1016/B978-0-12-821879-2.00014-4

Jiménez-Peñalver P, Gea T, Sánchez A, Font X (2016) Production of sophorolipids from winterization oil cake by solid-state fermentation: optimization, monitoring and effect of mixing. J Biochem Eng 115:93–100. https://doi.org/10.1016/j.bej.2016.08.006

Jiménez-Peñalver P, Rodríguez A, Daverey A, Font X, Gea T (2019) Use of wastes for sophorolipids production as a transition to circular economy: state of the art and perspectives. Reviews in Environmental Science and Bio/Technology, 18(3), 413–435.

Jimoh AA, Lin J (2019) Biosurfactant: a new frontier for greener technology and environmental sustainability. Ecotox Environ Safety 184:109607. https://doi.org/10.1016/j.ecoenv.2019.109607

Jisha VN, Benjamin S (2014) Solid-state fermentation for the concomitant production of δ-endotoxin and endospore from *Bacillus thuringiensis* subsp. kurstaki. Adv Biosci Biotechnol 5(10):797. https://doi.org/10.4236/abb.2014.510093

John I, Pola J, Thanabalan M, Appusamy A (2020) Bioethanol production from musambi peel by acid catalyzed steam pretreatment and enzymatic saccharification: optimization of delignification using Taguchi design. Waste Biomass Valor 11(6):2631–2643. https://doi.org/10.1007/s12649-018-00565-x

Jung J, Xing XH, Matsumoto K (2002) Recoverability of protease released from disrupted excess sludge and its potential application to enhanced hydrolysis of proteins in wastewater. J Biochem Eng 10(1):67–72. https://doi.org/10.1007/s12649-018-00565-x

Kam WYJ, Abas F, Hussain N, Mirhosseini H (2020) Comparison of crude extract from *Durio zibethinus* M.(durian) leaf waste via ultrasound-assisted extraction and accelerated solvent extraction: antioxidant activity and cytotoxicity. Nat Prod Res 34(13):1937–1941. https://doi.org/10.1080/14786419.2018.1564296

Kannah RY, Merrylin J, Devi TP, Kavitha S, Sivashanmugam P, Kumar G, Banu JR (2020) Food waste valorization: biofuels and value added product recovery. Biores Technol Rep 11:100524. https://doi.org/10.1016/j.biteb.2020.100524

Kanwar S, Kumar G, Sahgal M, Singh A (2012) Ethanol production through *Saccharomyces* based fermentation using apple pomace amended with molasses. Sugar Tech 14(3):304–311. https://doi.org/10.1007/s12355-012-0163-z

Karić N, Maia AS, Teodorović A, Atanasova N, Langergraber G, Crini G et al (2022) Bio-waste valorisation: agricultural wastes as biosorbents for removal of (in) organic pollutants in wastewater treatment. J Adv Chem Eng 9:100239. https://doi.org/10.1016/j.ceja.2021.100239

Karimi A, Tahmourespour A, Hoodaji M (2022) The formation of biocrust and improvement of soil properties by the exopolysaccharide-producing *Cyanobacterium*: a biogeotechnological study. Biomass Conv Bioref 1-11:15489. https://doi.org/10.1007/s13399-022-02336-0

Karmee SK, Linardi D, Lee J, Lin CSK (2015) Conversion of lipid from food waste to biodiesel. Waste Mgt 41:169–173. https://doi.org/10.1016/j.wasman.2015.03.025

Ke W, Li J, Mohammed F, Wang Y, Tou K, Liu X et al (2019) Therapeutic polymersome nanoreactors with tumor-specific activable cascade reactions for cooperative cancer therapy. ACS Nano 13(2):2357–2369. https://doi.org/10.1021/acsnano.8b09082

Khalid I, Ullah S, Umar IS (2022) The problem of solid waste: origins, composition, disposal, recycling, and reusing. Int J Adv Sci Comp Appl 1(1):27–40. https://doi.org/10.47679/ijasca.v1i1.6

Kim DM, Cho EJ, Kim JW, Lee YW, Chung HJ (2014) Production of cellulase by *Penicillium* sp. in a solid-state fermentation of oil palm empty fruit bunch. African. J Biotechnol 13(1):145–155. https://doi.org/10.5897/AJB12.2970

Kiran EU, Trzcinski AP, Liu Y (2014) Glucoamylase production from food waste by solid state fermentation and its evaluation in the hydrolysis of domestic food waste. J Biofuel Res 1(3):98–105. https://doi.org/10.18331/BRJ2015.1.3-7

Kumar A, Dutt D, Gautam A (2016) Production of crude enzyme from *aspergillus nidulans* AKB-25 using black gram residue as the substrate and its industrial applications. J Gen Eng Biotechnol 14(1):107–118. https://doi.org/10.1016/j.jgeb.2016.06.004

Kumar K, Yadav AN, Kumar V, Vyas P, Dhaliwal HS (2017a) Food waste: a potential bioresource for extraction of nutraceuticals and bioactive compounds. Biores Biopro 4(1):1–14. https://doi.org/10.1186/s40643-017-0148-6

Kumar V, Chhabra D, Shukla P (2017b) Xylanase production from *Thermomyces lanuginosus* VAPS-24 using low cost agro-industrial residues via hybrid optimization tools and its poten-

tial use for saccharification. Bioresour Technol 243:1009–1019. https://doi.org/10.1016/j.biortech.2017.07.094
Kumar A, Sadhya HK, Ahmad E, Dulawat S (2020a) Application of bio-enzyme in wastewater (greywater) treatment. IntJ Res Eng Technol 7(5):2886–2890
Kumar AN, Chatterjee S, Hemalatha M, Althuri A, Min B, Kim SH, Mohan SV (2020b) Deoiled algal biomass derived renewable sugars for bioethanol and biopolymer production in biorefinery framework. Bioresour Technol 296:122315. https://doi.org/10.1016/j.biortech.2019.122315
Lee YG, Cho EJ, Maskey S, Nguyen DT, Bae HJ (2023) Value-added products from coffee waste: a review. Mol 28(8):3562. https://doi.org/10.3390/molecules28083562
Li M, Buschle-Diller G (2017) Pectin-blended anionic polysaccharide films for cationic contaminant sorption from water. Int J Biol Macro- mol 101:481–489. https://doi.org/10.1016/j.ijbiomac.2017.03.091
Li S, Yang X (2016) Biofuel production from food wastes. In: Handbook biofuel prod. Woodhead Publishing, pp 617–653. https://doi.org/10.1016/B978-0-08-100455-5.00020-5
Li X, Wang H, Gan S, Jiang D, Tian G, Zhang Z (2013) Eco-stoichiometric alterations in paddy soil ecosystem driven by phosphorus application. PLoS One 8(5):e61141. https://doi.org/10.1371/annotation/a267ec0a-4ef6-4d9e-821f-5e818dac3d08
Li Y, Zhao P, Gong T, Wang H, Jiang X, Cheng H et al (2020a) Redox dyshomeostasis strategy for hypoxic tumor therapy based on DNAzyme-loaded electrophilic ZIFs. Int Angewandte Chem Edition 59(50):22537–22543. https://doi.org/10.1002/anie.202003653
Li Z, Sun Y, Yang Y, Han Y, Wang T, Chen J, Tsang DC (2020b) Biochar-supported nanoscale zero-valent iron as an efficient catalyst for organic degradation in groundwater. J Hazard Mater 383:121240. https://doi.org/10.1016/j.jhazmat.2019.121240
Li S, Tian Q, Meng T, Guan Z, Cai Y, Liao X (2022) Production, purification and activity evaluation of three novel antioxidant peptides obtained from grass carp (*Ctenopharyngodonidella*) scale waste by microbial protease BaApr1 hydrolysis. Syst Microbiol Biomanuf 1-12:568. https://doi.org/10.1007/s43393-022-00081-z
Liang X, Han L (2020) White peroxidase-mimicking nanozymes: colorimetric pesticide assay without interferences of O_2 and color. Adv Fun Mater 30(28):2001933. https://doi.org/10.1002/adfm.202001933
Liao R, Pon J, Chungyoun M, Nance E (2020) Enzymatic protection and biocompatibility screening of enzyme-loaded polymeric nanoparticles for neurotherapeutic applications. Biomaterials 257:120238. https://doi.org/10.1016/j.biomaterials.2020.120238
Lillford P, Hermansson AM (2021) Global missions and the critical needs of food science and technology. Trend Food Sci Technol 111:800–811. https://doi.org/10.1016/j.tifs.2020.04.009
Lima AC, Silva D, Silva V, Godoy M, Cammarota M, Gutarra M (2021) β-Mannanase production by *Penicillium citrinum* through solid-state fermentation using açaí residual biomass (*Euterpeoleracea*). J Chem Technol Biotechnol 96(10):2744–2754. https://doi.org/10.1002/jctb.6818
Litvinenko LV, Tishchenko AV, Ivshina IB (2019) Reduction of copper ion phytotoxicity using *Rhodococcus*-biosurfactants. Biol Bull 46(10):1333–1338. https://doi.org/10.1134/S1062359019100200
Liu B, Rajagopal D (2019) Life-cycle energy and climate benefits of energy recovery from wastes and biomass residues in the United States. Nat Ener 4(8):700–708. https://doi.org/10.1038/s41560-019-0430-2
Liu Z, Smith SR (2021) Enzyme recovery from biological wastewater treatment. Waste Biomass Valor 12(8):4185–4211. https://doi.org/10.1007/s12649-020-01251-7
Liu D, Zhang R, Yang X, Wu H, Xu D, Tang Z, Shen Q (2011) Thermostable cellulase production of *Aspergillus fumigatus* Z5 under solid-state fermentation and its application in degradation of agricultural wastes. IntBiodeter Biodegrad 65(5):717–725. https://doi.org/10.1016/j.ibiod.2011.04.005

Liu Y, Zhang Y, Xu J, Sun Y, Yuan Z, Xie J (2015) Consolidated bioprocess for bioethanol production with alkali-pretreated sugarcane bagasse. Appl Ener 157:517–522. https://doi.org/10.1016/j.apenergy.2015.05.004

Liu H, Chen D, Zhang R, Hang X, Li R, Shen Q (2016) Amino acids hydrolyzed from animal carcasses is a good additive for the production of bio-organic fertilizer. Front Microbiol 7:1290. https://doi.org/10.3389/fmicb.2016.01290

Liu H, Wang Y, Liang C, Yang Q, Wang S, Wang B et al (2020) Utilization of marigold (*Tagetes erecta*) flower fermentation wastewater as a fertilizer and its effect on microbial community structure in maize rhizosphere and non-rhizosphere soil. Biotechnol Equip 34(1):522–531. https://doi.org/10.1080/13102818.2020.1781548

Liu P, Li X, Xu X, Ye K, Wang L, Zhu H et al (2021) Integrating peroxidase-mimicking activity with photoluminescence into one framework structure for high-performance ratiometric fluorescent pesticide sensing. Sensor Actuator B: Chem 328:129024. https://doi.org/10.1016/j.snb.2020.129024

Liu Y, Zeng S, Ji W, Yao H, Lin L, Cui H et al (2022) Emerging theranostic nanomaterials in diabetes and its complications. Adv Sci 9(3):2102466. https://doi.org/10.1002/advs.202102466

López-Gómez JP, Venus J (2021) Potential role of sequential solid-state and submerged-liquid fermentations in a circular bioeconomy. Ferm 7(2):76. https://doi.org/10.3390/fermentation7020076

López-Prieto A, Martínez-Padrón H, Rodríguez-López L, Moldes AB, Cruz JM (2019) Isolation and characterization of a microorganism that produces biosurfactants in corn steep water. CyTA- J Food 17(1):509–516. https://doi.org/10.1080/19476337.2019.1607909

Lübeck M, Lübeck PS (2022) Fungal cell factories for efficient and sustainable production of proteins and peptides. Microorg 10(4):753. https://doi.org/10.3390/microorganisms10040753

Lucarini M, Durazzo A, Bernini R, Campo M, Vita C, Souto EB et al (2021) Fruit wastes as a valuable source of value-added compounds: a collaborative perspective. Mol 26(21):6338. https://doi.org/10.3390/molecules26216338

Luo J, Huang K, Zhou X, Xu Y (2020) Hybrid films based on holistic celery nanocellulose and lignin/hemicellulose with enhanced mechanical properties and dye removal. Int J Biol Macromol 147:699–705. https://doi.org/10.1016/j.ijbiomac.2020.01.102

Manasa P, Maitra S, Barman S (2020) Yield attributes, yield, competitive ability and economics of summer maize-legume intercropping system. International Journal of Agriculture, Environment and Biotechnology, 13(1), 33–38.

Mahapatra S, Ali MH, Samal K (2022) Assessment of compost maturity-stability indices and recent development of composting bin. Ener Nexus 100062. https://doi.org/10.1016/j.nexus.2022.100062

Mahawar MK, Jalgaonkar K, Bibwe B, Bhushan B, Meena VS, Sonkar RK (2020) Post-harvest processing and valorization of *Kinnow mandarin* (*citrus reticulate* L.): a review. J Food Sci Technol 57(3):799–815. https://doi.org/10.1007/s13197-019-04083-z

Mamimin C, Chanthong S, Leamdum C, Thong S-O, Prasertsan P (2021) Improvement of empty palm fruit bunches biodegradability and biogas production by integrating the straw mushroom cultivation as a pretreatment in the solid-state anaerobic digestion. Bioresour Technol 319:124227. https://doi.org/10.1016/j.biortech.2020.124227

Mamo M, Kassa H, Ingale L, Dondeyne S (2021) Evaluation of compost quality from municipal solid waste integrated with organic additive in Mizan–Aman town, Southwest Ethiopia. BMC Chem 15(1):1–11. https://doi.org/10.1186/s13065-021-00770-1

Mancini G, Papirio S, Lens PN, Esposito G (2018) Increased biogas production from wheat straw by chemical pretreatments. Renewab Ener 119:608–614. https://doi.org/10.1016/j.renene.2017.12.045

Markande AR, Patel D, Varjani S (2021) A review on biosurfactants: properties, applications and current developments. Bioresour Technol 330:124963. https://doi.org/10.1016/j.biortech.2021.124963

Masaphy S, Vanti GL, Zabari L (2022) Laccase enhancement and antifungal toxicity reduction: bidirectional influences between pomegranate peel extract and *Morchella conica* mycelium activity. Biores Technol Rep 17:100936. https://doi.org/10.1016/j.biteb.2021.100936

Medina-Torres N, Espinosa-Andrews H, Trombotto S, Ayora-Talavera T, Patrón-Vázquez J, González-Flores T et al (2019) Ultrasound-assisted extraction optimization of phenolic compounds from *Citrus latifolia* waste for chitosan bioactive nanoparticles development. Mol 24(19):3541. https://doi.org/10.3390/molecules24193541

Meftahi A, Samyn P, Geravand SA, Khajavi R, Alibkhshi S, Bechelany M, Barhoum A (2022) Nanocelluloses as skin biocompatible materials for skincare, cosmetics, and healthcare: formulations, regulations, and emerging applications. Carbohydr Poly 278:118956. https://doi.org/10.1016/j.carbpol.2021.118956

Mejias L, Estrada M, Barrena R, Gea T (2020) A novel two-stage aeration strategy for *Bacillus thuringiensis* biopesticide production from biowaste digestate through solid-state fermentation. J Biochem Eng 161:107644. https://doi.org/10.1016/j.bej.2020.107644

Mekonnen E, Kebede A, Tafesse T, Tafesse M (2020) Application of microbial bioenzymes in soil stabilization. Int J Microbiol 8:1725482. https://doi.org/10.1155/2020/1725482

Melendez JR, Mátyás B, Hena S, Lowy DA, El Salous A (2022) Perspectives in the production of bioethanol: a review of sustainable methods, technologies, and bioprocesses. Renewab Sustain Ener Rev 160:112260. https://doi.org/10.1016/j.rser.2022.112260

Melikoglu M, Lin CSK, Webb C (2015) Solid state fermentation of waste bread pieces by *Aspergillus awamori*: Analysing the effects of airflow rate on enzyme production in packed bed bioreactors. Food Bioprod Pro 95:63–75. https://doi.org/10.1016/j.fbp.2015.03.011

Melnichuk N, Braia MJ, Anselmi PA, Meini MR, Romanini D (2020) Valorization of two agroindustrial wastes to produce alpha-amylase enzyme from *Aspergillus oryzae* by solid-statefermentation. Waste Mgt 106:155–161. https://doi.org/10.1016/j.wasman.2020.03.025

Mesa J, Hinestroza-Córdoba LI, Barrera C, Seguí L, Betoret E, Betoret N (2020) High homogenization pressures to improve food quality, functionality and sustainability. Mol 25(14):3305. https://doi.org/10.3390/molecules25143305

Mishra S, Kumar P, Malik A (2016) Suitability of agricultural by-products as production medium for spore production by *Beauveria bassiana* HQ917687. Int J Recy Organic Waste Agr 5(2):179–184. https://doi.org/10.1007/s40093-016-0127-5

Mitchell MJ, Billingsley MM, Haley RM, Wechsler ME, Peppas NA, Langer R (2021) Engineering precision nanoparticles for drug delivery. Nat Rev Drug Discov 20(2):101–124. https://doi.org/10.1038/s41573-020-0090-8

Mitri S, Salameh SJ, Khelfa A, Leonard E, Maroun RG, Louka N, Koubaa M (2022) Valorization of brewers' spent grains: pretreatments and fermentation. A reviewFerm 8(2):50. https://doi.org/10.3390/fermentation8020050

Mohan SV, Varjani S, Pant D, Sauer M, Chang JS (2020) Circular bioeconomy approaches for sustainability. BioresTechnol 318:124084. https://doi.org/10.1016/j.biortech.2020.124084

Mojumdar A, Deka J (2019) Recycling agro-industrial waste to produce amylase and characterizing amylase–gold nanoparticle composite. Int J Recy Organic Waste Agr 8(1):263–269. https://doi.org/10.1007/s40093-019-00298-4

Molina-Peñate E, Sánchez A, Artola A (2022). Enzymatic hydrolysis of the organic fraction of municipal solid waste: optimization and valorization of the solid fraction for *Bacillus thuringiensis* biopesticide production through solid-state fermentation. Waste Mgt 137: 304–311. https://doi.org/10.1016/j.wasman.2021.11.014

Mupambwa HA, Nciizah AD, Nyambo P, Muchara B, Gabriel NN (eds) (2022) Food security for African smallholder farmers. Sustain Sci in Asia and Africa (pp. 89–112). Springer, Singapore. https://doi.org/10.1007/978-981-16-6771-8_6

Muthusamy S, Ajit S, Nath AV, Anupama Sekar J, Ramyaa Lakshmi TS (2022) Enzymes from genetically modified organisms and their current applications in food development and food chain. Novel Food Grade Enzymes: Applications in Food Processing and Preservation Industries 22:357–382. https://doi.org/10.1007/978-981-19-1288-7_13

Nadaf S, Kumbar VM, Killedar S, Torvi AI, Hoskeri JH, Shettar AK (2021) Microbial biosurfactants as cleaning and washing agents. In: Microbial Biosurf. Springer, Singapore, pp 293–314. https://doi.org/10.1007/978-981-15-6607-3_14

Nahak BK, Preetam S, Sharma D, Shukla SK, Syväjärvi M, Toncu DC, Tiwari A (2022) Advancements in net-zero pertinency of lignocellulosic biomass for climate neutral energy production. Renewab Sustain Ener Rev 161:112393. https://doi.org/10.1016/j.rser.2022.112393

Naik B, Kumar V, Rizwanuddin S, Chauhan M, Gupta AK, Rustagi S, Kumar V, Gupta S (2023) Agro-industrial waste: a cost-effective and eco-friendly substrate to produce amylase. Food Prod Proc Nutr 5(1):30. https://doi.org/10.1186/s43014-023-00143-2

Narisetty V, Maitra S, Tarafdar A, Alphy MP, Kumar AN, Madhavan A et al (2022) Waste-derived fuels and renewable chemicals for bioeconomy promotion: a sustainable approach. Bio Ener Res 1-17:16. https://doi.org/10.1007/s12155-022-10428-y

Nazim F, Meera V (2017) Comparison of treatment of greywater using garbage and citrus enzymes. Int J Innov Res Sci Eng Technol 6(4) 49-54.ISSN (Online): 2319–8753. www.ijirset.com

Negi A, Pare A, Manickam L, Rajamani M (2022) Effects of defect action level of *Tribolium castaneum* (Herbst)(*Coleoptera: Tenebrionidae*) fragments on quality of wheat flour. J Sci Food Agr 102:223–232. https://doi.org/10.1002/jsfa.11349

Nehad EA, Atalla SM (2020) Production and immobilization of invertase from *Penicillium* sp. using orange peel waste as substrate. Egyptian J Pharm 19(2):103. http://www.epj.eg.net/text.asp?2020/19/2/103/283855

Neupane K, Khadka R (2019) Production of garbage enzyme from different fruit and vegetable wastes and evaluation of its enzymatic and antimicrobial efficacy. Tribhuvan University Journal of Microbiology, 6, 113–118.

Nikku M, Deb A, Sermyagina E, Puro L (2019) Reactivity characterization of municipal solid waste and biomass. Fuel 254:115690. https://doi.org/10.1016/j.fuel.2019.115690

Novianto N (2022) Response of liquid organic fertilizer eco enzyme (EE) on growth and production of Shallot (*Allium ascalonicum.* L). J Agr Tanaman Trop 4(1):147–154. https://doi.org/10.36378/juatika.v4i1.1782

Nuhu A, Hussaini I, Gide S, Anas G, Madika A (2020) Production of laccase by fungi isolated from soil via submerged fermentation using corn cob as substrate. FUDMA J Sci 4(3):224–229. https://doi.org/10.33003/fjs-2020-0403-296

O'Connor J, Hoang SA, Bradney L, Dutta S, Xiong X, Tsang DC et al (2021) A review on the valorization of food waste as a nutrient source and soil amendment. Environ Pol 272:115985. https://doi.org/10.1016/j.envpol.2020.115985

O'shea N, Ktenioudaki A, Smyth TP, McLoughlin P, Doran L, Auty MAE et al (2015) Physicochemical assessment of two fruit by-products as functional ingredients: apple and orange pomace. J Food Eng 153:89–95. https://doi.org/10.1016/j.jfoodeng.2014.12.014

O'Connor J, Mickan B, Siddique KH, Rinklebe J, Kirkham MB, Bolan NS (2022) Physical, chemical, and microbial contaminants in food waste management for soil application: a review. Environ Pol 118860. https://doi.org/10.1016/j.envpol.2022.118860

Ortiz GE, Guitart ME, Cavalitto SF, Albertó EO, Fernández-Lahore M, Blasco M (2015) Characterization, optimization, and scale-up of cellulase production by *Trichoderma reesei* CBS 836.91 in solid-state fermentation using agro-industrial products. Bioprocess Biosyst Eng 38(11):2117–2128. https://doi.org/10.1007/s00449-015-1451-2

Pal P, Chew JW, Yen H, Lim J, Lam M, Show P (2019) Cultivation of oily microalgae for the production of third-generation biofuels. Sustain 11:5424. https://doi.org/10.3390/su11195424

Panda SK, Mishra SS, Kayitesi E, Ray RC (2016) Microbial-processing of fruit and vegetable wastes for production of vital enzymes and organic acids: biotechnology and scopes. Environ Res 146:161–172. https://doi.org/10.1016/j.envres.2015.12.035

Pandey P, Khan F, Ahmad V, Singh A, Shamshad T, Mishra R (2020) Combined efficacy of *Azadirachta indica* and *Moringa oleifera* leaves extract as a potential coagulant in ground water treatment. SN Appl Sci 2(7):1–8. https://doi.org/10.1007/s42452-020-3124-2

Pangsri P, Pangsri P (2017) Mannanase enzyme from *Bacillus subtilis* P2-5 with waste management. Ener Proced 138:343–347. https://doi.org/10.1016/j.egypro.2017.10.136

Panwar D, Saini A, Panesar PS, Chopra HK (2021) Unraveling the scientific per- spectives of citrus by-products utilization: Progress towards circular economy. Trend Food Sci Technol 111:549–562. https://doi.org/10.1016/j.tifs.2021.03.018

Pathak P, Bhardwaj NK, Singh AK (2014) Production of crude cellulase and xylanase from *Trichoderma harzianum* PPDDN10 NFCCI-2925 and its application in photocopier waste paper recycling. Appl Biochem Biotechnol 172(8):3776–3797. https://doi.org/10.1007/s12010-014-0758-9

Paul S, Joshi SR (2022) Industrial perspectives of fungi. In: Indust Microbiol Biotechnol. Springer, Singapore, pp 81–105. https://doi.org/10.1007/978-981-16-5214-1_3

Prabhu AA, Jayadeep A (2017) Optimization of enzyme-assisted improvement of polyphenols and free radical scavenging activity in red rice bran: a statistical and neural network-based approach. Prep Biochem Biotechnol 47(4):397–405. https://doi.org/10.1080/10826068.2016.1252926

Prabhu G, Bhat D, Bhat RM, Selvaraj S (2022) A critical look at bioproducts co-cultured under solid state fermentation and their challenges and industrial applications. Waste Biomass Valor 1-17:3095. https://doi.org/10.1007/s12649-022-01721-0

Qin X, Wu C, Niu D, Qin L, Wang X, Wang Q, Li Y (2021) Peroxisome inspired hybrid enzyme nanogels for chemodynamic and photodynamic therapy. Nat Commun 12(1):1-15.5243. https://doi.org/10.1038/s41467-021-25561-z

Qureshi N, Lin X, Liu S, Saha BC, Mariano AP, Polaina J et al (2020) Global view of biofuel butanol and economics of its production by fermentation from sweet sorghum bagasse, food waste, and yellow top presscake: application of novel technologies. Ferm 6(2):58. https://doi.org/10.3390/fermentation6020058

Rahayu MR, Situmeang YP (2021) Acceleration of production natural disinfectants from the combination of eco-enzyme domestic organic waste and frangipani flowers (*Plumeria alba*). Sustain Environ Agr Sci 5(1):15–21. https://doi.org/10.22225/seas.5.1.3165.15-21

Raheem BS, Oladiran GF, Oke DA, Musa SA (2020) Evaluation of strength properties of subgrade materials stabilized with bio-enzyme. Eur J Eng Technol Res 5(5):607–610. https://doi.org/10.24018/ejeng.2020.5.5.1329

Ramkumar A, Sivakumar N, Gujarathi AM, Victor R (2018) Production of thermotolerant, detergent stable alkaline protease using the gut waste of *Sardinella longiceps* as a substrate: optimization and characterization. Sci Rep 8(1):1–15. https://doi.org/10.1038/s41598-018-30155-9

Rani A, Negi S, Hussain A, Kumar S (2020) Treatment of urban municipal landfill leachate utilizing garbage enzyme. Bioresour Technol 297:122437. https://doi.org/10.1016/j.biortech.2019.122437

Rasit N, Kuan OC (2018) Investigation on the influence of bio-catalytic enzyme produced from fruit and vegetable waste on palm oil mill effluent. In IOP Conf series. Earth environ Sci 140(1): 012015. IOP Publishing. Doi:https://doi.org/10.1088/1755-1315/140/1/012015

Rasit N, Mohammad FS (2018) Production and characterization of bio catalytic enzyme produced from fermentation of fruit and vegetable wastes and its influence on aquaculture sludge.Int. J Sci Technol 4(2):12–26. https://doi.org/10.20319/mijst.2018.42.1226

Rasit N, Hwe Fern L, Ab Karim Ghani WAW (2019) Production and characterization of eco enzyme produced from tomato and orange wastes and its influence on the aquaculture sludge. Int J Civil Eng Technol 10(3):967–980. http://www.iaeme.com/IJCIET/index.asp

Ravindran B, Kumar AG, Bhavani PA, Sekaran G (2011) Solid-state fermentation for the production of alkaline protease by *Bacillus cereus* 1173900 using proteinaceous tannery solid waste. Cur Sci:726–730. https://www.jstor.org/stable/24075813

Rebelo R, Fernandes M, Fangueiro R (2017) Biopolymers in medical implants: a brief review. Proced Engin 200:236–243. https://doi.org/10.1016/j.proeng.2017.07.034

Ren X, Ghazani MS, Zhu H, Ao W, Zhang H, Moreside E et al (2022) Challenges and opportunities in microwave-assisted catalytic pyrolysis of biomass: a review. Appl Ener 315:118970. https://doi.org/10.1016/j.apenergy.2022.118970

Sala A, Artola A, Sánchez A, Barrena R (2020) Rice husk as a source for fungal biopesticide production by solid-state fermentation using *B. bassiana* and *T harzianum*. Biores Technol 296:122322. https://doi.org/10.1016/j.biortech.2019.122322

Salah WA, Abuhelwa M, Bashir MJK (2021) The key role of sustainable renewable energy technologies in facing shortage of energy supplies in. Current practice and future potential. Palestine J Clean Prod 293:125348. https://doi.org/10.1016/j.jclepro.2020.125348

Sampath P, Reddy KR, Reddy CV, Shetti NP, Kulkarni RV, Raghu AV (2020) Biohydrogen production from organic waste a review. Chem Eng Technol 43(7):1240–1248. https://doi.org/10.1002/ceat.201900400

Santis-Navarro A, Gea T, Barrena R, Sánchez A (2011) Production of lipases by solid state fermentation using vegetable oil-refining wastes. Bioresour Technol 102(21):10080–10084. https://doi.org/10.1016/j.biortech.2011.08.062

Saratale GD, Kshirsagar SD, Sampange VT, Saratale RG, Oh SE, Govindwar SP, Oh MK (2014) Cellulolytic enzymes production by utilizing agricultural wastes under solid state fermentation and its application for biohydrogen production. Appl Biochem Biotechnol 174(8):2801–2817. https://doi.org/10.1007/s12010-014-1227-1

Saxena R, Singh R (2011) Amylase production by solid-state fermentation of agro-industrial wastes using *Bacillus* sp. Brazilian J Microbiol 42(4):1334–1342. https://doi.org/10.1590/S1517-83822011000400014

Selvakumar P, Sivashanmugam P (2017) Optimization of lipase production from organic solid waste by anaerobic digestion and its application in biodiesel production. Fuel Pro Technol 165:1–8. https://doi.org/10.1016/j.fuproc.2017.04.020

Selvakumar P, Sivashanmugam P (2019) Ultrasound assisted oleaginous yeast lipid extraction and garbage lipase catalyzed transesterification for enhanced biodiesel production. Ener Conv Mgt 179:141–151. https://doi.org/10.1016/j.enconman.2018.10.051

Sengar AS, Rawson A, Muthiah M, Kalakandan SK (2020) Comparison of different ultrasound assisted extraction techniques for pectin from tomato processing waste. Ultrason Sonochem 61:104812. https://doi.org/10.1016/j.ultsonch.2019.104812

Sermyagina E, Martinez CLM, Nikku M, Vakkilainen E (2021) Spent coffee grounds and tea leaf residues: Characterization, evaluation of thermal reactivity and recovery of high-value compounds. Biomass and Bioenergy, 150, 106141.

Sethi B, Satpathy A, Tripathy S, Parida S, Singdevsachan SK, Behera B (2016) Production of ethanol and clarification of apple juice by pectinase enzyme produced from *Aspergillus terreus* NCFT 4269.10. Int. J Biol Res 4:67. https://doi.org/10.14419/ijbr.v4i1.6134

Shaddel S, Bakhtiary-Davijany H, Kabbe C, Dadgar F, Østerhus SW (2019) Sustainable sewage sludge management: from current practices to emerging nutrient recovery technologies. Sustain 11(12):3435. https://doi.org/10.3390/su11123435

Sharma P, Gaur VK, Kim SH, Pandey A (2020) Microbial strategies for bio-transforming food waste into resources. Bioresour Technol 299:122580. https://doi.org/10.1016/j.biortech.2019.122580

Sharma P, Gaur VK, Sirohi R, Varjani S, Kim SH, Wong JW (2021) Sustainable processing of food waste for production of bio-based products for circular bioeconomy. Bioresour Technol 325:124684. https://doi.org/10.1016/j.biortech.2021.124684

Sharma V, Tsai ML, Nargotra P, Chen CW, Kuo CH, Sun PP, Dong CD (2022) Agro-industrial food waste as a low-cost substrate for sustainable production of industrial enzymes: a critical review. Catalyst 12(11):1373. https://doi.org/10.3390/catal12111373

Shi H, Zhang M, Wang W, Devahastin S (2020) Solid-state fermentation with probiotics and mixed yeast on properties of Okara. Food Biosci 36:100610. https://doi.org/10.1016/j.fbio.2020.100610

Singhania RR, Ruiz HA, Awasthi MK, Dong CD, Chen CW, Patel AK (2021) Challenges in cellulase bioprocess for biofuel applications. Renewab Sustain Ener Rev 151:111622. https://doi.org/10.1016/j.rser.2021.111622

Singla G, Krishania M, Sandhu PP, Sangwan RS, Panesar PS (2019) Value additon of kinnow industry byproducts for the preparation of fiber enriched extruded products. J Food Sci Technol 56(3):1575–1582. https://doi.org/10.1007/s13197-019-03670-4

Sirajudheen P, Poovathumkuzhi NC, Vigneshwaran S, Chelaveettil BM, Meenakshi S (2021) Applications of chitin and chitosan based biomaterials for the adsorptive removal of textile dyes from water. A comprehensive review. Carbohydr Poly 273:118604. https://doi.org/10.1016/j.carbpol.2021.118604

Sivakumar D, Srikanth P, Ramteke PW, Nouri J (2022) Agricultural waste management generated by agro-based industries using biotechnology tools. Global J Environ Sci Mgt 8(2):281–296. https://doi.org/10.22034/gjesm.2022.02.10

Sivamani S, Baskar R (2018) Process design and optimization of bioethanol production from cassava bagasse using statistical design and genetic algorithm. Prep Biochem Biotechnol 48(9):834–841. https://doi.org/10.1080/10826068.2018.1514512

Sodhi AS, Sharma N, Bhatia S, Verma A, Soni S, Batra N (2022) Insights on sustainable approaches for production and applications of value added products. Chemosphe 286:131623. https://doi.org/10.1016/j.chemosphere.2021.131623

Sodhi PS, Ocean YK (2018) Stabilization of soil using alkaline bio-enzyme (Alkazyme). International Research Journal of Engineering and Technology, 5(8), 1681–1685.

Solanki YS, Agarwal M, Gupta AB, Gupta S, Shukla P (2022) Fluoride occurrences, health problems, detection, and remediation methods for drinking water: a comprehensive review. Sci Total Environ 807:150601. https://doi.org/10.1016/j.scitotenv.2021.150601

Song Y, Cho EJ, Park CS, Oh CH, Park BJ, Bae HJ (2019) A strategy for sequential fermentation by *Saccharomyces cerevisiae* and *Pichia stipitis* in bioethanol production from hardwoods. Renewab Ener 139:1281–1289. https://doi.org/10.1016/j.renene.2019.03.032

Souza MAD, Vilas-Boas IT, Leite-da-Silva JM, Abrahão PDN, Teixeira-Costa BE, Veiga-Junior VF (2022) Polysaccharides in agro-industrial biomass residues. Polysaccharid 3(1):95–120. https://doi.org/10.3390/polysaccharides3010005

Srimathi N, Subiksha M, Abarna J, Niranjana T (2020) Biological treatment of dairy wastewater using bio enzyme from citrus fruit peels. Int J Recent Technol Eng 9(1):292–295. https://doi.org/10.35940/ijrte.A1530.059120

Srivastava N, Shrivastav A, Singh R, Abohashrh M, Srivastava KR, Irfan S et al (2021) Advances in the structural composition of biomass: fundamental and bioenergy applications. J Renewab Mater 9(4):615–636. https://doi.org/10.32604/jrm.2021.014374

Streimikyte P, Viskelis P, Viskelis J (2022) Enzymes-assisted extraction of plants for sustainable and functional applications. Int J Mol Sci 23(4):2359. https://doi.org/10.3390/ijms23042359

Suri S, Singh A, Nema PK (2022) Current applications of citrus fruit processing waste: a scientific outlook. Appl Food Res 1:100050. https://doi.org/10.1016/j.afres.2022.100050

Swathy R, Rambabu K, Banat F, Ho SH, Chu DT, Show PL (2020) Production and optimization of high grade cellulase from waste date seeds by *Cellulomonas Uda* NCIM 2353 for biohydrogen production. Int J Hydrogen Ener 45(42):22260–22270. https://doi.org/10.1016/j.ijhydene.2019.06.171

Taokaew S, Kriangkrai W (2023) Chitinase-assisted bioconversion of chitinous waste for development of value-added Chito-oligosaccharides products. Biol 12(1):87. https://doi.org/10.3390/biology12010087

Thangaratham T, Manimegalai G (2014) Optimization and production of pectinase using agro waste by solid state and submerged fermentation. Int J Cur Microbiol Appl Sci 3(9):357–365. http://www.ijcmas.com/

Torres-León C, Chávez-González ML, Hernández-Almanza A, Martínez-Medina GA, Ramírez-Guzmán N, Londoño-Hernández L, Aguilar CN (2021) Recent advances on the microbiological and enzymatic processing for conversion of food wastes to valuable bioproducts. Curr Opinion Food Sci 38:40–45. https://doi.org/10.1016/j.cofs.2020.11.002

Van Dorst RM, Gårdmark A, Svanbäck R, Beier U, Weyhenmeyer GA, Huss M (2019) Warmer and browner waters decrease fish biomass production. Global Chang Biol 25(4):1395–1408. https://doi.org/10.1111/gcb.14551

Veana F, Martínez-Hernández JL, Aguilar CN, Rodríguez-Herrera R, Michelena G (2014) Utilization of molasses and sugar cane bagasse for production of fungal invertase in solid state fermentation using *Aspergillus Niger* GH1. Brazilian J Microbiol 45(2):373–377. https://doi.org/10.1590/S1517-83822014000200002

Velioglu Z, Urek RO (2016) Physicochemical and structural characterization of biosurfactant produced by *Pleurotus djamor* in solid-state fermentation. Biotechnol Bioprocess Eng 21(3):430–438. https://doi.org/10.1007/s12257-016-0139-z

Verduzco-Oliva R, Gutierrez-Uribe JA (2020) Beyond enzyme production: solid state fermentation (SSF) as an alternative approach to produce antioxidant polysaccharides. Sustain 12(2):495. https://doi.org/10.3390/su12020495

Vijayaraghavan P, Lazarus S, Vincent SGP (2014) De-hairing protease production by an isolated *Bacillus cereus* strain AT under solid-state fermentation using cow dung: biosynthesis and properties. Saudi. J Biol Sci 21(1):27–34. https://doi.org/10.1016/j.sjbs.2013.04.010

Wang C, Zhang Q, Wang X, Chang H, Zhang S, Tang Y et al (2017) Dynamic modulation of enzyme activity by near-infrared light. Angewandte Chem 129(24):6871–6876. https://doi.org/10.1002/ange.201700968

Wang T, Zhai Y, Li H, Zhu Y, Li S, Peng C et al (2018) Co-hydrothermal carbonization of food waste-woody biomass blend towards biofuel pellets production. Bioresour Technol 267:371–377. https://doi.org/10.1016/j.biortech.2018.07.059

Wiltschi B, Cernava T, Dennig A, Casas MG, Geier M, Gruber S et al (2020) Enzymes revolutionize the bioproduction of value-added compounds: from enzyme discovery to special applications. Biotechnol Adv 40:107520. https://doi.org/10.1016/j.biotechadv.2020.107520

Wohlgemuth R, Twardowski T, Aguilar A (2021) Bioeconomy moving forward step by step–a global journey. New Biotechnol 61:22–28. https://doi.org/10.1016/j.nbt.2020.11.006

Xie J, Zhang Y, Simpson B (2022) Food enzymes immobilization: novel carriers, techniques and applications. Curr Opinion Food Sci 43:27–35. https://doi.org/10.1016/j.cofs.2021.09.004

Yadav M, Paritosh K, Vivekanand V (2020) Lignocellulose to bio-hydrogen: An overview on recent developments. Int J Hydrogen Ener 45(36):18195–18210. https://doi.org/10.1016/j.ijhydene.2019.10.027

Yadav KD, Sharma D, Prasad R (2022) Challenges and opportunities for disposal of floral waste in developing countries by using composting method. In: Advanced organic waste management. Elsevier, pp 55–77. https://doi.org/10.1016/B978-0-323-85792-5.00018-6

Yafetto L (2022) Application of solid-state fermentation by microbial biotechnology for bioprocessing of agro-industrial wastes from 1970 to 2020: A review and bibliometric analysis. Heliyon 8:e09173. https://doi.org/10.1016/j.heliyon.2022.e09173

Yang G, Wang J (2018) Synergistic biohydrogen production from flower wastes and sewage sludge. Ener Fuel 32(6):6879–6886. https://doi.org/10.1021/acs.energyfuels.8b01122

Yankey R, Omoor IN, Karanja JK, Wang L, Urga RT, Fang CH et al (2022) Metabolic properties, gene functions, and biosafety analysis reveal the action of three rhizospheric plant growth-promoting bacteria of *Jujuncao* (*Pennisetum giganteum*). Environ Sci Pollut Res 1-15:38435. https://doi.org/10.1007/s11356-021-17854-z

Yazid NA, Barrena R, Sánchez A (2016) Assessment of protease activity in hydrolysed extracts from SSF of hair waste by and indigenous consortium of microorganisms. Waste Mgt 49:420–426. https://doi.org/10.1016/j.wasman.2016.01.045

Yong ZJ, Bashir MJ, Hassan MS (2020) Assessment of environmental, energy and economic prospective of anaerobic digestion of organic municipal solid waste in Malaysia. In IOP Conf Series: Earth Environ Sci 463(1):012054). IOP Publishing. https://doi.org/10.1088/1755-1315/463/1/012054

Yong EL, Halim KA, Liong VYF, Tee MLK, Yong ZY, See HH, Syafiuddin A (2021) Improving the water quality of iron-containing ponds using fermented kitchen wastes. Environ Quality Mgt 32:37. https://doi.org/10.1002/tqem.21821

Youssef AM, Hasanin MS, Abd El-Aziz ME, Darwesh OM (2019) Green, economic, and partially biodegradable wood plastic composites via enzymatic surface modification of lignocellulosic fibers. Heliyon 5(3):e01332. https://doi.org/10.1016/j.heliyon.2019.e01332

Zema DA, Calabrò PS, Folino A, Tamburino VINCENZO, Zappia G, Zimbone SM (2018) Valorisation of citrus processing waste: a review. Waste Mgt 80:252–273. https://doi.org/10.1016/j.wasman.2018.09.024

Zhang C, Ni D, Liu Y, Yao H, Bu W, Shi J (2017) Magnesium silicide nanoparticles as a deoxygenation agent for cancer starvation therapy. Nat Nanotechnol 12(4):378–386. https://doi.org/10.1038/nnano.2016.280

Zhang T, Jiang D, Zhang H, Lee DJ, Zhang Z, Zhang Q et al (2020) Effects of different pretreatment methods on the structural characteristics, enzymatic saccharification and photo-fermentative bio-hydrogen production performance of corn straw. Bioresour Technol 304:122999. https://doi.org/10.1016/j.biortech.2020.122999

Zhao S, Liang X, Hua D, Ma TS, Zhang H (2011) High-yield cellulase production in solid-state fermentation by *Trichoderma reesei* SEMCC-3.217 using water hyacinth (*Eichhornia crassipes*). Afr J Biotechnol 10(50):10178–10187. https://doi.org/10.5897/AJB10.748

Zhou H, Liu J, Chen X, Ying Z, Zhang Z, Wang M (2018) Fate of pharmaceutically active compounds in sewage sludge during anaerobic digestions integrated with enzymes and physicochemical treatments. Waste Mgt 78:911–916. https://doi.org/10.1016/j.wasman.2018.07.018

Zhou Y, Xu XY, Gan RY, Zheng J, LiY ZJJ et al (2019) Optimization of ultrasound-assisted extraction of antioxidant polyphenols from the seed coats of red sword bean (*Canavalia gladiate* (Jacq.) DC.). Antiox 8(7):200. https://doi.org/10.3390/antiox8070200

Zhu Y, Luan Y, Zhao Y, Liu J, Duan Z, Ruan R (2023) Current technologies and uses for fruit and vegetable wastes in a sustainable system: a review. FoodReview 12(10):1949. https://doi.org/10.3390/foods12101949

Chapter 6
Valorization of Fruit Processing Industry Waste into Value-Added Chemicals

Abas Siraj Hamda, Melkiyas Diriba Muleta, Mani Jayakumar, Selvakumar Periyasamy, and Baskar Gurunathan

Abstract Food production, processing, and preparation produce larger amounts of waste that endanger human health by polluting the environment. Approximately 25–30% of the final product, solid and liquid, is wasted in the fruit processing plants. Each year, waste fruits and vegetables amounting to over 100 billion tons are thrown. Fruit and vegetable wastes have a higher moisture content, making them unsuitable for burning. At the same time, there are so many organic components present in the stacking or landfilling practices that will generate many leachates, seriously polluting the environment. Understanding how to utilize these resources and convert waste fruits and vegetables into useful products is essential. Because of the generation of significant amounts of byproducts from fruits at various stages of the sequence, the recycling of waste from the industries that process fruits has emerged as one of the most challenging elements globally. This garbage can be utilized to generate a valuable product, which is a fresh step toward its sustainable

A. S. Hamda · M. D. Muleta
Department of Chemical Engineering, Haramaya Institute of Technology, Haramaya University, Dire Dawa, Ethiopia

M. Jayakumar (✉)
Department of Chemical Engineering, Haramaya Institute of Technology, Haramaya University, Dire Dawa, Ethiopia

Department of Biotechnology , Faculty of Engineering, Karpagam Academy of Higher Education, Coimbatore, Tamilnadu, India
e-mail: drjayakumarmani@haramaya.edu.et

S. Periyasamy
Department of Chemical Engineering, School of Mechanical, Chemical and Materials Engineering, Adama Science and Technology University, Adama, Ethiopia

Department of Biomaterials, Saveetha Dental College and Hospitals, SIMATS, Saveetha University, Chennai, Tamilnadu, India

B. Gurunathan
Department of Biotechnology, St. Joseph's College of Engineering, Chennai, Tamil Nadu, India

School of Engineering, Lebanese American University, Byblos, Lebanon

E. Cherian, B. Gurunathan (eds.), *Value Added Products From Food Waste*,
https://doi.org/10.1007/978-3-031-48143-7_6

application. This chapter offers an in-depth analysis of recent developments in the utilization of waste materials generated during fruit processing. In addition, the source of fruit waste, characteristics of fruit waste, byproducts of fruit processing industries, and opportunity challenges related to fruit waste are discussed.

Keywords Fruit waste · Food processing · Value-added products · Biomass valorization

6.1 Introduction

Fruits are the most widely consumed food products among horticulture crops. Fruit is a fundamental component of our meals and human life. Consequently, there is a higher demand for these fundamental food items as the world's population has increased and dietary preferences have changed (Sagar et al. 2018a). The estimates by the Food and Agriculture Organization of the United Nations, a total of 84.63 MMT of apples, 114.08 MMT of bananas, 124.73 million metric tons (MMT) of citrus, 74.49 MMT of grapes, 45.22 MMT of mangoes, mangosteens, and guavas, and 25.43 MMT of pineapples are harvested annually. A significant quantity of waste is generated during many steps of processing these fruits, including gathering, handling, shipping, and industrial processing, which accounts for about 40% of all food produced (Swetha et al. 2023). Fruits that have been processed produce a considerable number of leftover pods, peels, pulp, stones, and seeds (Nirmal et al. 2023).

Various byproducts of the fruit and fruit processing industries are an abundant supply of bioactive substances, including phenolic and antioxidant substances, that can increase the stability of meals by avoiding lipid peroxidation (Dos Santos et al. 2023). According to the location and processing method, the average quantity of solid wastes produced by the fruit processing industry is often very high (for example, mango waste can range from 30 to 50%, banana waste from 20 to 50%, and citrus waste from 30 to 50%) (Banerjee et al. 2017). These fruit effluents include a sizable amount of moisture as well as a wide range of biological components, some of which may be harmful due to their potential to generate phytotoxicity symptoms (Zema et al. 2019). These substances cause gastrointestinal problems in animals fed the wastes, contaminate aqueous media, degrade drinking water quality, interfere with plant growth, kill delicate marine species, and impede the germination of seeds. These wastes are disposed of in municipal dumpsters or allowed to degrade because the infrastructure is unable to handle such a big amount of biomass or because there is no known commercial use for them. Processors may incur additional costs when disposing of these wastes, and direct disposal in landfills or the soil may have detrimental environmental effects. However, the price of drying, storing, and shipping byproducts is an economic barrier (Bisht et al. 2020). Particularly in developing nations, there are financial, spatial, and sometimes strict governmental rules about trash disposal. Because the bulk of these businesses is small and informal, processing waste is viewed as having less value than processed fruit (Banerjee et al. 2017).

Utilizing trash or residues from the fruit processing sector has recently become a significant challenge for agro-processing businesses and has a negative influence on the environment (Jameel et al. 2023). The fruit processing industry produces a lot of waste dumped in landfills or rivers, threatening the ecosystem. Therefore, there is a need for such disposal techniques that recycle it, provide resources for livestock feed, or extract or create goods with added value (Jayakumar et al. 2023a). It is possible to lessen environmental degradation, increase energy security, and cut greenhouse gas emissions by turning wastes collected from the fruit processing industry into valuable products. It is believed that using fruit waste from processing can help the food industry recover value-added products and make operations commercially viable. In light of this, it is a promising field of research to use wastes from the processing of fruit to create goods with value added.

6.2 Sources and Characteristics of Waste from the Fruit Processing Industry

Fruit processing waste is divided into two categories: liquid waste (juice and washed water) and solid rubbish (peel, skin, seeds, stones, etc.). The seedling development, maturity, collection, listing, and processing produce these wastes. Fruit garbage is mostly produced in sectors of fruit transportation, manufacturing, and trash collection (Peng et al. 2019). The primary sources of waste from the industries that process fruit include waste from washing water, remaining fruit after sorting, and peel, seed, and pomace after juice extraction. Numerous types and quantities of waste produced by fruit processing activities are presented in Table. 6.1.

These wastes provide a plentiful source of essential elements like carbs, proteins, lipids, minerals, fibers, etc. (Sadh et al. 2018). Additionally, compared to agricultural wastes like maize stover, wheat straw, rice straw, etc., fruit-related wastes are

Table 6.1 Types and quantity of waste generated from fruit processing industries

Commodities	Percentage weight basis	Nature of waste
Apple	12–50	Peel, pomace, and seed
Pineapple	32–65	Peel, skin, core, and coarse solid
Mango	45–70	Peel, pulp, and stone
Peach	11–25	Stone
Pear	4–47	Peel, pomace
Citrus fruit	55–60	Peel, rag, and seed
Banana	–	Peel
Apricot	11–35	Stone

Source: Bisht et al. (2020)

higher in glucan, lignocellulose, pectin, and minerals. They are also lower in lignin than agricultural wastes like maize stover (Fierascu et al. 2020). Additionally, waste from the fruit processing industry is distinct from other wastes because it is organic and decomposes quickly. They are enormous in volume and have a high water content, which helps them degrade without causing pollution or return to the soil in the majority of cases. Despite their high volume, they have dispersed origins, which makes gathering and using them difficult and expensive; third, they have a tendency to degrade over time, which reduces the amount of time they can be stored, even when stored under the right conditions, such as low temperatures, controlled humidity, and dark, dry locations. These characteristics make the residue left over after digesting fruit into enzymes suitable substrates. However, it is unheard of to find commercial companies or pilot plants that exclusively use the fruit processing byproduct waste as a substrate when producing enzymes.

Although articles that have been published have documented the synthesis of industrially relevant enzymes utilizing fruit processing waste as a substrate (Sharma et al. 2016), BOD, COD, and other suspended solids are present in high concentrations in the waste produced by the fruit processing businesses (Okeke et al. 2022). Most of these wastes are not recovered or handled, which harms both the environment and the health of people and animals. Although these fruit wastes contained a lot of organic substances, this enabled the creation of a variety of products with better market values and reduced production costs (Esparza et al. 2020). Table 6.2 presents the major constituents of fruit processing wastes. The processed fruit has an effect on the chemical composition of the wastes, which varies. Mango seeds, for example, are high in lipids, proteins, carbohydrates, and minerals; melon, orange, and pumpkin seeds may also include lipids and minerals. These characteristics of the fruit processing waste have the potential to make them the ideal substrate for the production of enzymes and biochemical digestion procedures that result in useful products like biogas, bioethanol, and other readily marketable goods (Duan et al. 2021).

6.3 Pollution Prevention and Control in Fruit Processing Industries

High levels of moisture and a wide variety of biological substances that may be dangerous to humans can be found in the waste from the fruit processing businesses. Depending on this, various researchers suggested measures for preventing and controlling fruit processing wastes. These measures are

- Using cleansed raw fruit, reducing the concentration of undesirable components, and using organic insecticides, for example, in the effluents.
- For dry cleaning of raw fruit, use air jets or vibration. The organic load is reduced by up to 25% using dry peeling procedures, which also reduces effluent volume by up to 35%.

Table 6.2 Chemical constituents of industrial fruit waste (w/w)

S. no.	Waste source	Chemical constituents					
		Composition of cellulose	Composition of hemicellulose	Composition of lignin	Composition of ash	Composition of total solids	Composition of moisture
1.	PP	2.2%	–	–	7.7%	–	9.89
2.	OP	9.21%	10.5%	0.84%	3.5%	–	–
3.	PinPe	18.11%	–	1.37	–	93.6	91
4.	Cs	23.77 (g/100 g)	16.68 (g/100 g)	28.59 (g/100 g)	5.39 (g/100 g)	–	–

PP potato peel, *OP* orange peel, *PinPe* pineapple peel, *Cs* coffee skin

- Recirculate treated wastewater after separation.
- When washing is required, use countercurrent systems.
- To lessen the amount of wastewater that needs to be treated, use steam instead of hot water.
- Eliminate solid wastes without using water.
- Utilize solid waste and concentrated wastewater to produce byproducts.

For example, recycling processed water from fruit processing decreases the requirement by 95% and the organic burden by 75%. Similarly, it is feasible to decrease by 80% the liquid waste burden from the manufacturing of apple juice (measured in terms of BOD or biochemical oxygen demand).

6.4 Utilization of Fruit Processing Waste and Related Challenge

During fruit preparation, a lot of waste is produced, including leaves, pomace, peels, stones, crumbs, etc. The wastes are any pieces of the fruit that are left over after consumption or that are not used due to the morphological traits of the product, improper treatment, or just thrown away for various reasons. In addition, the unwanted produced by the fruit manufacturing industries, such as the manufacturing of fruit syrup and juice, contains a lot of moisture and a lot of biodegradable organic compounds, which directly causes environmental pollution during disposal, largely through the formation of leachate and odor emission (Oladzad et al. 2021). By encouraging the spread of bacteria, vermin, and mice, improper dumping of such material into the environment brings about inescapable ecological hazards and raises the danger of sickness. As a result, this waste could be further utilized to create other useful goods or to extract important components. Recycling these pollutants in different ways will not only lead to the creation of novel products but also contribute to the solution of the problem of environmental contamination (Sagar et al. 2018b).

Additionally, advanced fruit processing firms should concentrate on reducing waste and byproducts, using less energy, and making high-quality goods without contaminating the soil, air, or spas. The utilization of these wastes resulting from the disposal of fruit processing industry byproducts also necessitates collaborative efforts. There are numerous uses for the leftovers following the preparation of fruit. Due to the significant amount of sugar in waste products and the subsequent creation of alcohol, acetic acid, citric acid, and other substances, fermentation procedures may be used as one strategy. Landfilling and incineration are two more methods for disposing waste from fruit processing. In terms of solid waste disposal, landfilling is a common practice. Nevertheless, the influence of greenhouse gases is linked to landfilling. Because the waste from fruit processing enterprises has a large quantity of moisture, incinerating it is neither an effective nor desirable disposal method (Banerjee et al. 2017).

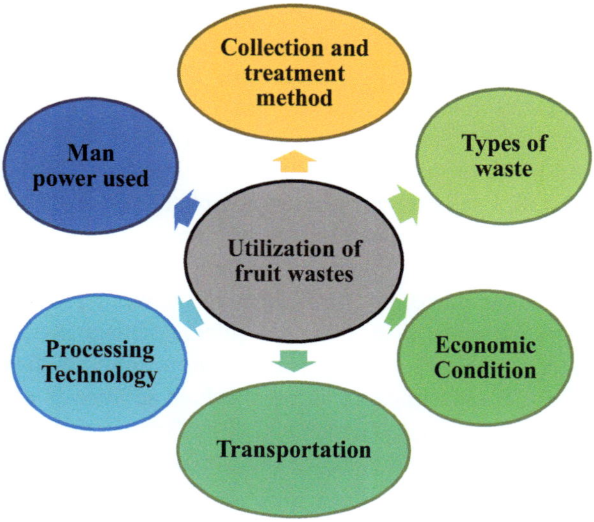

Fig. 6.1 Utilization of fruit wastes and its influencing factors

The usage of waste from the fruit processing industries is affected by several elements, such as collection, transportation, economics, technology, and waste types, as shown in Fig. 6.1. The usage of waste from the fruit processing industry is influenced by the economic situation. For instance, it was discovered that the average cost of transporting fruit trash in least-developed nations is approximately $11–15 per ton every trip, which could represent total landfilling expenses of more than $300 million (Banerjee et al. 2017). Technology for waste utilization is another aspect to take into account. Higher productivity yields were made possible by advancing technology, but care must be taken with how trash is used.

Typically, a byproduct of fruit processing, such as juice extraction, can be utilized to extract additional value commodities, such as fiber, seed oil, pectin, and vitamins, which are bioactive substances that can be utilized in the cosmetic, food processing, pharmaceutical, and textile processing sectors. Additionally, waste from the companies that process fruits can be used to make a variety of enzymes (such as cellulolytic, hemicellulose, proteases, lipases, pectinases, etc.) that are crucial for increasing the yield of juice by working on the fruit's cell walls (Jayasekara and Ratnayake 2019).

6.4.1 Opportunities for Utilizing Fruit Processing Waste

Wastes from the fruit-processing sector are geographically dispersed, plentiful, and of low nutritional value. As a result, the cost of harvesting, collecting, listing, moving, and processing byproducts may be greater than the cost of the items themselves. The distinction between the primary product and the byproduct is made by the degree of the products. In addition to manufacturing useable commodities from waste generated by the fruit processing industry, new scientific and technological procedures could be used to transform environmentally harmful byproducts into more valuable products than the primary products.

The process of turning fruit waste raw materials into commodities with added value is impacted by an array of factors. These consist of the kind and quality of fruit wastes, the kinds, standards, and traits of completed goods, processing technology, skill level, and operational state of the machinery and personnel (Dar et al. 2020). Utilizing fruit-processing industry waste helps to lessen environmental damage. Since organic waste makes up the majority of the waste produced by the fruit processing industry, improper disposal might quickly damage the ecosystem. As a result, these wastes can be properly disposed of using a variety of disposal methods, including landfilling, composting, and incineration, in addition to developing them into goods that bring value (Barbhuiya et al. 2021). On the other hand, fruit scraps left over after cleaning and seeding are fed to animals untreated.

6.4.2 Utilization Technology of Waste from the Fruit Processing Industry

Future environmental pollution reduction and technological adoption depend on the efficient use of waste fruits in industry. It is difficult to manage the leftovers from the preparation of fruits. In this situation, a partial substitution of industrial waste for conventional energy substrates is an alternative since it allows the utilization of residues while also lowering fixed operating costs (Osman et al. 2023). Fruit wastes tend to accumulate over the year, so using them to make valuable compounds can help cut down on overall waste output. In order to maintain a clean environment, waste treatment techniques, including burning and landfilling, have been used to remove significant components from wastes (Gupta et al. 2019).

The diverse sectors of the food processing industries are a part of producing a variety of processed items and different types of waste. If these wastes, whether solid or liquid, are not effectively managed, the manufacturer suffers a direct loss while also leading to environmental pollution. Among the most difficult aspects of the world today is how fruit processing companies use their waste. Wastes can be purified, extracted, and fermented using a variety of methods. These chemicals are rich in antioxidants, dietary fiber, enzymes, essential oils, food additives, organic acids, pectin, and other substances that can be utilized in industries for production activities. Discarded wastes contain higher amounts of antibacterial and antioxidant properties (Bisht et al. 2019).

To create clean and eco-friendly products, several scientists have sought fresh, practical, and efficient ways to recycle the trash from processed fruits and vegetables. Researchers are especially interested in characterization methods and processing extraction procedures (Wu et al. 2017). On the other hand, as they allow for the partial replacement of fossil fuel supplies and a reduction in the environmental impact, waste-to-energy disposal techniques are being researched. Anaerobic digestion is a process that creates biogas from solid or liquid organic waste, which can later be converted into usable energy (electricity and/or heat). Anaerobic digestion is the practice by which microbes transform organic waste materials into a gaseous mixture without the presence of oxygen, mostly made of methane (CH_4), carbon dioxide (CO_2), and nitrogen (N_2) (Gonçalves Neto et al. 2021). As a result, it is necessary to develop products using the right technologies to transform edible trash into food, helping to ease worries about overall food waste.

There are two primary categories of waste reuse from the fruit processing industries.

6.4.2.1 Conventional Utilization

Examples of conventional applications include adsorption, biochar, and energy from waste conversion (anaerobic digestion, composting, enzyme usage, and landfilling) (Patra et al. 2021).

6.4.2.2 Emerging Opportunities

Emerging applications are highlighted because they have a promising future, including nutraceuticals, encapsulation, flavoring agents, and sludge nanoparticles to produce value-added goods (Ganesh et al. 2022).

6.5 Byproducts Synthesized from Fruit Processing Wastes

The four main categories of agricultural residues are crop leftovers, agroindustry residues, livestock manure, and fruit wastes. Unrefined fruits, including mango, pineapple, tomato, jack fruit, bananas, and oranges, are included in fruit garbage, which makes up a sizeable amount of agricultural waste. Wholesale markets and companies that process food produce large amounts of fruit waste. These organic substances are a significant source of contamination and can be challenging to dispose of in landfills due to their high perishability. Furthermore, high fruit and vegetable waste production raises market operating costs (Pattanaik et al. 2019). Peels and seeds, the principal leftovers of fruit processing, make up about 40% of the weight of the fruit (de Moraes et al. 2020). Fruits are processed into peels, seeds, juices, and molasses that are high in carbohydrates, proteins, fiber, vitamins, and minerals but are also a large source of waste (Suri et al. 2021). The following are the principal sources of waste from the fruit business (Joshi 2020):

(i) Citrus fruit peel, rag, and seeds
(ii) Apple and pear pomace

(iii) Mangoes' peel and stones
(iv) Jack fruit seeds and rind
(v) Peel and core the guava
(vi) Grape seeds
(vii) Overripe and blemished fruit from canneries, etc.

6.5.1 Seed

The pulp and juice industries do not utilize passion fruit kernels and peels. If not recycled, these byproducts could pose a risk to the environment. The leftover cake obtained from the oil extraction methods can be utilized in soaps and other cosmetic formulations as a skin exfoliator (Goyal and Jerold 2021). A cosmetic and food-grade oil is also present in the seeds (Crini et al. 2020). Fruits like kiwi fruit and cantaloupes are frequently eaten with other fruit byproducts, like seeds. At the same time, larger seeds are often discarded after the flesh has been consumed since they cannot be digested. Fruit seeds have been used to create wholesome meals since they are rich in nutrients (Monika and Anna 2019).

6.5.2 Peel

The peel is the outer layer of a fruit that is thrown as trash during the processing process. Fruits, vegetables, and root crops account for 50% of overall food waste. The majority of these wastes comprise colorful peels that are high in iron/mineral content and contain a variety of accompanying ions. These characteristics will make them useful in the simultaneous extraction of natural pigments as well as bio-adsorbents (Gupta et al. 2019). Peels are the most common byproducts of fruit processing, and they have been demonstrated to be an excellent source of a variety of bioactive chemicals with a variety of therapeutic benefits. However, significant volumes of fruit peels, 20–30% for bananas and 30–50% for mangos, are discharged by the various fruit processing industries, leading to significant environmental issues (Joshi 2020). Fruit rinds or peels are two instances of leftovers from the manufacture or purchasing business that are commonly wasted. Fruits are safeguarded against the elements by their outer covering. Mango has a thin peel, but certain fruits, like the plum, have thick, rough skin. Because of their gritty texture and harsh aftertaste, most customers avoid eating fruit peels (Selvakumar and Sivashanmugam 2018; Lau et al. 2021).

Both fertilizer and animal feed can be made from the peels. The nutritional value of the skin floor of passion fruit has been the subject of numerous research, and pectin, a substance that can be used as a consistency agent, is also present in this region of the fruit (Silva et al. 2021). Pectin and flavonoids, including hesperidin, eriocitrin, and nobiletin, are also abundant in citrus peel (Liu et al. 2021).

6.5.3 Pomace

Fruit pomace, the solid material left over after fruit juice extract, includes seeds and skin, pulp, and stems for some fruits. For instance, even though the apple juice and cider industries might generate a lot of pomaces, pomace makes up 25% of apples. In Western nations like Germany and New Zealand, apple waste is particularly serious (Huettmann 2023).

6.6 Possible Products from Fruit Waste

Fruit processing companies generate around 0.5 billion tons of trash globally. Because of the abundant accessibility to this resource and the potential it still holds, researchers have done substantial research on the value-added potential for processing fruit wastes. As opposed to conventional food or other biomass-manufactured, it is noticed that waste from fruit processing is concentrated and separating in its nature. Pomace, peels, and seed fractions from waste from the processing of fruit might be helpful feedstock for the extraction of bioactive components as dietary fibers, flavonoids, lipids, pectin, and others (Banerjee et al. 2017).

Fruit wastes might be needed to increase the pilot scale of the financially viable biomass for the biofuel production process. Due to their organic nature, high availability, low lignin concentration, and low cost, these wastes may be of significant relevance for the synthesis of cellulase enzymes (Srivastava et al. 2021). Citrus wastes, grape pomace, apple pomace, banana peels, banana stalks, and pineapple peels are high in cellulose, hemicellulose, and reducing sugars. These wastes are created in great quantities each year throughout the world, yet they are still mostly undiscovered (Srivastava et al. 2021).

Fruit waste can be converted into biofuels (Dhande et al. 2021), bio-adsorbent, enzymes or catalysts (Srivastava et al. 2021), additives (Calderón-Oliver et al. 2016), bioactive (Duarte et al. 2017), and other goods. According to several studies on various Annonaceae fruits, peels and seeds are more abundant in phenolic compounds, antioxidants, and antibacterial activity than other edible parts. Various products synthesized from fruit processing wastes are illustrated in Fig. 6.2.

According to the findings of (Bisht et al. 2020; Dalal et al. 2020), the kind of waste and major potential byproducts recovered from fruit processing industry waste are presented in Table 6.3.

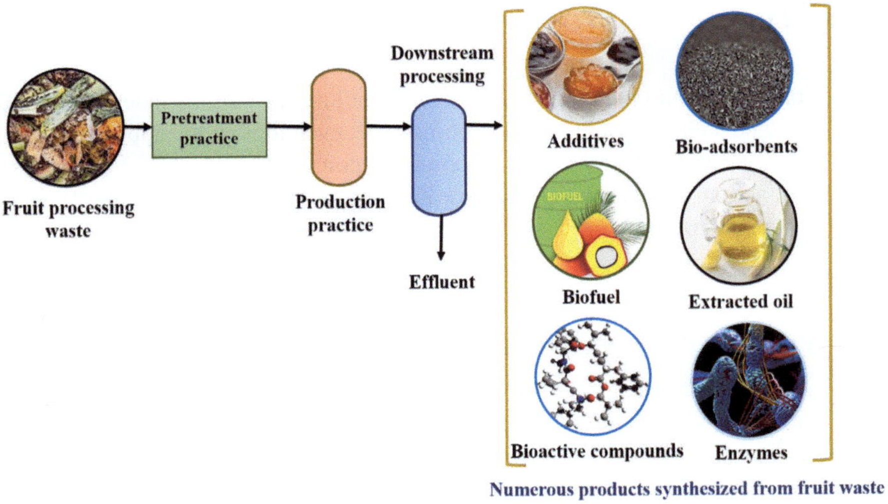

Fig. 6.2 Diagrammatic representation of several goods created from fruit processing wastes

Table 6.3 Potential byproducts from fruit processing wastes

S. no.	Fruits	Disposal (%)	Species of waste	Possible byproducts
1.	Apple	20–30	Pomace	Vinegar, pectin, wine, juice, and cow feed
2.	Orange	50	Pulp, seeds, and peel	Pectin, peel candy, animal feed, essential oils, etc.
3.	Lime	60	Seeds, peel, and pulp	
4.	Mango	40–60	Pulp and peel	Alcohol, beef feed, and pectin
5.	Grape	12–15	Skin, seed, and stem	Syrup and wine juice
6.	Pineapple	30–60	Peel, trim the core, and shred	Syrup, bromine, seed oil, juice wine, etc.

6.6.1 Biofuel

It is very feasible to replace fossil fuels with biomass for a number of reasons, including the fact that biomass is abundant in nature and, most importantly, renewable (Jayakumar et al. 2023b). Organic materials that are used as feedstocks can be divided into four categories (generations). Crops raised expressly for use in energy production are considered first-generation biomass feedstock. The emergence of the issue of food security competition led to the usage of second-generation feedstock produced from lignocellulosic materials such as forest leftovers and organic fractions from agricultural, industrial, and household wastes (Abo et al. 2023). The sole difference between third- and fourth-generation feedstock is that the latter is genetically engineered to boost production (Arpia et al. 2021).

Because of its properties, biofuel generation from fruit waste is rapidly developing at the moment. It is non-toxic and environmentally friendly. These residues from fruit processing could be used as feedstock for the manufacturing of bioethanol, which might be a better option than simply discharge to the environment. Few research publications address various practical applications of these fruit wastes and alcohol production. Examining its possible use in the production of bio-ethanol, banana, and mango fruit wastes (pulp and peels) underwent laboratory testing to determine their chemical composition (Joshi 2020).

One of the most consumed beverages in the world is coffee, and the production of it produces a lot of trash and toxins that, if not properly disposed of, could harm the environment (Ravaglio-pasquini et al. 2021). The most significant byproduct of wet-processing coffee is coffee pulp, which produces about 10 Mt worldwide annually. To develop biofuels and platform molecules, the coffee pulp is hydrothermally treated. The impacts of processing variables (temperature, pressure, reaction time, and solid/water ratio) are examined in the study. On the other hand, citrus trash is a great source of raw materials for making biological biofuels like ethanol and biogas because it contains a range of polymers of soluble and insoluble carbohydrates (Taghizadeh-alisaraei et al. 2016).

Organic wastes like fruit and vegetables offer the biological and chemical potential for bioethanol production (Sathendra et al. 2022). Some of the biological potentials include the possibility of indigenous microorganisms like *Candida* spp. On the other hand, fruit and vegetable wastes have chemical promise because lignocellulose contains significant amounts of complex saccharides (Jayakumar et al. 2022). A large amount of lignocellulose may be broken down into D-glucose and D-xylose, which microorganisms could subsequently convert into bioethanol (Utama et al. 2019).

6.6.2 Bio-adsorbents

For the treatment of wastewater, bio-adsorbents made from leftover biomass are environmentally safe, affordable, renewable, and sustainable (Gari et al. 2023). Because of the rise in demand for food and fruit globally, fruit-based biomass wastes are becoming more prevalent among other biomass waste sources (Hussain et al. 2021). Water that has been tainted has also been cleaned using fruit seeds and stones. Due to their stiff structure, which offers adsorbents with long life cycles, fruit shells and hulls also play a significant role as adsorbent materials. Lignin is abundant in the fruit's seeds and stones, and it can be used to create adsorbent materials with carbon concentrations of 45–50% (Hussain et al. 2021). With respect to their accessibility, low cost, and high biofiber content, the use of fruit wastes for the creation of bio-adsorbents (BAs) drew much interest (Hussain et al. 2021).

6.6.3 Catalysts

Trimmings and peels of horticulture leftovers may include a variety of enzymes with diverse uses in the food industry. Kinnow pulp and wheat bran can be utilized to boost cellulase activity in filter paper. Laccase, manganese peroxidase, and lignin can all be made from apple pomace. Pectinase can be made from citrus and sapodilla fruit peels (Bisht et al. 2021). It may be possible to address the global protein deficit by producing a cost-effective meal and feed by converting fruit wastes into biomass for individual cell protein synthesis. Making single-cell proteins from food processing waste could help cut down on contamination (Mondal et al. 2012). Cucumber and orange peels were used in the study of unicellular protein synthesis from submerged fermentation using *Saccharomyces cerevisiae* (Mondal et al. 2012).

6.6.4 Additives

The majority of additives are derived from various portions of fruits. Among the various additives, pectin is synthesized from both edible and peeling fruits. It can also be used in pharmaceuticals, foods, and preservatives. Many useful substances in the garbage are released into the environment. They are cutting-edge, inexpensive, and eco-friendly sources of protein, dietary fiber, antioxidants, and antimicrobials that can be employed as natural food components in the food business. New problems addressing their usage for further exploitation in producing high-nutritional food additives or supplements have piqued interest since they are high-value goods and their recovery may be financially alluring (Gowe 2015).

6.6.5 Bioactive Compounds

Due to their characteristics, fruits and vegetables are excellent biomass sources of bioactive constituents for the creation of nutraceuticals and functional foods, as well as their byproducts (Rodríguez et al. 2021). Identification of bioactive compounds is necessary to comprehend the basic processes of action and interactions of natural products in the human body (Albuquerque et al. 2016). Fruits and vegetables are high in bioactive chemicals, which help prevent a variety of degenerative disorders. These substances are frequently present in greater quantities as co-products of the preparation of fruits and vegetables. Because of this characteristic, these co-products are a desirable source for bioactive extraction, and the extraction method becomes a desirable value-added strategy for these co-products (Renard 2018).

Fruit and the residues of the fruit processing industry are important sources of bioactive chemicals including phenolic and antioxidant compounds that can be used to improve food stability by lowering lipid peroxidation. Various researchers have

Table 6.4 Various bio-active components derived from fruit processing industry waste

S. no.	Bio-actives	Source of waste	Concentration averages	Bioactivity
1.	Flavanols	Tobacco pomace, mango peels, pomegranate seeds, orange peels, and tamarind seeds	0.02–0.14% w/w on a moisture content	Oxidizing agent
2.	Phenolic acids (ferulic acid, vanillic acid, caffeic acid)	Orange peel, guava seeds, mango peels and kernels, leftover wheat germ, and apple pomace	0.02–0.2% w/w	Oxidizing agent
3.	Flavanones	Peels from citrus fruits, cucumbers, tomatoes, and citrus seeds	2–14% w/w	Oxidizing agent
4.	Anthocyanidins	Waste from grapes, berries, sapotas, corn cobs, Litchi pericarp, and seeds	0.8% w/w	Antioxidants, food coloring additives, and anticancer
5.	Carotenoids	Mango peel, papaya peel, tomato peel, and pumpkin seeds.	0.07–0.1% w/w	Extreme forager
6.	Glycosides	Banana stem, apple peel	0.02–1% w/w	Antioxidants, anticancer
7.	Bioactive lipids	Mango kernel, tomato seed, and pomegranate seed	10–30% w/w	provider of both required and optional fatty acids
8.	Dietary fibers	Carrot peel, tamarind seeds, flax seed cake, watermelon rind, and pomace	30–60% w/w	Stimulator of intestinal health, skin regeneration

Source: Banerjee et al. (2017)

found that the antioxidant activity of some fruits' leftovers is higher than that of fresh fruit pulp (Bisht et al. 2019). Waste from the fruit processing industry contains various bioactive components, which are illustrated in Table 6.4.

6.6.6 *Extracted Oil*

The oils found in various areas of discarded fruit processing are extracted and purified by various methods. Other traditional extraction techniques for making edible oils include cold pressing, hot pressing, maceration, distillation, Soxhlet or solvent extraction, and more modern extraction practices like 80 supercritical liquid extraction and techniques for pretreatment like enzyme, microwave, or pulsed electric field-assisted extractions and ultrasound (Paul and Radhakrishnan 2020). As a source of edible oil, fruit seed oil has received less attention than seed oil from field crops. Plant sources have come under scrutiny in studies to better understand their quality and functional attributes due to the rising demand for edible oil. Fruit seed is typically seen as garbage, making it a less expensive alternative source of edible oil. In addition to offering energy, seed oil also maintains bodily tissues, controls body temperature, and carries liposoluble vitamins. Food gains more texture, flavor, and palatability when seed oil is added. Consuming seed oil has been associated with a reduced risk of developing several degenerative and chronic diseases, including diabetes, cancer, and cardiovascular disease.

These beneficial effects are due to fatty acids that are stored and bioactive elements in the seed oil. As a result, seed oil is more widely used for nutraceutical and pharmaceutical purposes (Kaseke et al. 2020). After the fruit has been processed, passion fruit seeds, which make up 6–12% of the fruits weight and are a useful byproduct of juice manufacturing, are frequently thrown away. In the scientific community, it is common practice to use waste materials for research. Thus, it is possible to add value to this agricultural waste by pressing the delicious passion fruit's seeds to acquire their oil. According to the study reported by (Pereira et al. 2017), the yellow (*Passiflora edulis* Sims var. flavicarpa) passion fruit seeds contain about 30% oil by weight.

6.7 Future Prospects

Alternative lignocellulosic biomass has been used and researched recently for the creation of value-added goods. Biomass used in the fruit processing business is viewed in this context as redundant and renewable because it is thought to be a plentiful supply of carbon on earth. Future possibilities for the valorization of waste from the fruit waste processing sector into value-added goods are encouraging. This strategy entails using thermal or chemical procedures to transform organic waste resources, like leftover fruit, into valuable items. One of the main advantages of this strategy is that it can increase new revenue streams for companies while simultaneously assisting in reducing the quantity of waste that ends up in landfills or other disposal facilities. Biofuels, bioplastics, and fertilizers are just a few of the potential value-added goods that can be made from fruit waste. Numerous businesses, including industry, agriculture, and the energy sector, use these products in a variety of ways. For instance, bioplastics made from fruit waste can be utilized in packaging materials and other products, while biofuels made from fruit waste can be used in place of traditional fossil fuels. The volarization of fruit waste can assist the environment in addition to creating new cash streams. By keeping organic waste out of landfills, it is possible to lower greenhouse gas emissions while also enhancing soil health by creating organic fertilizers. In summary, it appears that the volarization of waste from the fruit waste processing industry into value-added goods has a bright future. The need for these items is probably going to grow as companies continue to look for more ecologically friendly and sustainable business methods. Additionally, new and inventive approaches to turning fruit waste into useful items are probably going to result from advances in technology and study.

6.8 Conclusion

The only method that appears to be effective for addressing the problems associated with the proper management of food processing wastes is waste valorization. The application of transdisciplinary strategies is necessary for the development of a sustainable bioeconomy. To significantly minimize waste creation, strict regulations must be put in place to limit waste at various stages of processing. Utilizing the leftovers from the food processing sector to make products with added value could be a better solution to this issue.

References

Abo LD et al (2023) Waste biomass utilization for production of bioenergy through gasification practice

Albuquerque TG, Santos F, Sanches-Silva A, Beatriz Oliveira M, Bento AC, Costa HS (2016) Nutritional and phytochemical composition of Annona cherimola Mill. fruits and by-products: potential health benefits. Food Chem 193:187–195. https://doi.org/10.1016/j.foodchem.2014.06.044

Arpia AA, Chen W, Shiung S, Rousset P, Daniel M, De Luna G (2021) Sustainable biofuel and bioenergy production from biomass waste residues using microwave-assisted heating: a comprehensive review. Chem Eng J 403(July):126233. https://doi.org/10.1016/j.cej.2020.126233

Banerjee J, Singh R, Vijayaraghavan R, Macfarlane D, Patti AF, Arora A (2017) Bioactives from fruit processing wastes: green approaches to valuable chemicals. Food Chem 225:10–22. https://doi.org/10.1016/j.foodchem.2016.12.093

Barbhuiya NH et al (2021) The future of flash graphene for the sustainable management of solid waste. ACS Nano 15(10):15461–15470

Bisht TS, Sharma SK, Rawat L (2019) A novel approach towards the fruit specific waste minimization and utilization: a review. 9:712–722

Bisht TS, Sharma SK, Rawat L, Chakraborty B, Yadav V (2020) A novel approach towards the fruit specific waste minimization and utilization: a review. J Pharmacogn Phytochem 9(1):712–722

Bisht B et al (2021) Food irradiation: effect of ionizing and non-ionizing radiations on preservation of fruits and vegetables – a review. Trends Food Sci Technol 114:372–385

Calderón-Oliver M, Escalona-Buendía HB, Medina-Campos ON, Pedraza-Chaverri J, Pedroza-Islas R, Ponce-Alquicira E (2016) Optimization of the antioxidant and antimicrobial response of the combined effect of nisin and avocado byproducts. LWT Food Sci Technol 65:46–52. https://doi.org/10.1016/j.lwt.2015.07.048

Crini G, Lichtfouse E, Chanet G, Morin-Crini N (2020) Applications of hemp in textiles, paper industry, insulation and building materials, horticulture, animal nutrition, food and beverages, nutraceuticals, cosmetics and hygiene, medicine, agrochemistry, energy production and environment: a review. Environ Chem Lett 18(5):1451–1476

Dalal N, Phogat N, Bisht V, Dhakar U (2020) Potential of fruit and vegetable waste as a source of pectin. Int J Chem Stud 8(1):3085–3090

Dar AH, Bashir O, Khan S, Wahid A, Makroo HA (2020) Fresh-cut products: processing operations and equipments. In: Fresh-cut fruits and vegetables. Elsevier, pp 77–97

de Moraes MR et al (2020) Bioactivity of atemoya fruits and by-products. Food Biosci 41(August):2021. https://doi.org/10.1016/j.fbio.2021.101036

Dhande DY, Nighot DV, Sinaga N, Dahe KB (2021) Extraction of bioethanol from waste pomegranate fruits as a potential feedstock and its blending effects on a performance of a single cylinder SI engine. Renew Sust Energ Rev 149(September):111349. https://doi.org/10.1016/j.rser.2021.111349

Dos Santos LF, Biduski B, Lopes ST, Bertolin TE, Dos Santos LR (2023) Brazilian native fruit pomace as a source of bioactive compounds on starch-based films: antimicrobial activities and food simulator release. Int J Biol Macromol:124900

Duan Y et al (2021) Apple orchard waste recycling and valorization of valuable product – a review. Bioengineered 12(1):476–495

Duarte PF, Chaves MA, Borges CD, Mendonça CRB (2017) Avocado: characteristics, health benefits, and uses. Int News Fats Oils Relat Mater 28(3):28–32. https://doi.org/10.1590/0103-8478cr20141516

Esparza I, Jiménez-Moreno N, Bimbela F, Ancín-Azpilicueta C, Gandía LM (2020) Fruit and vegetable waste management: conventional and emerging approaches. J Environ Manag 265:110510

Fierascu RC, Sieniawska E, Ortan A, Fierascu I, Xiao J (2020) Fruits by-products – a source of valuable active principles. A short review. 8(April):1–8. https://doi.org/10.3389/fbioe.2020.00319

Ganesh KS, Sridhar A, Vishali S (2022) Utilization of fruit and vegetable waste to produce value-added products: conventional utilization and emerging opportunities – a review. Chemosphere 287:132221. https://doi.org/10.1016/j.chemosphere.2021.132221

Gari MT et al (2023) Natural resources-based activated carbon synthesis

Gonçalves Neto J, Vidal Ozorio L, Campos de Abreu TC, Ferreira dos Santos B, Pradelle F (2021) Modeling of biogas production from food, fruits and vegetables wastes using artificial neural network (ANN). Fuel 285(August):119081. https://doi.org/10.1016/j.fuel.2020.119081

Gowe C (2015) Review on potential use of fruit and vegetables by-products as a valuable source of natural food additives. 45:47–61

Goyal N, Jerold F (2021) Biocosmetics: technological advances and future outlook. Environ Sci Pollut Res:1–22

Gupta N, Poddar K, Sarkar D, Kumari N, Padhan B, Sarkar A (2019) Fruit waste management by pigment production and utilization of residual as bioadsorbent. J Environ Manag 244(January):138–143. https://doi.org/10.1016/j.jenvman.2019.05.055

Huettmann F (2023) Betel nut, coconuts/copra, chocolate, strawberries, coffee, apples, spam and fish in Papua New Guinea: from ancient farming, highly bred species and sustainability concepts over diseases and dna into global market repercussions and wholesale (environmenta). In: Globalization and Papua New Guinea: ancient wilderness, paradise, introduced terror and hell. Springer, pp 253–275

Hussain N, Kumar J, Ali S, Ahmed S, Fatima N (2021) Development of fruit waste derived bio-adsorbents for wastewater treatment: a review. J Hazard Mater 416(January):125848. https://doi.org/10.1016/j.jhazmat.2021.125848

Jameel M et al (2023) Extraction of natural dyes from agro-industrial waste. In: Extraction of natural products from agro-industrial wastes. Elsevier, pp 197–216

Jayakumar M, Kuppusamy Vaithilingam S, Karmegam N, Jabesa A (2022) Bioethanol production from green biomass resources: emerging technologies. Encyclopedia Green Mater:1–12

Jayakumar M et al (2023a) A comprehensive outlook on topical processing methods for biofuel production and its thermal applications: current advances, sustainability and challenges. Fuel 349:128690

Jayakumar M et al (2023b) Bioethanol production from agricultural residues as lignocellulosic biomass feedstock's waste valorization approach: a comprehensive review. Sci Total Environ:163158

Jayasekara S, Ratnayake R (2019) Microbial cellulases: an overview and applications. Cellulose 22:92

Joshi VK (2020) Fruit and vegetable processing waste management – an overview. 10(December):30954. https://doi.org/10.30954/2277-9396.02.2020.4

Kaseke T, Opara UL, Fawole OA (2020) Fatty acid composition, bioactive phytochemicals, antioxidant properties and oxidative stability of edible fruit seed oil: effect of preharvest and processing factors. Heliyon 6(9):e04962. https://doi.org/10.1016/j.heliyon.2020.e04962

Lau KQ, Sabran MR, Shafie SR (2021) Utilization of vegetable and fruit by-products as functional ingredient and food. 8(June):1–12. https://doi.org/10.3389/fnut.2021.661693

Liu N et al (2021) A review of chemical constituents and health-promoting effects of citrus peels. Food Chem 365:130585

Mondal AK, Sengupta S, Bhowal J, Bhattacharya DK, Engineering B, Bengal W (2012) Utilization of fruit wastes in producing single cell. 1(5):430–438

Monika M, Anna K-D (2019) Nut oils and their dietetic and cosmetic significance: a review. J Oleo Sci 68(2):111–120

Nirmal NP et al (2023) Valorization of fruit waste for bioactive compounds and their applications in the food industry. Foods 12(3):556

Okeke ES et al (2022) Environmental and health impact of unrecovered API from pharmaceutical manufacturing wastes: a review of contemporary treatment, recycling and management strategies. Sustain Chem Pharm 30:100865

Oladzad S, Fallah N, Mahboubi A, Afsham N, Taherzadeh MJ (2021) Bioresource technology date fruit processing waste and approaches to its valorization: a review. Bioresour Technol 340(June):125625. https://doi.org/10.1016/j.biortech.2021.125625

Osman AI et al (2023) Materials, fuels, upgrading, economy, and life cycle assessment of the pyrolysis of algal and lignocellulosic biomass: a review. Environ Chem Lett 21(3):1419–1476

Patra BR, Mukherjee A, Nanda S, Dalai AK (2021) Biochar production, activation and adsorptive applications: a review. Environ Chem Lett 19(3):2237–2259. https://doi.org/10.1007/s10311-020-01165-9

Pattanaik L, Pattnaik F, Saxena DK, Naik SN (2019) Chapter 5. Biofuels from agricultural wastes. Elsevier

Paul A, Radhakrishnan M (2020) Pomegranate seed oil in food industry: extraction, characterization, and applications. Trends Food Sci Technol 105:273–283. https://doi.org/10.1016/j.tifs.2020.09.014

Peng W, Ma Q, Wang Z, Xie Z (2019) Research progress on comprehensive utilization of fruit and vegetable waste. 01106

Pereira MG, Hamerski F, Andrade EF, de Scheer AP, Corazza ML (2017) Assessment of subcritical propane, ultrasound-assisted and Soxhlet extraction of oil from sweet passion fruit (Passiflora alata Curtis) seeds. J Supercrit Fluids 128:338–348. https://doi.org/10.1016/j.supflu.2017.03.021

Ravaglio-pasquini F, Pedraza-segura L, Rem J, Pinilla L, Arcelus-arrillaga P, Suelves I (2021) Caffeinating the biofuels market: effect of the processing conditions during the production of biofuels and high-value chemicals by hydrothermal treatment of residual coffee pulp. 302. https://doi.org/10.1016/j.jclepro.2021.127008

Renard CMGC (2018) Extraction of bioactives from fruit and vegetables: state of the art and perspectives. LWT 93:390–395. https://doi.org/10.1016/j.lwt.2018.03.063

Rodríguez LGR, Gasga VMZ, Pescuma M, Van Nieuwenhove C, Mozzi F, Burgos JAS (2021) Fruits and fruit by-products as sources of bioactive compounds. Benefits and trends of lactic acid fermentation in the development of novel fruit-based functional beverages. Food Res Int 140:109854

Sadh PK, Duhan S, Duhan JS (2018) Agro-industrial wastes and their utilization using solid state fermentation: a review. Bioresour Bioprocess 5(1):1–15. https://doi.org/10.1186/s40643-017-0187-z

Sagar NA, Pareek S, Sharma S, Yahia EM, Lobo MG (2018a) Fruit and vegetable waste: bioactive compounds, their extraction, and possible utilization. Compr Rev Food Sci Food Saf 17:512–531

Sagar NA, Pareek S, Sharma S, Yahia EM, Lobo MG (2018b) Fruit and vegetable waste: bioactive compounds, their extraction, and possible utilization. 00(2014):1–20. https://doi.org/10.1111/1541-4337.12330
Sathendra ER, Praveenkumar R, Gurunathan B, Chozhavendhan S, Jayakumar M (2022) Generation biomass waste for bioethanol production, pp 87–110
Selvakumar P, Sivashanmugam P (2018) Multi-hydrolytic biocatalyst from organic solid waste and its application in municipal waste activated sludge pre-treatment towards energy recovery. Process Saf Environ Prot 117. https://doi.org/10.1016/j.psep.2018.03.036
Sharma R, Oberoi HS, Dhillon GS (2016) Fruit and vegetable processing waste: renewable feed stocks for enzyme production. Elsevier
Silva GC, Rodrigues RAF, Bottoli CBG (2021) Passion fruit seed extract enriched in piceatannol obtained by microwave-assisted extraction. Sustain Chem Pharm 22(June):3–9. https://doi.org/10.1016/j.scp.2021.100472
Srivastava N, Srivastava M, Alhazmi A, Kausar T, Haque S, Kumar V (2021) Technological advances for improving fungal cellulase production from fruit wastes for bioenergy application: a review☆. *Environ Pollut* 287(January):117370. https://doi.org/10.1016/j.envpol.2021.117370
Suri S, Singh A, Nema PK (2021) Recent advances in valorization of citrus fruits processing waste: a way forward towards environmental sustainability. Food Sci Biotechnol:1–26
Swetha TA et al (2023) A comprehensive review on polylactic acid (PLA)–Synthesis, processing and application in food packaging. Int J Biol Macromol:123715
Taghizadeh-alisaraei A, Hasan S, Ghobadian B, Motevali A (2016) Biofuel production from citrus wastes: a feasibility study in Iran. Renew Sust Energ Rev (September):1–13. https://doi.org/10.1016/j.rser.2016.09.102
Utama GL, Sidabutar FEE, Felina H, Wira DW, Balia L (2019) The utilization of fruit and vegetable wastes for bioethanol production with the inoculation of indigenous yeasts consortium. 25(2):264–270
Wu F, Jin Y, Xu X, Yang N (2017) Electrofluidic pretreatment for enhancing essential oil extraction from citrus fruit peel waste. J Clean Prod 159:85–94. https://doi.org/10.1016/j.jclepro.2017.05.010
Zema DA, Calabro PS, Folino A, Tamburino V, Zappia G, Zimbone SM (2019) Wastewater management in citrus processing industries: an overview of advantages and limits. Water 11(12):2481

Chapter 7
Wastes from Fruits and Vegetables Processing Industry for Value-Added Products

Abate Ayele, Chandran Masi, Ebrahim Mama Abda, and Gamachis Korsa

Abstract As the world's population grows, so does the demand for fruit and vegetable production and the processing sector that supports it, resulting in the production of large amounts of fruit and vegetable waste. Fruit and vegetable processing is one of today's most important industries. Nowadays, a large portion of fresh fruits and vegetables are processed. The industrial processing of fruits generates a large amount of waste or byproducts. Fruit and vegetable production has recently expanded dramatically as a result of the rising population and changing eating habits, with more people adopting vegetarian diets. Vegetables contain sugars, dietary and resistant fibers, and vitamins and minerals. Vegetable processing waste products such as peels, seeds, and stones can be successfully used as a source of phytochemicals and antioxidants. Vegetable tissue is rich in bioactive compounds such as phenolic compounds, carotenoids, and vitamins, and in most cases, waste byproducts can contain antioxidant and antibacterial substances that are comparable to or even greater than the end products. Because these are high-value products, and their recovery may be economically attractive, new aspects addressing the utilization of these wastes as byproducts for further exploitation in the formation of food additives or supplements with high nutritional value have captivated attention. The byproducts are a good source of sugars, minerals, organic acids, dietary fiber, and phenolics, all of which have anti-tumoral, antiviral, antibacterial, cardioprotective, and antimutagenic properties. However, due to a lack of awareness of the nutritional

A. Ayele · E. M. Abda · G. Korsa
Department of Biotechnology, College of Natural and Applied Sciences, Addis Ababa Science and Technology University, Addis Ababa, Ethiopia

Bioprocess and Biotechnology Center of Excellence, Addis Ababa Science and Technology University, Addis Ababa, Ethiopia

C. Masi (✉)
Department of Biotechnology, College of Natural and Applied Sciences, Addis Ababa Science and Technology University, Addis Ababa, Ethiopia

Bioprocess and Biotechnology Center of Excellence, Addis Ababa Science and Technology University, Addis Ababa, Ethiopia

Department of Food Technology, Dhanalakshmi Srinivasan Engineering College (Autonomous), Perambalur, Tamil Nadu, India.

E. Cherian, B. Gurunathan (eds.), *Value Added Products From Food Waste*,
https://doi.org/10.1007/978-3-031-48143-7_7

and economic worth of byproducts, their use is limited. The major goal of this review is to stimulate the production and processing of fruits and vegetables by emphasizing the possibility of extracting bioactive components from discarded fruits and vegetables and using them as natural food additives. All of these advantages will pave the way for the future use of fruit and vegetable waste in a variety of applications.

Keywords Bioactive compound · Fruits · Vegetables · Wastes · Bioactive compound · Value-added products

7.1 Introduction

Fruit and vegetables are one of the most consumed commodities in the world. A shift to healthier and more sustainable diets would have to include increasing fruit and vegetable consumption (Mason-D'Croz et al. 2019). Producing and processing horticulture crops, particularly fruits and vegetables, have increased significantly to meet the rising demand brought on by the increasing population and changing food requirements (Sagar et al. 2018). They can be consumed raw, cooked, or as a supplement to other foods. Nearly, one-third of all food produced for human consumption is lost or wasted each year, amounting to around 1.3 billion tons, with fruit and vegetable waste accounting for roughly 60% of all food waste (Ishangulyyev et al. 2019). This costs roughly 990 billion dollars and includes food losses and processing waste. Critical agricultural food commodities suffer huge losses and waste due to several variables, such as increased production, along with improper handling of fruits and vegetables, wasteful retail practices, and customer behavior (Spigarelli et al. 2018; Trigo et al. 2022; Seberini 2020). Additionally, population growth, along with technological developments, has created a demand–supply imbalance, resulting in increased food waste globally (Lin et al. 2013; Ganesh et al. 2022). Waste products from the processing of fruits and vegetables are challenging to manage. Management should be viewed in terms of its storage and, more importantly, disposal. The main issue is that the variety of waste produced in a given region is dependent on varied substrate specifications (Lipiński et al. 2018; Abdel-Shafy and Mansour 2018). The fruit and vegetable processing industry is one of the major producers of food byproducts with limited commercial potential, posing economic and environmental issues. These byproducts, on the other hand, contain a significant quantity of dietary fiber as well as bioactive chemicals with key biological functions such as antioxidant and antibacterial properties (Padam et al. 2014; Kasapidou et al. 2015; Trigo et al. 2022). Utilizing fruit and vegetable byproducts can also improve customer costs while increasing food sustainability. This review gives an overview of various types of value-added parts produced from fruit and vegetable waste. This review article explores and discusses the sources of fruit and vegetable waste as well as any potential valuable byproducts.

7.1.1 Sources of Fruit and Vegetable Wastes

Except for the edible part, remaining fragments such as leaves, peel, pomace, rind, stem, seeds, and spoiled fruits and vegetables are considered waste in the fruit and vegetable industry. These items are gathered throughout the washing, cleaning, and processing processes (Narasimmalu and Ramasamy 2020). Byproducts of vegetables and fruits are wastes from the agricultural, postharvest, processing, distribution, and consumption stages. Some of the byproducts have been used as fuel, construction material, or animal feed, while the majority has been discarded (Lau et al. 2021). Fruit and vegetable waste (FVW) is composed of from inedible food components such as the outer layers and extremities of fruits and vegetables that are removed during processing, primarily through peeling and pressing. During collection, handling, transportation, and processing, these parts are discarded (Roberta et al. 2014; Pérez-Marroquín et al. 2023). Fruit and vegetable waste includes pectin derived from apples, pineapples, and guavas, as well as ethanol-, tartrate-, and malate-rich grape pomace, and oil from tomato seeds and other vegetables. When a consumer rejects a product owing to unacceptable quality such as off-color, off-flavor, or damage sustained during shipping, the resources are termed waste in these industries (Narasimmalu and Ramasamy 2020; Cecchi and Carolis 2021). Plant tissue remnants generated from agricultural fields and also during processing are also considered as such waste (Singh et al. 2011; Mamma and Christakopoulos 2014; Sharma et al. 2017). By commodity, vegetable and fruit byproducts account for 44% of global food wastes, roots and tubers account for 20%, and cereal accounts for 19% (Lau et al. 2021). Fruit and vegetable processing industries can thus add value by producing 10–60% of raw material as waste (Sudha and Priyanka 2023). Potential quantities of waste generated from some fruit and vegetable in tones per year are illustrated in Fig. 7.1 (Gowe 2015).

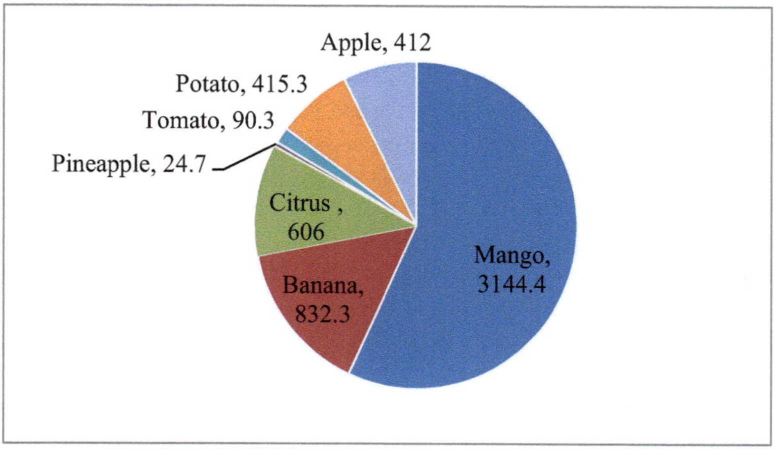

Fig. 7.1 Quantities of waste generated from fruit and vegetables (tones/year) (Gowe 2015)

7.1.2 Vegetable and Fruit Byproducts

Trimmings, peelings, stems, seeds, shells, bran, and residues left over for juice, oil, starch, and sugar extraction are all considered to be vegetable waste (Kumar et al. 2017). Vegetable byproducts are secondary products that are often discarded or wasted during food processing or manufacturing. In the preparation process, up to one-third of the vegetables may be wasted (Sagar et al. 2018; Lau et al. 2021). Surprisingly, several vegetable sections are voluntarily discarded due to their unpleasant flavor or texture. Vegetable pieces such as hulls, bagasse, and seeds, for example, are frequently thrown in the manufacturing process (Yusuf 2017, Lau et al. 2021).

Fruit waste is one of the main sources of municipal solid waste, which has posed ongoing threats to the environment. Any items dumped from residential, commercial, or industrial activity are classified as municipal solid waste (Pattnaik and Reddy 2010; Makwara and Snodia 2013). Fruit peels or rinds are an example of a byproduct that is frequently wasted throughout the consumption or production industry. It is the outer covering of fruits that protects them from the elements. Some fruits, like pomegranate, have thick and rough skin, while others, like mango, have a thin peel. Because of the hard texture and harsh aftertaste, most people avoid eating fruit skins and are consequently dumbed as waste (Dembitsky et al. 2011; Lau et al. 2021). Moreover, after pressing the juice or making other value-added products like jams, jellies, and marmalades, several wastes and byproducts from the fruit industry are generated.

7.2 Value-Added Products Derived from Fruit and Vegetable Wastes

The production of fruits and vegetables is rising significantly worldwide. As a result, the quantity of byproducts produced during the processing of fruits and vegetables in food factories has also significantly increased (Taghian Dinani and van der Goot 2022). Producing valuable products from waste is known as value addition (Narasimmalu and Ramasamy 2020). Fruit and vegetable waste is mostly made up of lignocellulosic biomass, which is composed of cellulose (30–50%), hemicellulose (15–35%), and lignin (10–20%), and is an excellent site of sugars and phenolic compounds (Gomes-Araújo et al. 2021). Hence, sugars, pectin, proteins, lipids, polysaccharides, fibers, polyphenols, vitamins (A and E), vital minerals, fatty acids, volatiles, anthocyanins, and pigments are the key components of value-added products extracted from fruit and vegetable byproducts or wastes (Fig. 7.2) (Sharma et al. 2017; Socas-Rodríguez et al. 2021). Bagasse, peels, trimmings, stems, shells, bran, and seeds, among other fruit and vegetable byproducts accumulated from industrial operations, make up more than half of all fresh fruit and have nutritional and functional values that are sometimes higher than the finished product (Mohd

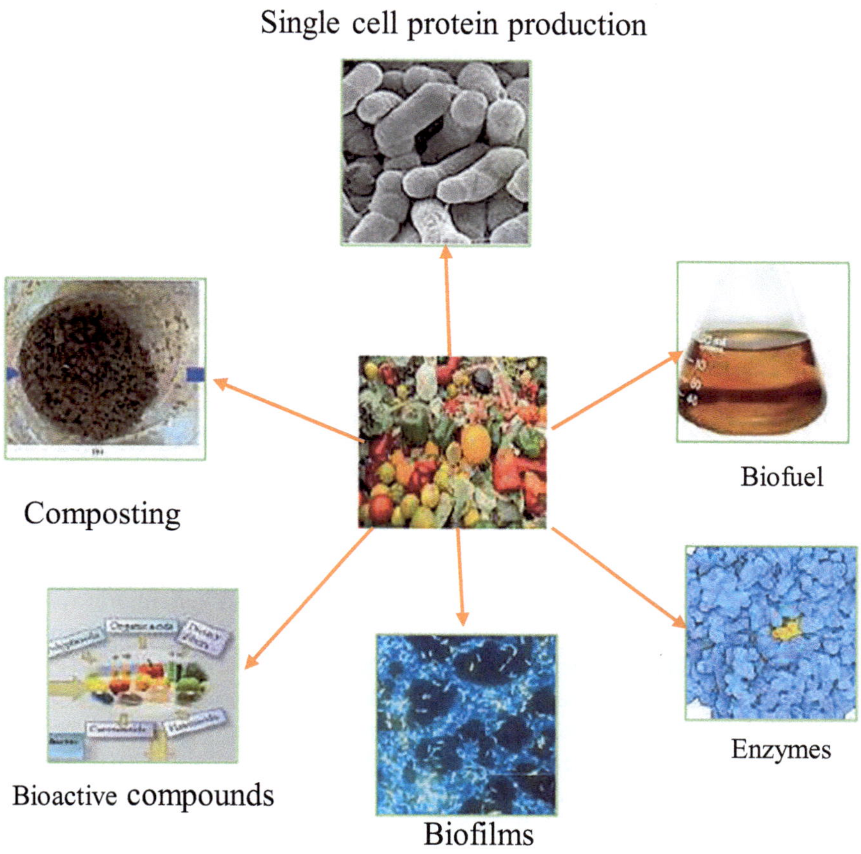

Fig. 7.2 Value-added products derived from fruit and vegetable wastes (Sánchez et al. 2021)

Basri et al. 2021). The food industry generates massive amounts of waste, primarily in the form of pomace (a mixture of pulp, skin, seeds, and stem), which typically contains much higher levels of bioactive compounds than the fruit juice itself. Peels from grapes, apples, citrus fruits, and avocado, jackfruit, and mango seeds have been found to have more than 15% more polyphenolic compounds than the pulp (Jiménez-Moreno et al. 2020; Coman et al. 2020; Socas-Rodríguez et al. 2021). Fruits and vegetables provide sugars, dietary and resistant fibers, and vitamins and minerals (Asif 2011; Slavin and Lloyd 2012; Rudra et al. 2015).

7.2.1 *Organic Acid Production*

Organic acids are biomolecules that are widely used in the food, cosmetics, and chemical industries. Organic acids have been found in fruit and vegetable wastes, and some researchers have employed these food industry wastes as a substrate to generate various organic acids (Upadhyay et al. 2010; Sánchez et al. 2021). The most important acids for the food and pharmaceutical industries are citric and lactic acids (Sagar et al. 2018). Lactic acid occupies a prominent position in the carboxylic acid family due to its use in both the food and non-food industries. It is used in the food industry as a preservative and acidulant. However, due to the high cost of the raw ingredients used, commercial production of lactic acid is expensive. Biological wastes can be used to reduce costs (Wadhwa et al. 2016; Iglesias et al. 2020).

7.2.2 *Biofuel*

Starch-rich byproducts of fruits and vegetables make great substrates for the production of biofuels. Even their leftovers from the extraction processes can be used to produce biofuels (Taghian Dinani and van der Goot 2022). For a long time, food waste—in particular, fruit and vegetable waste—has been viewed as an underutilized resource for bioenergy production. Bioenergy is described as energy derived from renewable resources, such as biodegradable waste and agricultural residues like fruit and vegetables. These energy sources have the lowest environmental impact, are less likely to deplete natural resources, and have the least negative impact on human health (Ganesh et al. 2022). Fruit and vegetable wastes are inexpensive and readily available renewable substrates that could be used to produce bioethanol, biogas, and other renewable energies (Wadhwa et al. 2016; Deressa et al. 2015). Fruit wastes contain low lignin content in addition to high cellulose and hemicellulose content, making them suitable for bioethanol production (Dhillon et al. 2013; John et al. 2017; Bhuvaneswari and Sivakumar 2019). Thereby, bananas, citrus fruit, apples, and pineapples are some of the most common fruit wastes used in bioethanol production (Zanivan et al. 2022).

7.2.3 *Polyhydroxybutrate Production (PHB)*

From glucose, cellulose, lignin, and oil, bioplastics are produced. Bagasse, agro wastes, fruit pulp, and peels are also used to produce biopolymers. It may be strategically beneficial and economically realistic to use agricultural and industrial waste as a source of carbon for the production of PHB to reduce production costs and energy consumption (Yusuf, 2017). Given that fruit and vegetable waste constitute a significant portion of food waste, poly-hydroxybutyrate from fruit waste is becoming more and more common nowadays (Sirohi et al. 2021). Biopolymers can also be made from fruit peels, fruit pulp, agron waste, and bagasse. Monomeric units of glucose make up cellulose-based biopolymers. Lignocellulosic waste from paper and pulp mills contains lignin, which is a good source of lignin for bioplastic production. Oil-based bioplastics such as poly-3-hydroxyalkanoates and polyamide11

(PA11) are also made from corn oil, soybean oil, and palm oil (Sanchez Vazquez 2014; Sirohi et al. 2021; Thulasisingh et al. 2021).

7.2.4 Single-Cell Protein (SCP) Production

Several researchers are interested in developing sustainable materials as an alternative to plastics by combining various compounds extracted from fruits and vegetables to make viable films for bio-packaging (Ludwicka et al. 2020). Bioplastic production from vegetable or fruit waste can be regarded as a sustainable process due to its biodegradability and carbon neutrality. Due to these environmental advantages, bioplastics could expand on the global market (Coppola et al. 2021; Ganesh et al. 2022). Proteins isolated from pure microbial cell culture are known as biomass, bioprotein, or microbial protein. Bacteria, fungi, algae, and yeasts are used to produce them (Mondal et al. 2021; Wadhwa et al. 2016; Bogale 2020). The term SCP refers to the dead, dried microbial cells or total protein extracted from the pure microbial culture of filamentous fungi, bacteria, algae, unicellular algae, and cyanobacteria that are produced on various sources of carbon and used as a protein supplement in human foods or animal feeds (Thiviya et al. 2022). Fruit and vegetable processing industries are estimated to produce large amounts of solid waste, such as seeds, which could have commercial use as protein sources. Utilization of these fruit and vegetable processing byproducts (usually seeds and skin), which are available at no additional cost, can help in generating value-added protein adjuncts while also reducing solid waste and, as a result, contributing to environmental sustainability (Sarkar and Kaul 2014; Gowe 2015). The single-cell protein is produced from microbial growth and biomass, and it can be used as a protein supplement in feed or food (Ravindra 2000; Joshi 2020). SCP may be produced in large quantities, and the process that results is very promising. Microorganisms that can be considered as single-cell protein (SCP), i.e., useful as protein additives and sources of important biomolecules, should have a range of traits, including rapid growth, low nucleic acid levels, and good biological values, as well as being non-pathogenic and non-toxic (Matos 2017; Gervasi et al. 2018).

7.2.5 Dietary Fiber

Dietary fiber (DF) is composed of cellulose, hemicellulose, pectin, B-glucans, gums, and lignin, which are nonstarch polysaccharides (Sharoba et al. 2013; Slavin 2013; Joshi 2020). Biomolecules resistant to digestion by gastrointestinal enzymes make up dietary fiber (DF). These include cellulose, hemicellulose, pectic substances, beta-glucans, resistant dextrin, inulin, gums, chitosan oligosaccharides, lignin, and other components (Malenica and Bhat 2020; Oyedepo and Kayode 2020). DF, also known as nonstarch polysaccharides, is abundant in fruits and vegetable byproducts. Because the human upper digestive tract lacks the enzymes required to break down polysaccharides, they reach the colon almost intact. DF is divided into two categories: water-soluble (pectin) and water-insoluble (cellulose and lignin) (Trigo et al. 2022). Fruit and vegetable fiber contains a higher proportion of soluble dietary fiber, whereas cereal fiber contains more insoluble cellulose and hemicellulose. Potato peels are a great source of fiber because they contain 9.7% to 68% weight (Elleuch et al. 2011; Sharoba et al. 2013). Fruit pomace from apples and berries has been proposed as an additive in bakery and dairy product formulations to enhance natural antioxidants and dietary fiber content (Sun-Waterhouse 2011; Socas-Rodríguez et al. 2021). Dietary fiber is made up of building blocks such as cellulose, hemicellulose, lignins, and pectins, as well as resins and waxes (Nawirska and Uklańska 2008; Zhu et al. 2016; Chavan et al. 2018). Dietary fibers are carbohydrate polymers including cellulose, hemicellulose, lignin, and pectin, which provide the plant cell wall structural rigidity. Dietary fibers are classified as soluble dietary fiber (SDF) or insoluble dietary fiber (IDF) based on their water solubility (Daou and Zhang 2014; Wu et al. 2020). Pectin (sugars from whole grains, legumes, and other sources), gums (sugar monomers from beans, legumes, as well as other sources), and mucilage are examples of dietary soluble fiber (aquatic plants, cactus, aloe vera, okra, as well as glycoproteins from food additives) (Hussain et al. 2020). Insoluble dietary fiber forms include cellulose (which provides glucose monomers and is found in fruits, root vegetables, and grains), hemicellulose (complex sugars found in cereal bran and grains), and lignin (aromatic alcohols found in plants) in Table 7.1 (Jalili et al. 2000; Hussain et al. 2020).

7.2.6 *Composting*

It is a technology that enables the handling of organic waste, making it suitable for food waste, especially for fruit and vegetable waste (Esparza et al. 2020; Neto et al. 2021). Over time, the weight of fruits and vegetables crushes them, damaging plant

Table 7.1 Dietary fiber contents in the waste of different fruits and vegetable wastes

S. no.	Vegetables/fruits	Soluble dietary fiber (SDF) (%)	Insoluble dietary fiber (IDF) (%)	References
1.	Apple pomace	15	36	Sudha et al. (2007)
2.	Apple	1.1	12	Bae et al. (2016)
3.	Cranberry pomace	6	66	White et al. (2010)
4.	Pomace	0.8	58	Gouw et al. (2017)
5.	Tommy Atkins	14.3	13.8	Larrauri et al. (1996)
6.	Orange	9	47.6	Chau et al. (2003)
7.	Kiwi fruit	10	19	Soquetta et al. (2016)
8.	Pomegranate peels	2	12	Colantuono et al. (2017)
9.	Carrot	14	50	Chau et al. (2004)

Table 7.2 Compositions of vegetable waste and dried compost product (Chang et al. 2006)

Composition	Vegetable waste	Dried compost product
Moisture (%)	92–94	12–14
Ash (%)	0.6–0.7	16–20
Volatile matter (%)	5.3–7.4	70–74
Carbon (%)	37–39	43–45
Hydrogen (%)	6.0–7.0	5.5–6.3
Nitrogen (%)	2.5–3.0	0.7–0.8
Oxygen (%)	51–53	49–51
C/N (solid)	12–13	52–60
pH value	6.2–6.4	8.1–8.5

cells that provide structural stiffness and strength, allowing water to seep into the pores (Sall et al. 2016; Padayachee et al. 2017). With nitrogen recovery of 6 to 22%, composting vegetable and fruit wastes can be utilized to replace a significant amount of nitrogen fertilizer, as shown in Table 7.2. Owing to the increase in the more resistant carbon components caused by long-term vegetable and fruit compost (VFC) use, carbon accumulated on soils (Tits et al. 2014; Wadhwa et al. 2016).

7.2.7 Enzymes Production

Enzymes are proteins that, under mild conditions, act as active and highly selective catalysts for particular chemical reactions. These biomolecules have important applications in a variety of industries, including food, cosmetics, medicines, textiles, chemicals, and fuels (Esparza et al. 2020). Enzymes are essential in the industrial sector because they ensure sustainability by lowering pollution, reducing the chemical load of processes, and removing harmful substances. The most widely utilized enzymes in beverages, processed meat, dairy, fruits, and vegetables are lipases, carbohydrases, proteases, and polymerases (Osorio et al. 2021). Enzymes are high-value bioactive compounds that can be produced by the biotransformation of FVWs (Sánchez et al. 2021). Fruit and vegetable wastes/residues are used as a substrate for microbial fermentation to produce enzymes, which has a wide range of uses in Table 7.3 (Wadhwa et al. 2016).

7.2.8 Bioactive Compounds

Bioactive compounds or components are food or dietary supplements that, in addition to meeting basic nutritional requirements, have a beneficial health effect on the host (Laparra and Sanz 2010; Saini et al. 2019). These are known to have antioxidant, antibacterial, or anti-inflammatory effects on the host; however, these effects

Table 7.3 Production of enzymes using fruit and vegetable waste (FVW)

Enzymes	Purposes	FVW used for enzyme production	References
Amylase	Food industries for the production of fruit juices, cheese, chocolate cakes, and syrups	Banana peel, orange peel, fruit juices, starch syrup, moist cakes, chocolate cakes	Sagar et al. (2018)
Invertase	Used to produce invert sugar	Fruit peel	Sagar et al. (2018)
Pectinase	Pectinases degrade pectic compounds, which are important fruit and vegetable structural components in cell walls	Pineapple peel, lemon peel, orange peel, banana peel, wheat bran, rice bran, and sugarcane bagasse	Sagar et al. (2018), Namasivayam et al. (2011) and Garzón and Hours (1992)
Cellulase	Key enzymes of potential use for industrial saccharification of cellulosic materials into simple sugars	Banana waste, orange peels, pineapple, olive pomace, wheat bran, oil palm trunk, and corncob residue	Amaeze et al. (2015) and Tian et al. (2018)
Xylanase	To hydrolyze xylan, which is an abundant natural polysaccharide	Peel, olive pomace, wheat bran, oil palm trunk, and corncob residue	Martins et al. (2018) and Tian et al. (2018)

fluctuate depending on their bioactivity, chemical structure, dosage, and other factors (Saini et al. 2019; Leyva-López et al. 2020). Because of their biological properties, bioactive compounds can affect human health (antioxidant, antimicrobial, anti-inflammatory, anticancer, etc.). In general, they are less abundant in the edible organs of plants than in their byproducts. Vitamins C and E, carotenoids, phenolic compounds, and dietary fiber are the most common bioactive chemical compounds found in fruits and vegetables (Gowe 2015; Chamorro et al. 2022). Due to their richness in several bioactive components, fruits and vegetables are among the most basic types of functional foods (Gemechu 2020, Kumar et al. 2017). Fruit and vegetable waste is rich in many bioactive components; therefore, rather than being seen as garbage, these materials can be used to recover useful compounds, furthering the zero-waste concept. A zero-waste concept is a useful approach that enables efficient valorization of the generated agro-industrial wastes into value-added products, which have numerous applications in the food industry as coloring agents, antioxidants, and preservatives (Saini et al. 2019). Bioactive fractions of fruit and vegetable wastes usually include carbohydrates, proteins, lipids, and secondary metabolites (Okino Delgado and Fleuri 2016; Sánchez et al. 2021).

Bioactive peptides and proteins can be found in abundance in animal industry byproducts (Ben-Othman et al. 2020; Vilas-Boas et al. 2021). Phenolics and carotenoids are the bioactive compounds found in Fruit and Vegetable Waste (FVW) that are most widely distributed (Bailão et al. 2015; Trigo et al. 2022). In most cases, the waste byproducts may have similar or even higher antioxidant and antimicrobial content than the end product because fruits and vegetables' entire tissue is rich in bioactive compounds such as phenolic compounds, carotenoids, and vitamins

Table 7.4 Bioactive compounds from fruits and vegetable waste

S. no.	Sources	Residue	Metabolites/compounds	References
1.	Pineapple	Husk	Phenolic compounds	Alias and Abbas (2017)
2.	Papaya	Seeds	Sulforaphane and phenolic compounds	Briones-Labarca et al. (2015)
3.	Grape	Peel	Anthocyanins	Corrales et al. (2008)
4.	Tomatoes	Wastes	Sterols, tocopherols, carotenes, terpenes, and polyphenols	Kalogeropoulos et al. (2012)
5.	Purple eggplant	Peel and pulp	Anthocyanins and phenolic compounds	Ferarsa et al. (2018)
6.	Elderberry	Stalks	Phenolic and lipophilic compounds	Patinha et al. (2021)
7.	Apples	Seeds	Phenolic compounds	Gunes et al. (2019)
8.	Garlic	Husk	Phenolic acids, flavonoids	Kallel et al. (2014)
9.	Carrots	Peel	Polyphenols, flavonoids, carotenoids, Vitamins	Nguyen and Scarlett(2016)
10.	Beetroots	Leaf	Polyphenols, betalain	Calvo et al. (2018)
11.	Mango	Peel	Flavonol glycosides	Maisuthisakul and Gordon (2009)
12.	Pomegranate	Seeds and peel	Anthocyanins, ellagic acid	Akhtar et al. (2015)
13.	Acerola	Peel	Anthocyanins, phenolics, and vitamin C	Sancho et al. (2015)
14.	Quinoa	Leaves	Flavonoids, polyphenols, anthocyanins, and vitamin C	Świeca et al. (2014)

(Ayala-Zavala et al. 2010; Rudra et al. 2015). Additionally rich in value-added components, fruits and vegetables contain pectin and polyphenols (bioactive compounds) that can be extracted before fermentation and used to produce industrial chemicals (Mahato et al. 2019; Shehu et al. 2019). Fruit and vegetable wastes have been investigated for their ability to extract phenolic compounds, dietary fibers, and other active compounds, as shown in Table 7.4 (Sagar et al. 2018; Mohd Basri et al. 2021). Pomace, a mixture of pulp, skin, seeds, and stem, is the main type of waste produced by the food industry, and it frequently has levels of bioactive compounds that are far higher than those found in fruit juice (Ben-Othman et al. 2020). Researchers have discovered that the peels of grapes, apples, citrus fruits, avocado, jackfruit, and mango seeds contain more polyphenolic compounds than pulp fibers by more than 15% (Socas-Rodríguez et al. 2021; Bhardwaj et al. 2022).

7.2.9 Nutraceutical Compounds

A nutraceutical may be defined as a food (or part of food) that provides medical or health benefits, including the prevention and/or treatment of a disease (Chauhan et al. 2013; Wadhwa et al. 2016). They define nutraceutical as "a product isolated or purified from foods that are generally sold in medicinal forms not usually associated with food" (Bishnoi and Mudgil 2022). Findings in the literature showed that vegetable and fruit byproducts have high nutritional value (O'Shea et al. 2012; Lau et al. 2021).

7.2.10 Functional Foods

Functional food is defined as "a food that resembles or may resemble a conventional food, that is consumed as part of a regular diet, and that has been shown to offer physiological advantages and/or lower the risk of chronic diseases beyond basic nutritional functions" (Salim et al. 2017). Functional foods contain bioactive ingredients that provide benefits in addition to nutritional effects (Pérez-Marroquín et al. 2023).

7.2.11 Biofilms

In biotic or abiotic surfaces and interfaces, biofilms are structurally complex assemblages of microorganisms that are characterized by population interactions. They possess exopolymeric substances (EPS) and survive as self-organized, three-dimensional structures with modified phenotypic and genotypic characters (Gemechu 2020, Jahid and Ha 2012). Thin layers of edible coatings are applied to food surfaces, increasing food shelf life and maintaining food qualities, properties, and functionality for a low cost (Hussain et al. 2022). The production of biodegradable films using biopolymers and glycerol solutions is one of the useful applications of polysaccharides derived from tomato processing industry wastes and granadilla peels (Poli et al. 2011; Rudra et al. 2015). Electrospinning and dip-coating are the processes used to produce the biofilm. Fruits and vegetables with a coating have been discovered to have a longer shelf life (Singh and Packirisamy 2022; Polat and Aygun 2022). Nanotechnology and various technologies such as dip coating, spray coating, and electrospun nanofibers coating are used to apply thin edible coatings on perishable fruits and vegetables (Miteluţ et al. 2021; Singh and Packirisamy 2022).

7.3 Future Prospects of Value-Added Products from Fruit and Vegetable Waste

The conversion of fruit and vegetable waste into other value-added products has been investigated in order to reduce waste management burdens, reduce resource and energy consumption, and protect the environment (Ingrao et al. 2021). The growing recognition of the importance of sustainable practices, as well as the need to reduce fruit and vegetable waste, has motivated the development of novel and innovative bioactive compound extraction technologies. The use of cutting-edge technologies has many advantages, including waste reduction, the creation of new economic opportunities, the promotion of a circular economy, and the development of functional food ingredients, cosmetics, and dietary supplements. It will be critical in the coming years to develop specific extraction methods for various fruit and vegetable waste sources or compounds of interest (Liu et al. 2023). In the current scenario, FVW management has become critical for sustainable growth. As a result, there is an urgent need to investigate alternative solutions to fully utilize FVW in order to gain economic, environmental, and social benefits from these waste materials (Bhardwaj et al. 2022). Green solvents should be used as much as possible to reduce environmental and human health impacts. Setting specific standards for value-added products resulting from fruit and vegetable waste conversion may be difficult but worthwhile, given the lack of a uniform standard for fruit and vegetable waste sources (Liu et al. 2023). Furthermore, efforts should be made to bridge the quality gap between value-added products derived from fruit and vegetable waste and the original product. Effective food waste management, particularly of fruit and vegetable waste, is a critical issue that must be addressed immediately on a global scale. These food items account for a significant portion of wastage and have an important economic value (Ganesh et al. 2022).

7.4 Conclusion

As agro-industrial wastes, many low-cost raw materials are available and valuable sources of value-added products. As hot-spot agri-food commodities, fruits and vegetables are produced in large quantities, have a significant potential for waste reuse in line with the circular economy concept, and have a positive significant impact on quality of life. Their wastes hold great potential for the next generation of eco-materials used in the pharmaceutics, cosmetics, energy, environment, and biomedical industries. The literature that is currently available makes it clear that wastes and byproducts from fruits and vegetables offer several opportunities for the isolation of naturally occurring value added products with potential uses in the food, pharmaceutical, and cosmetic industries. This will help to completely utilize the industrial waste, providing the industries extra compensation through the sale of residues, and it will also help to eliminate environmental pollution caused by improperly disposing of industrial fruit and vegetable waste. The urgent need is to improve and optimize the isolation, extraction, processing, and production processes of fruit and vegetable wastes and byproducts using a sustainable approach.

References

Abdel-Shafy HI, Mansour MS (2018) Solid waste issue: sources, composition, disposal, recycling, and valorization. Egypt J Pet 27:1275–1290. https://doi.org/10.1016/j.ejpe.2018.07.003

Akhtar S, Ismail T, Fraternale D, Sestili P (2015) Pomegranate peel and peel extracts: chemistry and food features. Food Chem 174:417–425. https://doi.org/10.1016/j.foodchem.2014.11.035

Alias NH, Abbas Z (2017) Preliminary investigation on the total phenolic content and antioxidant activity of pineapple wastes via microwave-assisted extraction at fixed microwave power. Chem Eng Trans 56:1675–1680. https://doi.org/10.3303/CET1756280

Amaeze NJ, Okoliegbe IN, Francis ME (2015) Cellulase production by *Aspergillus niger* and *Saccharomyces cerevisiae* using fruit wastes as substrates. Int J Appl Microbiol Biotechnol 3:36–44. http://bluepenjournals.org/ijambr/pdf/2015/May/Amaeze_et_al.pdf

Asif M (2011) The role of fruits, vegetables, and spices in diabetes. Int J Nutr Pharmacol Neurol Dis 1:27. https://www.ijnpnd.com/text.asp?2011/1/1/27/77527

Ayala-Zavala JF, Rosas-Domínguez C, Vega-Vega V, González-Aguilar GA (2010) Antioxidant enrichment and antimicrobial protection of fresh-cut fruits using their own byproducts: looking for integral exploitation. J Food Sci 75:175–181. https://doi.org/10.1111/j.1750-3841.2010.01792.x

Bae IY, Jun Y, Lee S, Lee HG (2016) Characterization of apple dietary fibers influencing the in vitro starch digestibility of wheat flour gel. LWT Food Sci Technol 65:158–163. https://doi.org/10.1016/j.lwt.2015.07.071

Bailão EF, Devilla IA, Da Conceição EC, Borges LL (2015) Bioactive compounds found in Brazilian Cerrado fruits. Int J Mol Sci 16:23760–23783. https://doi.org/10.3390/ijms161023760

Ben-Othman S, Jõudu I, Bhat R (2020) Bioactives from agri-food wastes: present insights and future challenges. Molecules 25:510. https://doi.org/10.3390/molecules25030510

Bhardwaj K, Najda A, Sharma R, Nurzyńska-Wierdak R, Dhanjal DS, Sharma R, Manickam S, Kabra A, Kuča K, Bhardwaj P (2022) Fruit and vegetable peel-enriched functional foods: potential avenues and health perspectives. Evid Based Complement Alternat Med 2022:14. https://doi.org/10.1155/2022/8543881

Bhuvaneswari M, Sivakumar N (2019) Bioethanol production from fruit and vegetable wastes. Bioprocess Biomol Prod 18:417–427. https://doi.org/10.1002/9781119434436.ch20

Bishnoi S, Mudgil D (2022) Current concepts and prospects of herbal nutraceutical. In: Handbook of nutraceuticals and natural products: biological, medicinal, and nutritional properties and applications, vol 1, pp 189–204. https://doi.org/10.1002/9781119746843.ch10

Bogale TT (2020) Microbial protein production from agro-industrial wastes as food and feed. Am J Life Sci 8:121–126. https://doi.org/10.11648/j.ajls.20200805.16

Briones-Labarca V, Plaza-Morales M, Giovagnoli-Vicuña C, Jamett F (2015) High hydrostatic pressure and ultrasound extractions of antioxidant compounds, sulforaphane and fatty acids from Chilean papaya (*Vasconcelleapubescens*) seeds: effects of extraction conditions and methods. LWT Food SciTechnol 60:525–534. https://doi.org/10.1016/j.lwt.2014.07.057

Calvo TR, Perullini M, Santagapita PR (2018) Encapsulation of betacyanins and polyphenols extracted from leaves and stems of beetroot in Ca (II)-alginate beads: a structural study. J Food Eng 235:32–40. https://doi.org/10.1016/j.jfoodeng.2018.04.015

Cecchi T, Carolis CD (2021) Food processing industries, food waste classification and handling, target compounds. In: Biobased products from food sector waste. Springer, Cham, pp 17–78. https://doi.org/10.1007/978-3-030-63436-0_2

Chamorro F, Carpena M, Fraga-Corral M, Echave J, Rajoka MS, Barba FJ, Cao H, Xiao J, Prieto MA, Simal-Gandara J (2022) Valorization of kiwi agricultural waste and industry by-products by recovering bioactive compounds and applications as food additives: a circular economy model. Food Chem 370:131315. https://doi.org/10.1016/j.foodchem.2021.131315

Chang JI, Tsai JJ, Wu KH (2006) Composting of vegetable waste. Waste Manag Res 24:354–362. https://doi.org/10.1177/0734242X06065727

Chau, C. F., Huang, Y. L., & Lee, M. H. (2003). In vitro hypoglycemic effects of different insoluble fiber-rich fractions prepared from the peel of Citrus sinensis L. cv. Liucheng. Journal of Agricultural and Food Chemistry, 51(22), 6623–6626.

Chau, C. F., Chen, C. H., & Lin, C. Y. (2004). Insoluble fiber-rich fractions derived from Averrhoa carambola: hypoglycemic effects determined by in vitro methods. LWT-Food Science and Technology, 37(3), 331–335.

Chauhan B, Kumar G, Kalam N, Ansari SH (2013) Current concepts and prospects of herbal nutraceutical: a review. J Adv Pharm Technol 4:4. https://doi.org/10.4103/2231-4040.107494

Chavan P, Singh AK, Kaur G (2018) Recent progress in the utilization of industrial waste and by-products of citrus fruits: a review. J Food Process Eng 41:e12895. https://doi.org/10.1111/jfpe.12895

Colantuono A., Vitaglione P., Ferracane R., Campanella O. H., Hamaker B. R. (2017). Development and functional characterization of new antioxidant dietary fibers from pomegranate, olive and artichoke by-products. Food Res Int 101:155–164.

Coman V, Teleky BE, Mitrea L, Martău GA, Szabo K, Călinoiu LF, Vodnar DC (2020) Bioactive potential of fruit and vegetable wastes. Adv Food Nutr Res 91:157–225. https://doi.org/10.1016/bs.afnr.2019.07.001

Coppola G, Gaudio MT, Lopresto CG, Calabro V, Curcio S, Chakraborty S (2021) Bioplastic from renewable biomass: a facile solution for a greener environment. Earth Syst Environ 5:231–251. https://doi.org/10.1007/s41748-021-00208-7

Corrales M, Toepfl S, Butz P, Knorr D, Tauscher B (2008) Extraction of anthocyanins from grape by-products assisted by ultrasonics, high hydrostatic pressure or pulsed electric fields: a comparison. Innov Food Sci Emerg Technol 9:85–91. https://doi.org/10.1016/j.ifset.2007.06.002

Daou C, Zhang H (2014) Functional and physiological properties of total, soluble, and insoluble dietary fibres derived from defatted rice bran. J Food Sci Technol 51:3878–3885. https://doi.org/10.1007/s13197-013-0925-y

Dembitsky VM, Poovarodom S, Leontowicz H, Leontowicz M, Vearasilp S, Trakhtenberg S, Gorinstein S (2011) The multiple nutrition properties of some exotic fruits: biological activity and active metabolites. Int Food Res J 44:1671–1701. https://doi.org/10.1016/j.foodres.2011.03.003

Deressa L, Libsu S, Chavan RB, Manaye D, Dabassa A (2015) Production of biogas from fruit and vegetable wastes mixed with different wastes. Environ Ecol Res 3:65–71. https://doi.org/10.13189/eer.2015.030303

Dhillon GS, Kaur S, Brar SK (2013) Perspective of apple processing wastes as low-cost substrates for bioproduction of high value products: a review. Renew Sust Energ Rev 27:789–805. https://doi.org/10.1016/j.rser.2013.06.046

Elleuch M, Bedigian D, Roiseux O, Besbes S, Blecker C, Attia H (2011) Dietaryfibre and fibre-rich by-products of food processing: characterisation, technological functionality and commercial applications: a review. Food Chem 124:411–421. https://doi.org/10.1016/j.foodchem.2010.06.077

Esparza I, Jiménez-Moreno N, Bimbela F, Ancín-Azpilicueta C, Gandía LM (2020) Fruit and vegetable waste management: conventional and emerging approaches. J Environ Manag 265:110510. https://doi.org/10.1016/j.jenvman.2020.110510

Ferarsa S, Zhang W, Moulai-Mostefa N, Ding L, Jaffrin MY, Grimi N (2018) Recovery of anthocyanins and other phenolic compounds from purple eggplant peels and pulps using ultrasonic-assisted extraction. Food Bioprod Process 109:19–28. https://doi.org/10.1016/j.fbp.2018.02.006

Ganesh KS, Sridhar A, Vishali S (2022) Utilization of fruit and vegetable waste to produce value-added products: conventional utilization and emerging opportunities – a review. Chemosphere 287:132221. https://doi.org/10.1016/j.chemosphere.2021

Garzón CG, Hours RA (1992) Citrus waste: an alternative substrate for pectinase production in solid-state culture. Bioresour Technol 39:93–95. https://doi.org/10.1016/0960-8524(92)90061-2

Gemechu FG (2020) Embracing nutritional qualities, biological activities and technological properties of coffee byproducts in functional food formulation. Trends Food Sci Technol 104:235–261. https://doi.org/10.1016/j.tifs.2020.08.005

Gervasi T, Pellizzeri V, Calabrese G, Di Bella G, Cicero N, Dugo G (2018) Production of single cell protein (SCP) from food and agricultural waste by using *Saccharomyces cerevisiae*. Nat Prod Res 32:648–653. https://doi.org/10.1080/14786419.2017.1332617

Gomes-Araújo R, Martínez-Vázquez DG, Charles-Rodríguez AV, Rangel-Ortega S, Robledo-Olivo A (2021) Bioactive compounds from agricultural residues, their obtaining techniques, and the antimicrobial effect as postharvest additives. Int J Food Sci 2021. https://doi.org/10.1155/2021/9936722

Gouw VP, Jung J, Zhao Y (2017) Functional properties, bioactive compounds, and in vitro gastrointestinal digestion study of dried fruit pomace powders as functional food ingredients. LWT 80:136–144. https://doi.org/10.1016/j.lwt.2017.02.015

Gowe C (2015) Review on potential use of fruit and vegetables by-products as a valuable source of natural food additives. Food Sci Qual Manag 45:47–61

Gunes R, Palabiyik I, Toker OS, Konar N, Kurultay S (2019) Incorporation of defatted apple seeds in chewing gum system and phloridzin dissolution kinetics. J Food Eng 255:9–14. https://doi.org/10.1016/j.jfoodeng.2019.03.010

Hussain S, Jõudu I, Bhat R (2020) Dietary fiber from underutilized plant resources – a positive approach for valorization of fruit and vegetable wastes. Sustainability 12:5401. https://doi.org/10.3390/su12135401

Hussain H, Mamadalieva NZ, Hussain A, Hassan U, Rabnawaz A, Ahmed I, Green IR (2022) Fruit peels: food waste as a valuable source of bioactive natural products for drug discovery. Curr Issues Mol Biol 44:1960–1994. https://doi.org/10.3390/cimb44050134

Iglesias J, Martínez-Salazar I, Maireles-Torres P, Alonso DM, Mariscal R, Granados ML (2020) Advances in catalytic routes for the production of carboxylic acids from biomass: a step forward for sustainable polymers. Chem Soc Rev 49:5704–5771. https://doi.org/10.1039/D0CS00177E

Ingrao C, Arcidiacono C, Siracusa V, Niero M, Traverso M (2021) Life cycle sustainability analysis of resource recovery from waste management systems in a circular economy perspective key findings from this special issue. Resources 10:32. https://doi.org/10.3390/resources10040032

Ishangulyyev R, Kim S, Lee SH (2019) Understanding food loss and waste – why are we losing and wasting food? Foods 8:297. https://doi.org/10.3390/foods8080297

Jahid IK, Ha SD (2012) A review of microbial biofilms of produce: future challenge to food safety. Food Sci Biotechnol 21:299–316. https://doi.org/10.1007/s10068-012-0041-1

Jalili T, Wildman RE, Medeiros DM (2000) Nutraceutical roles of dietary fiber. J Nutraceut Funct Med Foods 2:19–34. https://doi.org/10.1300/J133v02n04_03

Jiménez-Moreno N, Esparza I, Bimbela F, Gandía LM, Ancín-Azpilicueta C (2020) Valorization of selected fruit and vegetable wastes as bioactive compounds: opportunities and challenges. Crit Rev Environ Sci Technol 50:2061–2108. https://doi.org/10.1080/10643389.2019.1694819

John I, Muthukumar K, Arunagiri A (2017) A review on the potential of citrus waste for D-Limonene, pectin, and bioethanol production. Int J Green Energy 14:599–612. https://doi.org/10.1080/15435075.2017.1307753

Joshi VK (2020) Value addition to fruit and vegetable processing waste – an appraisal. Int J Food Ferment Technol 10:35–58. https://doi.org/10.30954/2277-9396.02.2020.2

Kallel F, Driss D, Chaari F, Belghith L, Bouaziz F, Ghorbel R, Chaabouni SE (2014) Garlic (*Allium sativum* L.) husk waste as a potential source of phenolic compounds: influence of extracting solvents on its antimicrobial and antioxidant properties. Ind Crop Prod 62:34–41. https://doi.org/10.1016/j.indcrop.2014.07.047

Kalogeropoulos N, Chiou A, Pyriochou V, Peristeraki A, Karathanos VT (2012) Bioactive phytochemicals in industrial tomatoes and their processing byproducts. LWT Food SciTechnol 49:213–216. https://doi.org/10.1016/j.lwt.2011.12.036

Kasapidou E, Sossidou E, Mitlianga P (2015) Fruit and vegetable co-products as functional feed ingredients in farm animal nutrition for improved product quality. Agriculture 5:1020–1034. https://doi.org/10.3390/agriculture5041020

Kumar K, Yadav AN, Kumar V, Vyas P, Dhaliwal HS (2017) Food waste: a potential bioresource for extraction of nutraceuticals and bioactive compounds. Bioresour Bioprocess 4:1–4. https://doi.org/10.1186/s40643-017-0148-6

Laparra JM, Sanz Y (2010) Interactions of gut microbiota with functional food components and nutraceuticals. Pharmacol Res 61:219–225. https://doi.org/10.1016/j.phrs.2009.11.001
Larrauri, J. A., Rupérez, P., Borroto, B., & Saura-Calixto, F. (1996). Mango peels as a new tropical fibre: preparation and characterization. LWT-Food Science and Technology, 29(8), 729–733.
Lau KQ, Sabran MR, Shafie SR (2021) Utilization of vegetable and fruit by-products as functional ingredient and food. Front Nutr 8:261. https://doi.org/10.3389/fnut.2021.661693
Leyva-López N, Lizárraga-Velázquez CE, Hernández C, Sánchez-Gutiérrez EY (2020) Exploitation of agro-industrial waste as potential source of bioactive compounds for aquaculture. Food 9:843. https://doi.org/10.3390/foods9070843
Lin CS, Pfaltzgraff LA, Herrero-Davila L, Mubofu EB, Abderrahim S, Clark JH, Koutinas AA, Kopsahelis N, Stamatelatou K, Dickson F, Thankappan S (2013) Food waste as a valuable resource for the production of chemicals, materials and fuels. Current situation and global perspective. Energy Environ Sci 6:426–464. https://doi.org/10.1039/C2EE23440H
Lipiński AJ, Lipiński S, Kowalkowski P (2018) Utilization of post-production waste from fruit processing for energetic purposes: analysis of Polish potential and case study. J Mater Cycles Waste Manag 20:1878–1883. https://doi.org/10.1007/s10163-018-0729-2
Liu Z, de Souza TS, Holland B, Dunshea F, Barrow C, Suleria HA (2023) Valorization of food waste to produce value-added products based on its bioactive compounds. Processes 11(3):840. https://doi.org/10.3390/foods12030556
Ludwicka K, Kaczmarek M, Białkowska A (2020) Bacterial nanocellulose – a biobased polymer for active and intelligent food packaging applications: recent advances and developments. Polymers 12:2209. https://doi.org/10.3390/polym12102209
Mahato N, Sinha M, Sharma K, Koteswararao R, Cho MH (2019) Modern extraction and purification techniques for obtaining high purity food-grade bioactive compounds and value-added co-products from citrus wastes. Foods 8:523. https://doi.org/10.3390/foods8110523
Maisuthisakul P, Gordon MH (2009) Antioxidant and tyrosinase inhibitory activity of mango seed kernel by product. Food Chem 117:332–341. https://doi.org/10.1016/j.foodchem.2009.04.010
Makwara EC, Snodia S (2013) Confronting the reckless gambling with people's health and lives: urban solid waste management in Zimbabwe. Eur J Sustain Dev 2:67–67. https://doi.org/10.14207/ejsd.2013.v2n1p67
Malenica D, Bhat R (2020) Current research trends in fruit and vegetables wastes and by-products management-Scope and opportunities in the Estonian context. Agron Res 18:1760–1795. https://doi.org/10.15159/ar.20.086
Mamma D, Christakopoulos P (2014) Biotransformation of citrus by-products into value added products. Waste Biomass Valor 5:529–549. https://doi.org/10.1007/s12649-013-9250-y
Martins MD, Guimarães MW, de Lima VA, Gaglioti AL, Da-Silva PR, Kadowaki MK, Knob A (2018) Valorization of passion fruit peel by-product: xylanase production and its potential as bleaching agent for kraft pulp. Biocatal Agric Biotechnol 16:172–180. https://doi.org/10.1016/j.bcab.2018.07.033
Mason-D'Croz D, Bogard JR, Sulser TB, Cenacchi N, Dunston S, Herrero M, Wiebe K (2019) Gaps between fruit and vegetable production, demand, and recommended consumption at global and national levels: an integrated modelling study. Lancet Planet Health 3:e318–e329. https://doi.org/10.1016/S2542-5196(19)30095-6
Matos ÂP (2017) The impact of microalgae in food science and technology. J Am Oil Chem Soc 94:1333–1350. https://doi.org/10.1007/s11746-017-3050-7
Miteluţ AC, Popa EE, Drăghici MC, Popescu PA, Popa VI, Bujor OC, Ion VA, Popa ME (2021) Latest developments in edible coatings on minimally processed fruits and vegetables: a review. Foods 10:2821. https://doi.org/10.3390/foods10112821
Mohd Basri MS, Abdul Karim Shah NN, Sulaiman A, Mohamed Amin Tawakkal IS, Mohd Nor MZ, Ariffin SH, Abdul Ghani NH, Mohd Salleh FS (2021) Progress in the valorization of fruit and vegetable wastes: active packaging, biocomposites, by-products, and innovative technologies used for bioactive compound extraction. Polymers 13:3503. https://doi.org/10.3390/polym13203503

Mondal AK, Sengupta S, Bhowal J, Bhattacharya DK (2021) Utilization of fruit wastes in producing single cell protein. Int J Sci Environ Technol 1:430–438

Namasivayam E, Ravindar JD, Mariappan K, Akhil J, Mukesh K, Jayaraj R (2011) Production of extracellular pectinase by *Bacillus cereus* isolated from market solid waste. J Bioanal Biomed 3:70–75. https://doi.org/10.4172/1948-593X.1000046

Narasimmalu A, Ramasamy R (2020) Food processing industry waste and circular economy. IOP Conf Ser Mater Sci Eng 955:012089. https://iopscience.iop.org/article/10.1088/1757-899X/955/1/012089/meta

Nawirska A, Uklańska C (2008) Waste products from fruit and vegetable processing as potential sources for food enrichment in dietary fibre. Acta Sci Pol Technol Aliment 7:35–42. https://www.food.actapol.net/pub/3_2_2008.pdf

Neto JG, Ozorio LV, de Abreu TC, Dos Santos BF, Pradelle F (2021) Modeling of biogas production from food, fruits and vegetables wastes using artificial neural network (ANN). Fuel 285:119081. https://doi.org/10.1016/j.fuel.2020.119081

Nguyen VT, Scarlett CJ (2016) Mass proportion, bioactive compounds and antioxidant capacity of carrot peel as affected by various solvents. Technologies 4:36. https://doi.org/10.3390/technologies4040036

O'Shea N, Arendt EK, Gallagher E (2012) Dietary fibre and phytochemical characteristics of fruit and vegetable by-products and their recent applications as novel ingredients in food products. Innov Food Sci Emerg Technol 16:1–10. https://doi.org/10.1016/j.ifset.2012.06.002

Okino Delgado CH, Fleuri LF (2016) Orange and mango by-products: agro-industrial waste as source of bioactive compounds and botanical versus commercial description – a review. Food Rev Int 32:1–4. https://doi.org/10.1080/87559129.2015.1041183

Osorio LL, Flórez-López E, Grande-Tovar CD (2021) The potential of selected agri-food loss and waste to contribute to a circular economy: applications in the food, cosmetic and pharmaceutical industries. Molecules 26:515. https://doi.org/10.3390/molecules26020515

Oyedepo TA, Kayode AA (2020) Bioactive carbohydrates, biological activities, and sources. In: Egbuna C, Dable Tupas G (eds) Functional foods and nutraceuticals. Springer, Cham. https://doi.org/10.1007/978-3-030-42319-3_4

Padam BS, Tin HS, Chye FY, Abdullah MI (2014) Banana by-products: an under-utilized renewable food biomass with great potential. J Food Sci Technol 51:3527–3545. https://doi.org/10.1007/s13197-012-0861-2

Padayachee A, Day L, Howell K, Gidley MJ (2017) Complexity and health functionality of plant cell wall fibers from fruits and vegetables. Crit Rev Food Sci Nutr 57:59–81. https://doi.org/10.1080/10408398.2013.850652

Patinha S, Murteira JV, Costa CP, Salvador ÂC, Santos SA, Silvestre AJ, Rocha SM (2021) Elderberry stalks as a source of high-value phytochemical: essential minerals and lipophilic compounds. Appl Sci 12:382. https://doi.org/10.3390/app12010382

Pattnaik S, Reddy MV (2010) Assessment of municipal solid waste management in Puducherry (Pondicherry), India. Resour Conserv Recycl 54:512–520. https://doi.org/10.1016/j.resconrec.2009.10.008

Pérez-Marroquín XA, Estrada-Fernández AG, García-Ceja A, Aguirre-Álvarez G, León-López A (2023) Agro-food waste as an ingredient in functional beverage processing: sources, functionality, market and regulation. Foods 12:1583. https://doi.org/10.3390/foods12081583

Polat S, Aygun A (2022) Nanoencapsulation of essential oil-based packaging for shelf-life extension of foods. In: Nanotechnology interventions in food packaging and shelf life. CRC Press, pp 217–244

Poli A, Anzelmo G, Fiorentino G, Nicolaus B, Tommonaro G, Di Donato P (2011) Polysaccharides from wastes of vegetable industrial processing: new opportunities for their eco-friendly re-use. Biotechnol Biopol 33:56

Ravindra P (2000) Value-added food: single cell protein. Biotechnol Adv 18:459–479. https://doi.org/10.1016/S0734-9750(00)00045-8

Roberta MS, Mariana SL, Édira CB (2014) Functional capacity of flour obtained from residues of fruit and vegetables. Int Food Res J 2:1675. https://agris.fao.org/agris-search/search.do?recordID=MY2021001226

Rudra SG, Nishad J, Jakhar N, Kaur C (2015) Food industry waste: mine of nutraceuticals. Int J Sci Environ Technol 4:205–229. ISSN:2278-3687 (O)

Sagar NA, Pareek S, Sharma S, Yahia EM, Lobo MG (2018) Fruit and vegetable waste: bioactive compounds, their extraction, and possible utilization. Compr Rev Food Sci Food Saf 17:512–531. https://doi.org/10.1111/1541-4337.12330

Saini A, Panesar PS, Bera MB (2019) Valorization of fruits and vegetables waste through green extraction of bioactive compounds and their nanoemulsions-based delivery system. Bioresour Bioprocess 6:1–12. https://doi.org/10.1186/s40643-019-0261-9

Salim NS, Singh A, Raghavan V (2017) Potential utilization of fruit and vegetable wastes for food through drying or extraction techniques. Nov Tech Nutr Food Sci 1:1–2

Sall PM, Antoun H, Chalifour FP, Beauchamp CJ (2016) On farm composting of fruit and vegetable waste from grocery stores: a case under cold climatic conditions of Eastern Canada. In: Proceedings of the SUM2016, third symposium on urban mining (23–25). http://www.bashanfoundation.org/contributions/Antoun-H/2016.-Hani-SUMCE.pdf

Sanchez Vazquez SA (2014) Polymers from food wastes. Doctoral dissertation, UCL (University College London)

Sánchez M, Laca A, Laca A, Díaz M (2021) Value-added products from fruit and vegetable wastes: a review. CLEAN Soil Air Water 49:2000376. https://doi.org/10.1002/clen.202000376

Sancho SD, da Silva AR, Dantas AN, Magalhães TA, Lopes GS, Rodrigues S, da Costa JM, Fernandes FA, Silva MG (2015) Characterization of the industrial residues of seven fruits and prospection of their potential application as food supplements. J Chem 2015. https://doi.org/10.1155/2015/264284

Sarkar A, Kaul P (2014) Evaluation of tomato processing by-products: a comparative study in a pilot scale setup. J Food Process Eng 37:299–307. https://doi.org/10.1111/jfpe.12086

Seberini A (2020) Economic, social and environmental world impacts of food waste on society and zero waste as a global approach to their elimination. SHS Web Conf 74:03010. https://doi.org/10.1051/shsconf/20207403010

Sharma K, Mahato N, Cho MH, Lee YR (2017) Converting citrus wastes into value-added products: economic and environmently friendly approaches. Nutrition 34:29–46. https://doi.org/10.1016/j.nut.2016.09.006

Sharoba AM, Farrag MA, El-Salam A (2013) Utilization of some fruits and vegetables wastes as a source of dietary fibers in cake making. J Food Dairy Sci 4:433–453. https://doi.org/10.21608/jfds.2013.72084

Shehu I, Akanbi TO, Wyatt V, Aryee AN (2019) Fruit, nut, cereal, and vegetable waste valorization to produce biofuel. In: Byproducts from agriculture and fisheries: adding value for food, feed, pharma, and fuels, vol 14, pp 665–684. https://doi.org/10.1002/9781119383956.ch30

Singh DP, Packirisamy G (2022) Biopolymer based edible coating for enhancing the shelf life of horticulture products. Food Chem Mol Sci 4:100085. https://doi.org/10.1016/j.fochms.2022.100085

Singh A, Kuila A, Adak S, Bishai M, Banerjee R (2011) Use of fermentation technology on vegetable residues for value added product development: a concept of zero waste utilization. Int J Food Ferment Technol 1:173–184

Sirohi R, Gaur VK, Pandey AK, Sim SJ, Kumar S (2021) Harnessing fruit waste for poly-3-hydroxybutyrate production: a review. Bioresour Technol 326:124734. https://doi.org/10.1016/j.biortech.2021.124734

Slavin J (2013) Fiber and prebiotics: mechanisms and health benefits. Nutrients 5:1417–1435. https://doi.org/10.3390/nu5041417

Slavin JL, Lloyd B (2012) Health benefits of fruits and vegetables. Adv Nutr 3:506–516. https://doi.org/10.3945/an.112.002154

Socas-Rodríguez B, Álvarez-Rivera G, Valdés A, Ibáñez E, Cifuentes A (2021) Food by-products and food wastes: are they safe enough for their valorization? Trends Food Sci Technol 114:133–147. https://doi.org/10.1016/j.tifs.2021.05.002

Soquetta, M. B., Stefanello, F. S., da Mota Huerta, K., Monteiro, S. S., da Rosa, C. S., & Terra, N. N. (2016). Characterization of physiochemical and microbiological properties, and bioactive compounds, of flour made from the skin and bagasse of kiwi fruit (Actinidia deliciosa). Food Chemistry, 199, 471–478.

Spigarelli F, Natali L, Compagnucci L, Cavicchi A (2018) A multi-stakeholder attempt to address food waste: the case of Wellfood Action EU project, pp 503–528. https://www.torrossa.com/it/resources/an/4447336

Sudha ML, Priyanka R (2023) Fruit and vegetable waste (by-product) utilization in bakery products – a review. Nov Tech Nutr Food Sci 6(NTNF):000650. https://doi.org/10.31031/NTNF.2023.06.000650

Sudha ML, Baskaran V, Leelavathi K (2007) Apple pomace as a source of dietary fiber and polyphenols and its effect on the rheological characteristics and cake making. Food Chem 104:686–692. https://doi.org/10.1016/j.foodchem.2006.12.016

Sun-Waterhouse D (2011) The development of fruit-based functional foods targeting the health and wellness market: a review. Int J Food Sci Technol 46:899–920. https://doi.org/10.1111/j.1365-2621.2010.02499.x

Świeca M, Sęczyk Ł, Gawlik-Dziki U, Dziki D (2014) Bread enriched with quinoa leaves–The influence of protein–phenolics interactions on the nutritional and antioxidant quality. Food Chem 162:54–62. https://doi.org/10.1016/j.foodchem.2014.04.044

Taghian Dinani S, van der Goot AJ (2022) Challenges and solutions of extracting value-added ingredients from fruit and vegetable by-products: a review. Crit Rev Food Sci Nutr 4:1–23. https://doi.org/10.1080/10408398.2022.2049692

Thiviya P, Gamage A, Kapilan R, Merah O, Madhujith T (2022) Single cell protein production using different fruit waste: a review. Separations 9:178. https://doi.org/10.3390/separations9070178

Thulasisingh A, Kumar K, Yamunadevi B, Poojitha N, SuhailMadharHanif S, Kannaiyan S (2021) Biodegradable packaging materials. Polym Bull 6:1–30. https://doi.org/10.1007/s00289-021-03767-x

Tian M, Wai A, Guha TK, Hausner G, Yuan Q (2018) Production of endoglucanase and xylanase using food waste by solid-state fermentation. Waste Biomass Valor 9:2391–2398. https://doi.org/10.1007/s12649-017-0192-7

Tits M, Elsen A, Bries J, Vandendriessche H (2014) Short-term and long-term effects of vegetable, fruit and garden waste compost applications in an arable crop rotation in Flanders. Plant Soil 376:43–59. https://doi.org/10.1007/s11104-012-1318-0

Trigo JP, Alexandre EM, Saraiva JA, Pintado ME (2022) High value-added compounds from fruit and vegetable by-products – characterization, bioactivities, and application in the development of novel food products. Crit Rev Food Sci Nutr 60:1388–1416. https://doi.org/10.1080/10408398.2019.1572588

Upadhyay A, Lama JP, Tawata S (2010) Utilization of pineapple waste: a review. J Food Sci Technol Nepal 6:10–18. https://doi.org/10.3126/jfstn.v6i0.8255

Vilas-Boas AA, Pintado M, Oliveira AL (2021) Natural bioactive compounds from food waste: toxicity and safety concerns. Foods 10:1564. https://doi.org/10.3390/foods10071564

Wadhwa M, Bakshi MP, Makkar HP (2016) Wastes to worth: value added products from fruit and vegetable wastes. CABI Rev 10:1–25. https://doi.org/10.1079/PAVSNNR201510043

White BL, Howard LR, Prior RL (2010) Proximate and polyphenolic characterization of cranberry pomace. J Agric Food Chem 58:4030–4036. https://doi.org/10.1021/jf902829g

Wu W, Hu J, Gao H, Chen H, Fang X, Mu H, Han Y, Liu R (2020) The potential cholesterol-lowering and prebiotic effects of bamboo shoot dietary fibers and their structural characteristics. Food Chem 332:127372. https://doi.org/10.1016/j.foodchem.2020.127372

Yusuf M (2017) Agro-industrial waste materials and their recycled value-added applications. Handb Ecomater 1:1–1

Zanivan J, Bonatto C, Scapini T, Dalastra C, Bazoti SF, Júnior SL, Fongaro G, Treichel H (2022) Evaluation of bioethanol production from a mixed fruit waste by *Wickerhamomyces* sp. UFFS-CE-3.1. 2. Bionerg Res 15:175–182. https://doi.org/10.1007/s12155-021-10273-5

Zhu H, Luo W, Ciesielski PN, Fang Z, Zhu JY, Henriksson G, Himmel ME, Hu L (2016) Wood-derived materials for green electronics, biological devices, and energy applications. Chem Rev 116:9305–9374. https://doi.org/10.1021/acs.chemrev.6b00225

Chapter 8
Commercial Products Derived from Vegetable Processing Industrial Wastes and Their Recent Conversion Studies

Desta Getachew Gizaw, Selvakumar Periyasamy, Zinnabu Tassew Redda, Mani Jayakumar, and S. Kavitha

Abstract Demand for the vegetable cultivation and processing industry increases due to the growing global population, resulting in increased waste generation. Organic waste generation is unavoidable, particularly during pre-consumption or processing. Slow progress in developing appropriate waste management approaches as well as methods for adequate waste treatment and removal has exacerbated the situation. The vegetable wastes are enriched in complex carbohydrates, proteins, lipids, and nutraceuticals, all of which might be utilized for making value-added products. Several novel technologies are being developed for producing valuable

D. G. Gizaw
Department of Chemical Engineering, School of Mechanical, Chemical and Materials Engineering, Adama Science and Technology University, Adama, Ethiopia

S. Periyasamy (✉)
Department of Chemical Engineering, School of Mechanical, Chemical and Materials Engineering, Adama Science and Technology University, Adama, Ethiopia

Department of Biomaterials, Saveetha Dental College and Hospitals, SIMATS, Saveetha University, Chennai, India

Z. T. Redda
Faculty 1, University of Applied Sciences (HTW) Berlin, Berlin, Germany

School of Chemical and Bio Engineering, Addis Ababa Institute of Technology, Addis Ababa University, Addis Ababa, Ethiopia

M. Jayakumar
Department of Chemical Engineering, Haramaya Institute of Technology, Haramaya University,
Dire Dawa, Ethiopia

Department of Biotechnology, Faculty of Engineering, Karpagam Academy of Higher Education, Coimbatore, India

S. Kavitha
Department of Biotechnology, Adhiyamaan College of Engineering, Hosur, Tamil Nadu, India

E. Cherian, B. Gurunathan (eds.), *Value Added Products From Food Waste*,
https://doi.org/10.1007/978-3-031-48143-7_8

products such as biofuels, enzymes, bioactive compounds, biodegradable polymers, and nanoparticles from vegetable processing industrial waste. The conversion technologies that have the least detrimental impact on the environment and biochemical processing procedures to recover biologically active chemicals provide economically appealing food waste prospects. Recent valorization studies and commercial products derived from vegetable processing industrial wastes are discussed in this chapter.

Keywords Vegetable waste · Bioactive compounds · Biofuel · Nanoparticles · Biodegradable polymer

8.1 Introduction

Food consumption has been increasing with the human population globally, leading to enormous agricultural product production. Globally, 1800 million tons of foods are produced annually to meet the demand. Not all the parts of vegetables are consumed and remain left as waste. This phenomenon can increase the massive vegetable waste generation. These biological wastes quickly decompose due to their high moisture content, which causes a foul odor and spreads diseases (Periyasamy et al. 2023). Most vegetable waste has risen due to postharvesting leftovers in the field and commercial establishments. In this connection, waste generation is categorized into production waste (20%), processing waste (1%), distribution waste (19%), and consumer and household waste (60%). The zero-waste notions are handled using various tactics to reduce waste and increase its reusability.

The zero-waste concepts have created a new revolution to avoid waste generation. The designing approach of this process has increased waste reusability for converting waste to products with added value (Du et al. 2018; Karthik et al. 2022). The generated food wastes are leaf, stem, root, seed, peel, and roots. The waste made from leftover vegetables has various characteristics that vary depending on the plant species. The wastes possess phytochemical, antioxidant, and antidiabetic properties and are utilized to develop multiple value-added products such as flavoring agents, nutraceuticals, biopolymers, and therapeutic enzymes (Mohamed et al. 2022; Velusamy et al. 2022; Zena et al. 2023). Therefore, developing different commercial products from vegetable wastes using the zero-waste concept is an effective waste valorization to minimize waste accumulation and increase the environmental protection and economy of the country (Sridhar et al. 2021; Mohamed et al. 2022).

Waste creation is inevitable, particularly during the processing at the commercial level and pre-consumption phase. Most vegetable waste generated from the vegetable processing industry has constituted higher carbohydrates, proteins, lipids, and nutraceuticals (Selvakumar and Sivashanmugam 2017a). This nutrient-rich waste may be used to make valuable products such as biofuels, enzymes, bioactive compounds, biodegradable polymers, and nanoparticles. In addition, these wastes are a renewable source of industrially essential compounds and raw materials for commercial manufacturing. The waste conversion technologies that majorly impact the environment, biochemical processing, and extraction processes for biologically active chemicals provide economically appealing food waste prospects. However, vegetable waste could be a sustainable source for producing different commercially important products only by equipping upgraded conversion routes to meet future demand for food products. This chapter focuses on the physicochemical characteristics of vegetable waste and traditional conventional and emerging waste conversion techniques. Besides, recent commercial products developed from different types of vegetable wastes as suitable ingredients are well explained.

8.2 General Conversion Routes of Vegetable Processing Waste

Due to the increased waste produced by the world's expanding population, the demand for vegetable production and the processing sector that supports it surged. The problem has worsened due to the slow establishment of efficient waste management strategies and procedures for adequate waste treatment and removal. Vegetable waste can be used to create numerous useful products since it is an extensive supply of complex carbohydrates, proteins, lipids, and nutraceuticals. Generally, different conversion routes, such as landfill, composting, and fermentation, are used to convert vegetable waste. Landfilling is the traditional disposal method for solid wastes such as vegetables, fruit, or agro-residual. Economically viable and low processing costs are more beneficial for producing landfill gas than other conventional waste disposal methods. This method of waste utilization may affect wildlife, emit foul odors, and challenge maintaining optimal process conditions. The worldwide shortage of land for disposal is the great concern of not prioritizing landfilling route of disposal (Karthik et al. 2021; Sridhar et al. 2021; Calbry-Muzyka et al. 2022).

Composting is the most important natural or biological method of solid waste treatment to generate fertilizer or manure. This has been achieved by utilizing a microbial community. The microorganisms lead to catabolism and oxidation of carbonaceous substances in vegetable wastes, producing CO_2 and methane, and finally, compost developed can be effectively used as a biofertilizer for agricultural practices. Composting of vegetable waste is enhanced by physicochemical properties such as pH, temperature, carbon-to-nitrogen ratio, moisture, and aeration (aerobic catabolism). Some microorganisms such as *Eiseni fetida*, *Eudrilus eugeniae*, *Perionyx excavates*, *Megascolex mauritii*, and *Lumbricus rubellus* are commonly used for the composting of vegetable wastes to produce valuable products. This process has increased water retention capacity, soil aggregate stability, soil aeration, and water infiltration. However, this method has drawbacks such as groundwater contamination due to infiltrated leachate and emitting bad odor (Jain et al. 2019; Asaithambi et al. 2020; Perumal et al. 2023; Sun and Yu 2023).

Fermentation is a traditional vegetable or agro-residual waste method. This process decomposes organic matter in the solid waste into liquid or gas phase by the activity of aerobic or anaerobic microorganisms under the optimized environment.

The substrate components in the solid waste such as macronutrients and micronutrients influence product yield. Wild-type bacteria have generally been used for the anaerobic degradation of solid waste, consisting of different biochemical stages such as hydrolysis, acidogenesis, acetogenesis, and methanogenesis (Periyasamy et al. 2022b). Throughout the process, the mass and volume of solid waste are reduced, and the organic components are converted to biogas through the microbial communities. During this process, the organic matter in the vegetable wastes is hydrolyzed into mono sugars, fatty acid, and amino acids. Further, the hydrolysates turned into various products, such as biogas, organic acids, and short-chain alcohols, according to the process conditions (Arhoun et al. 2019). Besides, vegetable wastes are commercially utilized as potential adsorbents after pyrolyzing to remove the contaminants in water bodies. The vegetable wastes such as peel, bark, leaf, and stem constitute increased porosity and surface area, which facilitates improved adsorption of contaminants (Pradhan et al. 2020). Therefore, the way of waste processing and products developments are unique with each method. Based on the overall performance and environmental concern, each conversion route has advantages and disadvantages, as indicated in Table 8.1.

Table 8.1 Merits, demerits, and products of various vegetable waste conversion routes

Conversion routes	Merits	Demerits	Products	References
Landfills	Economically viable Low process cost Suitable for solid waste utilization	Difficult to maintain controlled conditions Creates unpleasant odors and generates leachates Larger area is required and decreases soil quality Affects groundwater quality and wildlife	Landfill gases and biofertilizers	Sridhar et al. (2021)
Composting	High water retention capacity Enhances microbial activity Improves soil nutritional and aggregate stability Increases soil aeration and water infiltration	Creates infiltrated leachate Causes groundwater contamination Emits unpleasant odor	Fertilizer and compost	Jain et al. (2019), and Sun and Yu (2023)
Fermentation	Generates minimal residue Emits less odor Increases sludge value and used as fertilizer	Expensive procedure High consumption Temperature sensitive	Biogas, organic acids, short-chain alcohols, and digestate	Arhoun et al. (2019), and Periyasamy et al. (2022b)

8.3 Recent Valorization Studies on Vegetable Industry Wastes

Each year, industries involved in food processing, such as those that produce juice, chips, meat, confectionery, and fruit, produce enormous amounts of by-products and related effluents of organic waste. Various wastes are formed when vegetables and fruits are grown, processed, and prepared. The waste can be in the part of leaves or straws, the garbage generated during harvests, debris from the processing industry, or waste from the processing itself. Because of the great diversity of the fruits and vegetables used, the large group of operations, and the diversity of the output, the waste generated by the fruit processing sector is quite diversified.

However, the massive volume of waste in these industries, both liquid and solid, includes various reusable substances of significant importance with immense economic possibilities. Wastes such as peels, seeds, and wasted flesh produced at various stages of production chains can still lead to environmental issues if it is not appropriately utilized. Unfortunately, this waste material, released into the environment, is a treasure trove of useful chemicals. The buildup of such wastes at the disposal sites causes natural decay to take longer than it should, emits a bad smell, and raises potential health hazards. Because of their sluggish biodegradability, acidic pH, and pungent smell in landfills, volatile solids, excess moisture, and few lignocellulosic materials in these fruit and vegetable wastes pose severe ecological risks (Zhou and Wen 2019). A wide variety of commercially valuable products such as biofuels, nanoparticles enzymes, biofertilizers, bioactive compounds, and biodegradable plastics has been produced by the bioconversion of the vegetable processing industry, as shown in Fig. 8.1. This section presents the most current developments in turning vegetable industrial wastes into value-added products, which may greatly aid in the transition to a holistic, cyclic paradigm of "reuse, recycle, and regenerate." It is essential in establishing a sustainable agricultural economy (Khaksar et al. 2022). In addition, it could aid in the growth of a circular economy, which is an economic model based on the efficient use of resources through waste reduction, long-term value retention, primary resource minimization, and closed production cycles within the constraints and guidelines of socio-environmental policy initiatives (Hofmann 2019).

8.3.1 Biofuels

The need for more energy has compelled society to transition from a linear economy based on fossil fuels to a sustainable and circular economy (Selvakumar et al. 2022). Biofuels offer advantages and disadvantages regarding social, economic, and environmental sustainability. The major worldwide drivers of biofuels are, on the one hand, the mitigation of carbon emissions, security of energy, and rural development (Kavitha et al. 2022). On the contrary, there are issues with enhancing the

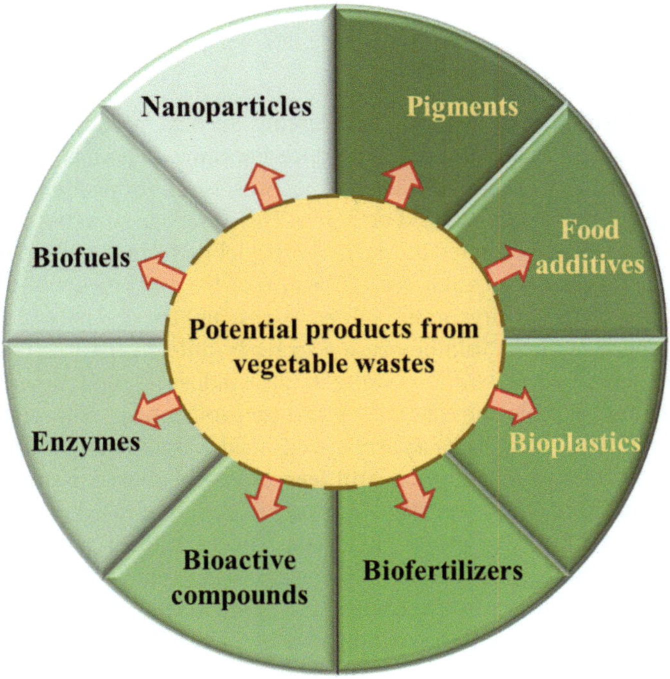

Fig. 8.1 Value-added products from vegetable wastes

levels of biofuels such as stress on food cost, the possibility of a rise in greenhouse gas emissions, as well as the dangers of ecosystem degradation and the deterioration of lands, forests, water resources, and other resources (Subramaniam et al. 2020; Dhandayuthapani et al. 2021, 2022).

However, this is not the case with fruits and vegetable wastes, which are particularly significant because they are not competing with food and feed crops as biomasses for anaerobic digestion (AD). Potato peel wastes, apple pomace, orange peel, carrot leftovers, and other biofuel wastes are examples that can be converted into biofuel. Converting such wastes for biofuel synthesis generates useful products and lowers waste treatment expenses. A diverse range of products is included under the umbrella term "biofuel" such as bioethanol, biodiesel, biohydrogen, biobutanol, biogas, and syngas. The investigations that have been done on using vegetable wastes to make different kinds of biofuels are presented in this subsection (Periyasamy et al. 2022a).

8.3.1.1 Bioethanol

As cellulose, hemicellulose, and lignin are polysaccharide-rich residues products in the vegetable processing industry, they can be treated for solid-state fermentation for making ethanol, which has different applications, including as a solvent and a liquid fuel additive used in various sectors. Because of its abundant accessibility, significant starch and cellulose content, and lack of competition with the food supply chain, vegetable wastes may provide a possible source of fuel for bioethanol synthesis (Singh et al. 2012). The vegetable waste composition varies depending on the kind of waste, as presented in Table 8.2. Whatever content is available, pretreatment, saccharification, and fermentation are the main steps in turning vegetable waste into biofuel, as shown in Fig. 8.2. Several studies have been performed on synthesizing bioethanol utilizing various vegetable wastes such as banana waste, potato peel, apple pomace, banana peel, and pineapple waste. High-reducing sugars found in fruit waste are used to make bioethanol. Using the fermentation process with yeast *Saccharomyces cerevisiae*, bioethanol is synthesized from common vegetable leftovers such as onion peel, potato peel, and carrot peel (Mushimiyimana and Tallapragada 2017; Kavitha et al. 2022).

Recently, the potato peel waste's conversion into bioethanol was performed with hydrolytic enzymes such as alpha-amylase, cellulase, termamyl, and amyloglucosidase (Awogbemi et al. 2022). Ingale et al. (2014) synthesized bioethanol utilizing pseudo banana stem as substrate after pretreating it with *Aspergillus fumigatus* and *A. ellipticus*. Similarly, several vegetable wastes such as pineapple waste, potato mash, and peels are effectively valorized for bioethanol production using suitable yeast organisms, namely *Aspergillus niger* and *Saccharomyces cerevisiae* (Chintagunta et al. 2016; Gil and Maupoey 2018).

8.3.1.2 Biodiesel

Alkyl esters with higher fatty acid content and lower aliphatic alcohol content make up biodiesel. Since this biofuel emits zero net CO_2, biodiesel can be regarded as "carbon neutral." The best materials for biodiesel are those with high lipid content such as used or fresh vegetable oils, animal fats, and microbial lipids (Selvakumar and Sivashanmugam 2017b, 2018a). The transesterification technique, frequently used for making biodiesel, only needs low pressure and temperatures and can provide a 98% conversion yield (Muniraj et al. 2015). Triglycerides and alcohol react with a suitable transesterification catalyst to enhance biodiesel and crude glycerol. On the other hand, it is also possible to use supercritical fluid extraction techniques to produce biodiesel from oilseed (Syimir Fizal et al. 2022).

For biodiesel synthesis, lipid- and oil-rich vegetable wastes, such as tomato seed, rapeseed, palm, and canola, are used. Tomato seeds, for instance, may be utilized for making biodiesel. Tomato seed oil may be converted through trans-esterification into fatty acid methyl esters (Karami et al. 2023). This biodiesel is capable of being used as fuel alone or in combination with petroleum diesel (Selvakumar and Sivashanmugam 2017a, 2018b). Despite the large amount of tomato waste produced and the fact that tomato seeds contain roughly 24% oil, quite a few researchers have examined biodiesel production (Sivasubramanian et al. 2020).

Table 8.2 Chemical and physical compositions of vegetable wastes

Type of waste	Starch (%)	Cellulose (%)	Hemicellulose (%)	Lignin (%)	Protein (%)	Moisture (%)	Ash (%)	Total solid (%)	References
Potato waste	30–40	17–25	10–15		3–5	85–87	6–12	1.7–19	Afifi (2011)
Tomato waste	10–18	30–32	5–18		17–22	85–90	3.1–5.3	7–22	Saev et al. (2009)
Onion waste	–	–	–	–	8.5	82–93	4.7	91	Benítez et al. (2011)
Sugar beet	–	–	–	–	–	85	3.8–8.8	7–11	Hampannavar and Shivayogimath (2010)
Carrot	1–2	13–52	12–19	–	5–8	–	–	–	Sagar et al. (2018)
Orange peel	–	9.2	10.5	0.8	–	11.8	3.5	–	Rivas et al. (2008)
Pineapple peel	–	18	–	1.4		91	–	94	Paepatung et al. (2009)

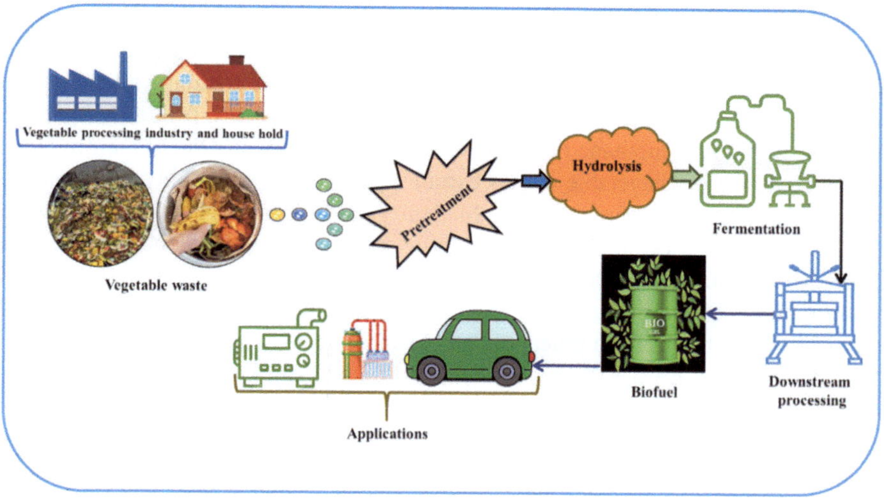

Fig. 8.2 Schematic representation of vegetable waste generation, processing, and end applications

8.3.1.3 Biohydrogen

The hydrogen (H_2) economy offers one of the most potential ecologically sound and long-term energy options. Currently, 95% of H_2 is generated via coal gasification or steam reforming of natural gas, emitting CO_2 (IRENA 2020). Furthermore, due to its ease of accessibility, environmental friendliness, and greater energy density, biohydrogen has the potential to be a biofuel. Since it emits no CO_2 and tends to replace exhausted oil supplies, it is one of the cleanest fuels (Magama et al. 2022). The fermentation method that uses light-dependent and dark fermentative processes to produce hydrogen offers several benefits over other approaches (Scotto di Perta et al. 2022). Because of its organic composition and biodegradability, fruit and vegetable waste (FVW) have great potential for biohydrogen production (Mishra et al. 2023). Additionally, FVW satisfies the primary criteria taken into account when choosing waste for biohydrogen generation, namely availability and affordability, carbohydrate content, and biodegradability (Moussa et al. 2022). With the addition of sewage, it was demonstrated a rise of up to 55% in hydrogen synthesis using vegetable wastes (Mohanakrishna et al. 2010). A substrate for biohydrogen generation can be made from various wastes, including potato, pumpkin, fruit wastes, fennel, and olive pomace wastes, as well as green vegetables such as cabbage, water celery, and cauliflower (Kalita and Sit 2022).

8.3.1.4 Biomethane

Anaerobic digestion of biogenic wastes by different microorganisms can result in the production of biomethane, an inexpensive source of bioenergy. The method of using vegetable waste to produce biomethane reduces wood fuel dependency, solves residual disposal, and reduces soil contamination and air and water pollution (Koul et al. 2022). Vegetable waste with a high organic content can be treated anaerobically to generate high methane yield with changing organic loading and a digested slurry is utilized for soil conditioner. As previously mentioned, fruit and vegetable wastes can produce methane through anaerobic fermentation. Due to their reduced lignin composition, vegetable wastes exhibit greater biodegradability and methane yields than lignocellulosic residues. Methane yields from vegetable wastes were 390, 320, 291, and 198 mL CH_4/g of volatile solids (VS) in the wastes of onion (Ji et al. 2017), potato (Parawira et al. 2005), carrot pomace (Garcia et al. 2011), and watermelon (Scapini et al. 2023), respectively. Similarly, the maximum observed methane yields found from pineapple and kiwi waste were 413 and 371 mL CH_4/g VS (Menardo and Balsari 2012). Bananas and fluted pumpkin peels yielded significantly lower methane, with 164 and 188 mL CH_4/g VS (Dahunsi et al. 2016). According to another study, out of 13 fruit and vegetable wastes, carrot waste produced the higher methane (417 mL/g volatile solids, with an organic loading rate (OLR) of 0.8–0.9 g VS/L/day) (Velmurugan 2011). Many researchers have recently been engaged to increase methane yield by incorporating vegetable waste alongside other organic wastes such as food wastes, secondary sludge, and cow dung (Agrawal et al. 2023; Shi and Xu 2023).

8.3.2 *Biofertilizer*

Significant economic and environmental threats are related to the widespread use of irrational disposal techniques for fruit and vegetable wastes (FVWs). One strategy for minimizing these environmental issues and enhancing the nutrition of soils is to produce biofertilizers from leftover fruit and vegetable using microbial degradation (Chakravarty and Mandavgane 2021; Bala et al. 2023). The biological conversion of vegetable residues into useful biofertilizers and biogas is a potentially effective approach. The microorganism that produces methane reaches the stationary phase earlier by using the sugar that is accessible to the development by the buildup of the hazardous components in a single aerobic digestion (AD) (Fu et al. 2017), where pH begins to decrease and biomass approaches early steady state with longer retention period (Weimer 2019). In several steps, fermentation avoids toxin buildup, accelerates biomass fall, and aids in pH regulation, all of which can be used to resolve these difficulties of slow decomposition efficiencies (Chakravarty and Mandavgane 2021).

Utilizing organic fertilizers enhances the physical, chemical, and biological properties of soil by enabling bacteria to degrade organic substances. Vegetable wastes are rich in soil nutrients such as nitrogen, phosphorus, and potassium, which can improve soil fertility and crop output (Beesigamukama et al. 2021). When it comes to reducing environmental problems and managing sustainable agricultural land, they are an improved substitute for chemical fertilizers. By utilizing different microorganisms to decompose in anoxic circumstances, such as bacteria and archaea, the decomposition of intricate organic materials in fruit and vegetable waste (FVWs) is aided by anaerobic digestion. It produces biogas as a substitute for other biofuels (Mirmohamadsadeghi et al. 2019). Anaerobic digesters' end effluent could be used as a biofertilizer. In a tomato cropping system, compost based on tomatoes found nutrient and bio-stimulation effects responsible for the enhanced response in productivity (Pane et al. 2015). Additionally, it has been noted that continuous applications of these compost wastes enhance the soil's nitrogen conditions over time (Tits et al. 2014).

8.3.3 Industrial Enzymes

Solid-state fermentation using vegetable wastes has produced enzymes such as cellulases, amylases, proteases, lipases, and pectinases. Enzymes are used in diverse fields to manufacture a huge range of goods. For example, tannases are used to reduce the quantity of tannic acid in effluents, pectinases are utilized in food, and the biofuel industry utilizes cellulases. Moreover, roughly 30% of the operating expenses are attributable to the raw material required to produce various enzymes (Ravindran and Jaiswal 2016). When using microorganisms to produce different enzymes, it is important to note that the synthesis for enzymes differs according to the substrates the microbes are growing on and the fermentation techniques (Shrestha et al. 2021).

Several studies have been done on how to make enzymes from vegetable wastes. Using solid-state fermentation with *Pleurotus ostreatus*, produced ligninolytic enzymes from potato peel waste such as manganese peroxidase, laccase, lignin peroxidase, and aryl alcohol oxidase (Ergun and Urek 2017). Numerous FVWs such as banana waste, date waste, and potato peel have been utilized in synthesizing amylase utilizing both submerged fermentation (SmF) and solid-state fermentation. Amylases are frequently used in baking, brewing, and producing digestive aids such as when making sugar, paper, and other products (Mehta and Satyanarayana 2016). Lipase was synthesized using *Yarrowia lipolytica* by SmF utilizing discarded mango peel and seed from industrial processing (Pereira et al. 2019). At the optimum temperature (27.9 °C), pH (5.0), and substrate concentration resulted in a lipase output as high as 3500 U/L of extracellular lipase. Furthermore, by utilizing *Trichoderma harzianum*, wide variety of fruit and vegetable wastes are used for the synthesis of xylanase (Rodríguez Couto 2008), and grape pomace is used as a substrate by *Aspergillus niger* to produce xylanase (Teles et al. 2019).

8.3.4 Nanoparticles

The synthesis of nanomaterials from vegetable and fruit processing residues is a relatively recent topic of study. Useful bioactive components found in vegetables act as reducing agents for producing metal nanoparticles (Ghosh et al. 2017; Sithara et al. 2017). Several metal nanoparticles have been produced using fruit and vegetable residues. Because of exceptional characteristics such as size, shape, surface area, alteration, and optical, magnetic, and electrical properties, the advancement of nanomaterials offers multiple fields of work and its relevance to a broad array of uses ranging from traditional chemical processes to environmental to medicine technologies (Aswathi et al. 2022). There are two general ways used in traditional techniques for producing nanoparticles. Chemical and physical methods are the two options. A wide range of nanoparticles has been produced using both methods.

Nonetheless, these methods frequently use hazardous solvents, nonbiodegradable stabilizing chemicals, and unsustainable procedures that pose several human and ecological risks. However, several other biological techniques have recently been researched for producing a wide range of nanoparticles. Utilization of hazardous solvents as well as chemicals can be avoided, and a green chemistry-based approach that uses biological processes is environmentally benign (Nasrollahzadeh et al. 2019). Alternative options to standard production processes that are appealing and environmentally benign include green chemistry-based techniques using vegetable and fruit wastes such as orange peel, banana peel, tomato waste, potato waste, and others. Recently, the ability of a broad spectrum of vegetable and fruit waste to synthesize nanoparticles has been investigated. For instance, using green synthesis, Mythili et al. (2018) produced silver nanoparticles (Ag-NPs) from vegetable waste (vegetable peels and discarded vegetables). Based on the data, Ag-NPs could be produced a 43 nm particle size and a range of 10–90 nm (Mythili et al. 2018). In another study, Skiba and Vorobyova (2019) prepared an extract of orange peel using a plasma solvent extraction approach, and silver nanoparticles were created by degrading methylene blue under solar irradiation (Skiba and Vorobyova 2019). Ahmad and Sharma (2012) also found that an extract from the fruit *Ananas comosus* resulted in spherical silver 5- to 35-nm-sized nanoparticles (Ahmad and Sharma 2012). Many researchers have utilized vegetable wastes for nanoparticle synthesis, with low-cost and environmentally friendly for various applications.

8.3.5 Biodegradable Plastics

The most current investigation into bioplastic production focuses on waste components and by-products of the vegetable industries of processing. Because those waste disposal sites produce negative outcomes, including pollution of groundwater and the release of greenhouse gases, their valorization by employing the development of bioplastics may provide a way to solve their disposal issue using sustainable and renewable technologies (Gong et al. 2023). Vegetables and fruit residues are rich in cellulose and starch that can be used to produce bio-based polymers. Peels of several vegetables have been investigated for use in bio-based polymer production. For instance, the synthesis of starch-based bioplastics using cassava peels has been studied. To produce high-quality plastics, cassava peel starch was enhanced with cellulose and sorbitol. As stated by Saridewi and Malik (2019), there are significant uses for polymers made from cassava peel in food packaging (Saridewi and Malik 2019). It has also been stated that waste potato peels can be used to produce plastics. These peels, produced in vast amounts by industrial potato processing, are high in starch, cellulose, and hemicelluloses. The synthesis of starch-based plastic using potato peels and a review of its biodegradability were reported recently (ARIKAN and Bilgen 2019). In recent years, various biopolymers, such as cellulose, polyhydroxyalkanoates, corn starch, and chitosan/pectin, have been extracted from several sources of vegetable waste for manufacturing bioplastics (Aslam et al. 2023; Merino et al. 2023).

8.3.6 Bioactive Compounds

Significant wastes or by-products are created during vegetable and fruit processing, making up between 25% and 30% of the whole commodity category. The waste comprises seeds, skin, and pomace, a significant supply of bioactive substances with potential value (Sagar et al. 2018). Carotenoids, polyphenols, dietary fibers, vitamins, and enzymes are a few examples of these components, which are utilized in a variety of industries such as the food industry to create enriched or functional foods, in the health industry to create medications and medicines, and in the textile industry (Selvakumar et al. 2021). An important step toward sustainable development is using waste to manufacture important bioactive substances.

Phenolic compounds are one of the most numerous types of bioactive chemicals, which have diverse and significant biological roles. Fruit and vegetable rind, peel, and seeds contain high levels of phenolic chemicals. The key beneficial constituents, such as total polyphenols, were extracted significantly from pineapple wastes. According to Hernández-Carranza et al. (2016), apple pomace, orange, and banana peels contain a substantial quantity of flavonoids, phenolic substances, and vitamin C (Hernández-Carranza et al. 2016). Orange had the highest phenolic compounds and vitamin C (729 mg of GAE/100 g DW and 96 mg of ascorbic acid/100 g DW, respectively). In contrast, banana peel had the highest concentration of flavonoids (752 mg of catechin/100 g DW). Citrus peels, grape seeds and skins, apple peels and pomace, grape skins and seeds, banana peels and seeds, and mango kernels have all been utilized successfully to extract phenolic chemicals for nutraceuticals (Varzakas et al. 2016). In addition, vegetable waste is a rich source of dietary fiber, a vital bioactive substance.

8.3.7 Food Additives and Essential Oils

Food additives serve various purposes such as minimizing food spoilage and microbial degradation, improving food qualities, adding color or flavor, and lowering acidity. They also help reduce food waste and increase the population's access to a wider variety of foods. Due to studies demonstrating the negative consequences of synthetic additives, interest in and demand for natural foods have surged in recent years. The bioactive molecule-rich by-products and biowaste from the vegetable processing sector amount to tons of raw resources. For instance, orange peels contain sugars that are soluble, cellulose, hemicellulose, pectin, and important oils, making them suitable for flavorings, sweeteners, and antioxidants (Dávila-Aviña et al. 2018). The color of food is critical in determining its approval since it influences the user's sense of its quality. As a result, using pigments and coloring agents is becoming a strategy for drawing in customers. Natural colorants such as carotenoids, anthocyanins, betanin, and chlorophyll have health benefits in addition to their decorative impacts (Pant et al. 2023).

Since ancient times, essential oils have been used in healing and fragrance. Essential oils are now one of the high-value agricultural commodities and as a result of the growing consumer desire for natural products. Several methods, such as steam distillation and hydrodistillation, extract essential oils from plant source materials. Essential oils are extensively employed and distinguished by their numerous advantages for usage in healthcare, pharmaceutical, and food industries, as well as in the cosmetics and fragrance industries. Essential oils also have a variety of biological impacts (Hikal et al. 2021). Organic aromatic or bactericidal components were reliably present in the essential oil generated.

8.3.8 Pigments

Petrochemical-derived synthetic pigments have been widely used in a variety of processed foods. However, these pigments' detrimental impacts on public health have forced the scientific community to look into far safer, natural, and environmentally friendly pigments. In this sense, it is important to utilize the potential of vegetable wastes, which may be collected primarily through environmentally friendly processing and extraction techniques. Vegetable residues and their by-products might be valued to meet the industrial demands of natural pigment manufacturing for prospective uses in food, medicine, and cosmetics (Sharma et al. 2021). Natural pigments such as anthocyanins, betalains, carotenoids, and chlorophyll are abundant in these wastes. These natural pigments are believed to have a substantial effect on the production of functional foods and offer a variety of biotherapeutic possibilities (Malabadi et al. 2022).

According to several studies, carotenoids can be extracted from vegetal wastes. Elik et al. (2020) recently used microwave-aided extraction to recover carotenoids

(77.48%) from waste produced during the processing of carrot juice under ideal conditions (165 W of microwave power, 9.39 min of extraction time, and 8.06:1 g/g of oil-to-waste ratio) (Elik et al. 2020). Kehili et al. (2016) proposed a biorefinery cascade approach to turn tomato wastes into value-added goods. The researchers extracted carotenoids from the oil portion of tomato peels and seeds using supercritical CO_2 technology (Kehili et al. 2016). Prokopov et al. (2017) showed that an enzymatic blend (cellulase 100 U g and hemicellulase 400 U g) improved the quantity of carotenoids recovered from tomato skins. Similarly, Kehili et al. (2017) reported that they recovered between 32% and 61% of lycopene and 28.38–58.8% of β-carotene utilizing a supercritical CO_2 extraction from tomato by-products (Kehili et al. 2017). Carotenoids are recovered through supercritical fluid extraction from different vegetal wastes. Recovered carotenoids from peel waste of sweet potato, tomato, apricot, pumpkin, peach, and pepper waste were 165.1, 253.5, 285.1, 142.0, 59.5, and 109.2 g/g dry weight basis, respectively (de Andrade Lima et al. 2018).

8.3.9 Animal Feed

By-products from fruits and vegetables have low hemicellulose and cellulose content in lignin. Consequently, they can be a suitable feed source for animals, particularly ruminants. Extracting some value-added elements first, followed by utilizing the remaining waste for feed or fertilizer, can be an economically sound method of using vegetable residues. However, such a reusing technique also has significant limitations. These wastes are vulnerable to microbial contamination because the higher moisture level frequently exceeds 80%. Hence, partial drying is frequently necessary. Additionally, low protein content and a high concentration of chemicals that cannot be digested do not always lead to good animal feed (Oliveira et al. 2017). Furthermore, the seasonal variations in vegetable product composition force producers to frequently alter feed compositions. Next, fermentation may be used following the process of extracting valuable components to further enhance the quality of feeds. To improve the nutritional content of the feed, functional ingredients such as phenolic and protein extracts obtained from various vegetable residues may occasionally be added (Kowalska et al. 2017). By-products from the industrial processing of tomatoes could be used as livestock feed to lower the price of animal feed and can also be utilized as a dietary supplement for humans by isolating the key chemicals.

8.4 Future Opportunities

To enhance the long-term environmental viability of the vegetable processing industries, wastes/by-products should be utilized or recycled appropriately. Reusing such waste is still in its early stages because of the requirement of novel recovery techniques and/or new production lines, whose prices are still higher than already-in-place landfill taxes and composting charges. The food industry's only goal is to lower energy and water use and, to a lesser extent, recover energy from waste. Increased research–industry collaboration, the adaptation and use of already available technology, and the adoption of scale economies necessary for treating large amounts of biomass with low specific economic values can all be advantageous for using vegetable wastes. To boost the reuse of the individual components found in vegetable wastes, cutting-edge extraction techniques must be used. These techniques can reduce the need for solvents and so improve the overall eco-sustainability of the food life cycle.

8.5 Conclusions

Population growth and increased fruit and vegetable intake due to increased nutrition knowledge are causing massive amounts of fruit and vegetable waste. However, it has become crucial to establish a sustainable method of handling these vegetable residues in the current circumstances. As a result, it necessitates establishing a strategy that can effectively exploit the possibilities of these waste materials and assist in gaining benefits from these wastes in the social, environmental, and economic spheres. Additionally, using leftover fruit and vegetable material to make products with added value, such as biofuels, enzymes, nanoparticles, biofertilizers, bioactive compounds, and biodegradable plastics, could be a green and sustainable method of developing innovative business prospects and functionalizing this waste. However, most conversion approaches are still in their infancy, with few technological improvements and achievements. As a result, there is a strong necessity to form consortiums of researchers and manufacturers to enhance the economic viability of these precious vegetable wastes through initial investment.

References

Afifi MM (2011) Enhancement of lactic acid production by utilizing liquid potato. Int J Biol Chem 5:91–102

Agrawal AV, Chaudhari PK, Ghosh P (2023) Effect of mixing ratio on biomethane potential of anaerobic co-digestion of fruit and vegetable waste and food waste. Biomass Convers Biorefin:1–10. https://doi.org/10.1007/s13399-023-03737-5

Ahmad N, Sharma S (2012) Green synthesis of silver nanoparticles using extracts of Ananas comosus. Green Sustain Chem 2(4):141–147

Arhoun B, Villen-Guzman MD, Vereda-Alonso C et al (2019) Anaerobic co-digestion of municipal sewage sludge and fruit/vegetable waste: effect of different mixtures on digester stability and methane yield. J Environ Sci Health A 54:628–634

Arikan EB, Bilgen HD (2019) Production of bioplastic from potato peel waste and investigation of its biodegradability. Int Adv Res Eng J 3:93–97

Asaithambi P, Govindarajan R, Busier Yesuf M et al (2020) Enhanced treatment of landfill leachate wastewater using sono(US)-ozone(O3)–electrocoagulation(EC) process: role of process parameters on color, COD and electrical energy consumption. Process Saf Environ Prot 142:212–218. https://doi.org/10.1016/j.psep.2020.06.024

Aslam M, Nadeem H, Azeem F et al (2023) Applications of bioplastics in disposable products. In: Handbook of bioplastics and biocomposites engineering applications, pp 445–455. https://doi.org/10.1002/9781119160182.ch20

Aswathi VP, Meera S, Maria CGA, Nidhin M (2022) Green synthesis of nanoparticles from biodegradable waste extracts and their applications: a critical review. Nanotechnol Environ Eng 8:377–397

Awogbemi O, Von Kallon DV, Owoputi AO (2022) Biofuel generation from potato peel waste: current state and prospects. Recycling 7:23

Bala S, Garg D, Sridhar K et al (2023) Transformation of agro-waste into value-added bioproducts and bioactive compounds: micro/nano formulations and application in the agri-food-pharma sector. Bioengineering 10:152

Beesigamukama D, Mochoge B, Korir NK et al (2021) Low-cost technology for recycling agro-industrial waste into nutrient-rich organic fertilizer using black soldier fly. Waste Manag 119:183–194

Benítez V, Mollá E, Martín-Cabrejas MA et al (2011) Characterization of industrial onion wastes (Allium cepa L.): dietary fibre and bioactive compounds. Plant Foods Hum Nutr 66:48–57

Calbry-Muzyka A, Madi H, Rüsch-Pfund F et al (2022) Biogas composition from agricultural sources and organic fraction of municipal solid waste. Renew Energy 181:1000–1007. https://doi.org/10.1016/J.RENENE.2021.09.100

Chakravarty I, Mandavgane SA (2021) Valorization of fruit and vegetable waste for biofertilizer and biogas. J Food Process Eng 44:e13512

Chintagunta AD, Jacob S, Banerjee R (2016) Integrated bioethanol and biomanure production from potato waste. Waste Manag 49:320–325

Dahunsi SO, Oranusi S, Owolabi JB, Efeovbokhan VE (2016) Comparative biogas generation from fruit peels of fluted pumpkin (Telfairia occidentalis) and its optimization. Bioresour Technol 221:517–525

Dávila-Aviña JE, Zoellner C, Solís-Soto L et al (2018) Economic and environmental benefits of utilizing plant food by-products. In: Plant food by-products. Apple Academic Press, Toronto, pp 25–53

de Andrade Lima M, Charalampopoulos D, Chatzifragkou A (2018) Optimisation and modelling of supercritical CO2 extraction process of carotenoids from carrot peels. J Supercrit Fluids 133:94–102

Dhandayuthapani K, Sarumathi V, Selvakumar P et al (2021) Study on the ethanol production from hydrolysate derived by ultrasonic pretreated defatted biomass of chlorella sorokiniana NITTS3. Chem Data Collect 31:100641. https://doi.org/10.1016/j.cdc.2020.100641

Dhandayuthapani K, Kumar PS, Chia WY et al (2022) Bioethanol from hydrolysate of ultrasonic processed robust microalgal biomass cultivated in dairy wastewater under optimal strategy. Energy 244:122604. https://doi.org/10.1016/j.energy.2021.122604

Du C, Abdullah JJ, Greetham D et al (2018) Valorization of food waste into biofertiliser and its field application. J Clean Prod 187:273–284

Elik A, Yanık DK, Göğüş F (2020) Microwave-assisted extraction of carotenoids from carrot juice processing waste using flaxseed oil as a solvent. Lwt 123:109100

Ergun SO, Urek RO (2017) Production of ligninolytic enzymes by solid state fermentation using Pleurotus ostreatus. Ann Agrar Sci 15:273–277
Fu S-F, Xu X-H, Dai M et al (2017) Hydrogen and methane production from vinasse using two-stage anaerobic digestion. Process Saf Environ Prot 107:81–86
Garcia SL, Jangid K, Whitman WB, Das KC (2011) Transition of microbial communities during the adaption to anaerobic digestion of carrot waste. Bioresour Technol 102:7249–7256
Ghosh PR, Fawcett D, Sharma SB, Poinern GEJ (2017) Production of high-value nanoparticles via biogenic processes using aquacultural and horticultural food waste. Materials (Basel) 10:852
Gil LS, Maupoey PF (2018) An integrated approach for pineapple waste valorisation. Bioethanol production and bromelain extraction from pineapple residues. J Clean Prod 172:1224–1231
Gong L, Passari AK, Yin C et al (2023) Sustainable utilization of fruit and vegetable waste bioresources for bioplastics production. Crit Rev Biotechnol:1–19. https://doi.org/10.1080/07388551.2022.2157241
Hampannavar US, Shivayogimath CB (2010) Anaerobic treatment of sugar industry wastewater by upflow anaerobic sludge blanket reactor at ambient temperature. Int J Environ Sci 1:631–639
Hernández-Carranza P, Ávila-Sosa R, Guerrero-Beltrán JA et al (2016) Optimization of antioxidant compounds extraction from fruit by-products: apple pomace, orange and banana peel. J Food Process Preserv 40:103–115
Hikal WM, Said-Al Ahl HAH, Tkachenko KG et al (2021) Sustainable and environmentally friendly essential oils extracted from pineapple waste. Biointerface Res Appl Chem 12:6833–6844
Hofmann F (2019) Circular business models: business approach as driver or obstructer of sustainability transitions? J Clean Prod 224:361–374. https://doi.org/10.1016/J.JCLEPRO.2019.03.115
Ingale S, Joshi SJ, Gupte A (2014) Production of bioethanol using agricultural waste: banana pseudo stem. Braz J Microbiol 45:885–892
IRENA RES (2020) International renewable energy agency, Abu Dhabi
Jain MS, Daga M, Kalamdhad AS (2019) Variation in the key indicators during composting of municipal solid organic wastes. Sustain Environ Res 29:1–8
Ji C, Kong C-X, Mei Z-L, Li J (2017) A review of the anaerobic digestion of fruit and vegetable waste. Appl Biochem Biotechnol 183:906–922
Kalita B, Sit N (2022) Biohydrogen from fruit and vegetable industry wastes. In: Organic waste to biohydrogen. Springer, Singapore, pp 69–91
Karami R, Rasul MG, Khan MMK (2023) An empirical and computational fluid dynamics analysis of combustion performance of a diesel engine fueled with tomato seed oil biodiesel. J Energy Resour Technol 145:41302
Karthik V, Periyasamy S, Isabel B, Temesgen T (2021) Restoration of contaminated agricultural soils. In: Biochar and its application in bioremediation. Springer, Singapore
Karthik V, Karuna B, Jeyanthi J, Periyasamy S (2022) Biochar production from Manilkara zapota seeds, activation and characterization for effective removal of Cu2+ ions in polluted drinking water. Biomass Convers Biorefin. https://doi.org/10.1007/s13399-022-03627-2
Kavitha S, Gajendran T, Saranya K et al (2022) An insight – a statistical investigation of consolidated bioprocessing of Allium ascalonicum leaves to ethanol using Hangateiclostridium thermocellum KSMK1203 and synthetic consortium. Renew Energy 187:403–416. https://doi.org/10.1016/j.renene.2022.01.047
Kehili M, Schmidt LM, Reynolds W et al (2016) Biorefinery cascade processing for creating added value on tomato industrial by-products from Tunisia. Biotechnol Biofuels 9:1–12
Kehili M, Kammlott M, Choura S et al (2017) Supercritical CO2 extraction and antioxidant activity of lycopene and β-carotene-enriched oleoresin from tomato (Lycopersicum esculentum L.) peels by-product of a Tunisian industry. Food Bioprod Process 102:340–349
Khaksar G, Sirijan M, Suntichaikamolkul N, Sirikantaramas S (2022) Metabolomics for agricultural waste valorization: shifting toward a sustainable bioeconomy. Front Plant Sci 13:938480
Koul B, Yakoob M, Shah MP (2022) Agricultural waste management strategies for environmental sustainability. Environ Res 206:112285

Kowalska H, Czajkowska K, Cichowska J, Lenart A (2017) What's new in biopotential of fruit and vegetable by-products applied in the food processing industry. Trends Food Sci Technol 67:150–159

Magama P, Chiyanzu I, Mulopo J (2022) A systematic review of sustainable fruit and vegetable waste recycling alternatives and possibilities for anaerobic biorefinery. Bioresour Technol Rep 18:101031

Malabadi RB, Kolkar KP, Chalannavar RK (2022) Plant natural pigment colorants-health benefits: toxicity of synthetic or artificial food colorants. Int J Innov Sci Res Rev 4:3418–3429

Mehta D, Satyanarayana T (2016) Bacterial and archaeal α-amylases: diversity and amelioration of the desirable characteristics for industrial applications. Front Microbiol 7:1129

Menardo S, Balsari P (2012) An analysis of the energy potential of anaerobic digestion of agricultural by-products and organic waste. Bioenergy Res 5:759–767

Merino D, Bellassi P, Paul UC et al (2023) Assessment of chitosan/pectin-rich vegetable waste composites for the active packaging of dry foods. Food Hydrocoll 139:108580

Mirmohamadsadeghi S, Karimi K, Tabatabaei M, Aghbashlo M (2019) Biogas production from food wastes: a review on recent developments and future perspectives. Bioresour Technol Rep 7:100202

Mishra P, Hai T, Zain JM et al (2023) Co-digestion of domestic kitchen food waste and palm oil mill effluent for biohydrogen production. Sustain Energy Technol Assess 55:102965

Mohamed BA, Bilal M, Salama ES et al (2022) Phenolic-rich bio-oil production by microwave catalytic pyrolysis of switchgrass: experimental study, life cycle assessment, and economic analysis. J Clean Prod 366:132668. https://doi.org/10.1016/j.jclepro.2022.132668

Mohanakrishna G, Goud RK, Mohan SV, Sarma PN (2010) Enhancing biohydrogen production through sewage supplementation of composite vegetable based market waste. Int J Hydrog Energy 35:533–541

Moussa RN, Moussa N, Dionisi D (2022) Hydrogen production from biomass and organic waste using dark fermentation: an analysis of literature data on the effect of operating parameters on process performance. Processes 10:156

Muniraj IK, Uthandi SK, Hu Z et al (2015) Microbial lipid production from renewable and waste materials for second-generation biodiesel feedstock. Environ Technol Rev 4:1–16

Mushimiyimana I, Tallapragada P (2017) Bioethanol production from agro wastes by acid hydrolysis and fermentation process. J Sci Ind Res 75(6):383–388

Mythili R, Selvankumar T, Kamala-Kannan S et al (2018) Utilization of market vegetable waste for silver nanoparticle synthesis and its antibacterial activity. Mater Lett 225:101–104

Nasrollahzadeh M, Sajjadi M, Sajadi SM, Issaabadi Z (2019) Green nanotechnology. In: Interface science and technology. Elsevier, pp 145–198. https://doi.org/10.1016/B978-0-12-813586-0.00005-5

Oliveira DM, Minuceli F, Ribeiro M et al (2017) Production lycopene dye São Caetano melon (Momordica charantia L.) for food application. Chem Eng Trans 57:1951–1956

Paepatung N, Nopharatana A, Songkasiri W (2009) Bio-methane potential of biological solid materials and agricultural wastes. Asian J Energy Environ 10:19–27

Pane C, Celano G, Piccolo A et al (2015) Effects of on-farm composted tomato residues on soil biological activity and yields in a tomato cropping system. Chem Biol Technol Agric 2:1–13

Pant A, Xin Ni PL, Chua CK, Tan U-X (2023) Valorisation of vegetable food waste utilising three-dimensional food printing. Virtual Phys Prototyp 18:e2146593

Parawira W, Murto M, Read JS, Mattiasson B (2005) Profile of hydrolases and biogas production during two-stage mesophilic anaerobic digestion of solid potato waste. Process Biochem 40:2945–2952

Pereira A d S, Fontes-Sant'Ana GC, Amaral PFF (2019) Mango agro-industrial wastes for lipase production from Yarrowia lipolytica and the potential of the fermented solid as a biocatalyst. Food Bioprod Process 115:68–77

Periyasamy S, Karthik V, Senthil Kumar P et al (2022a) Chemical, physical and biological methods to convert lignocellulosic waste into value-added products. A review. Environ Chem Lett 20:1129–1152. https://doi.org/10.1007/s10311-021-01374-w

Periyasamy S, Kavitha S, Beula Isabel J et al (2022b) Biogas recovery from sludge. In: Clean energy resource recovery wastewater treat plants as biorefineries, vol 2, pp 381–394. https://doi.org/10.1016/B978-0-323-90178-9.00008-1

Periyasamy S, Beula Isabel J, Kavitha S et al (2023) Recent advances in consolidated bioprocessing for conversion of lignocellulosic biomass into bioethanol – a review. Chem Eng J 453:139783. https://doi.org/10.1016/j.cej.2022.139783

Perumal A, Yesuf MB, Govindarajan R et al (2023) Photo-alternating current-electro-Fenton process for pollutant removal and energy usage from industrial wastewater: response surface approach optimization of operational parameters. ChemElectroChem 10:e202300086

Pradhan S, Abdelaal AH, Mroue K et al (2020) Biochar from vegetable wastes: agro-environmental characterization. Biochar 2:439–453

Prokopov T, Nikolova M, Dobrev G, Taneva D (2017) Enzyme-assisted extraction of carotenoids from Bulgariantomato peels. Acta Alimentaria, 46(1):84–91. https://doi.org/10.1556/066.2017.46.1.11

Ravindran R, Jaiswal AK (2016) Exploitation of food industry waste for high-value products. Trends Biotechnol 34:58–69

Rivas B, Torrado A, Torre P et al (2008) Submerged citric acid fermentation on orange peel autohydrolysate. J Agric Food Chem 56:2380–2387

Rodríguez Couto S (2008) Exploitation of biological wastes for the production of value-added products under solid-state fermentation conditions. Biotechnol J Healthc Nutr Technol 3:859–870

Saev M, Koumanova B, Simeonov IV (2009) Anaerobic co-digestion of wasted tomatoes and cattle dung for biogas production. J Univ Chem Technol Metall 44:55–60

Sagar NA, Pareek S, Sharma S et al (2018) Fruit and vegetable waste: bioactive compounds, their extraction, and possible utilization. Compr Rev Food Sci Food Saf 17:512–531. https://doi.org/10.1111/1541-4337.12330

Saridewi N, Malik M (2019) Food packaging development of bioplastic from basic waste of cassava peel (manihot uttilisima) and shrimp shell. IOP Conf Ser: Mater Sci Eng 602:12053. IOP Publishing

Scapini T, Bonatto C, Dalastra C et al (2023) Bioethanol and biomethane production from watermelon waste: a circular economy strategy. Biomass Bioenergy 170:106719

Scotto di Perta E, Cesaro A, Pindozzi S et al (2022) Assessment of hydrogen and volatile fatty acid production from fruit and vegetable waste: a case study of Mediterranean markets. Energies 15:5032

Selvakumar P, Sivashanmugam P (2017a) Optimization of lipase production from organic solid waste by anaerobic digestion and its application in biodiesel production. Fuel Process Technol 165:1–8. https://doi.org/10.1016/j.fuproc.2017.04.020

Selvakumar P, Sivashanmugam P (2017b) Thermo-chemo-sonic pre-digestion of waste activated sludge for yeast cultivation to extract lipids for biodiesel production. J Environ Manag 198:90–98. https://doi.org/10.1016/j.jenvman.2017.04.064

Selvakumar P, Sivashanmugam P (2018a) Study on lipid accumulation in novel oleaginous yeast Naganishia liquefaciens NITTS2 utilizing pre-digested municipal waste activated sludge: a low-cost feedstock for biodiesel production. Appl Biochem Biotechnol 186:731–749. https://doi.org/10.1007/s12010-018-2777-4

Selvakumar P, Sivashanmugam P (2018b) Multi-hydrolytic biocatalyst from organic solid waste and its application in municipal waste activated sludge pre-treatment towards energy recovery. Process Saf Environ Prot 117. https://doi.org/10.1016/j.psep.2018.03.036

Selvakumar P, Karthik V, Kumar PS et al (2021) Enhancement of ultrasound assisted aqueous extraction of polyphenols from waste fruit peel using dimethyl sulfoxide as surfac-

tant: assessment of kinetic models. Chemosphere 263:128071. https://doi.org/10.1016/j.chemosphere.2020.128071
Selvakumar P, Adane AA, Zelalem T et al (2022) Optimization of binary acids pretreatment of corncob biomass for enhanced recovery of cellulose to produce bioethanol. Fuel 321:124060. https://doi.org/10.1016/j.fuel.2022.124060
Sharma M, Usmani Z, Gupta VK, Bhat R (2021) Valorization of fruits and vegetable wastes and by-products to produce natural pigments. Crit Rev Biotechnol 41:535–563
Shi Y, Xu J (2023) A multi-objective approach to kitchen waste and excess sludge co-digestion for biomethane production with anaerobic digestion. Energy 262:125243
Shrestha S, Khatiwada JR, Sharma HK, Qin W (2021) Bioconversion of fruits and vegetables wastes into value-added products. In: Sustain bioconversion waste to value added products. Springer, Cham, pp 145–163
Singh A, Kuila A, Adak S et al (2012) Utilization of vegetable wastes for bioenergy generation. Agric Res 1:213–222
Sithara R, Selvakumar P, Arun C et al (2017) Economical synthesis of silver nanoparticles using leaf extract of Acalypha hispida and its application in the detection of Mn(II) ions. J Adv Res 8:561–568. https://doi.org/10.1016/j.jare.2017.07.001
Sivasubramanian H, Sundaresan V, Ramasubramaniam SK et al (2020) Investigation of biodiesel obtained from tomato seed as a potential fuel alternative in a CI engine. Biofuels 11:57–65
Skiba MI, Vorobyova VI (2019) Synthesis of silver nanoparticles using orange peel extract prepared by plasmochemical extraction method and degradation of methylene blue under solar irradiation. Adv Mater Sci Eng 2019:1–8
Sridhar A, Kapoor A, Kumar PS et al (2021) Conversion of food waste to energy: a focus on sustainability and life cycle assessment. Fuel 302:121069
Subramaniam Y, Masron TA, Azman NHN (2020) Biofuels, environmental sustainability, and food security: a review of 51 countries. Energy Res Soc Sci 68:101549
Sun F-S, Yu G-H (2023) Fate of bio-contaminants in organic wastes during composting and vermicomposting processes. In: Fate of biological contaminants during recycling of organic wastes, pp 143–156. https://doi.org/10.1016/B978-0-323-95998-8.00004-2
Syimir Fizal AN, Hossain MS, Zulkifli M et al (2022) Implementation of the supercritical CO2 technology for the extraction of candlenut oil as a promising feedstock for biodiesel production: potential and limitations. Int J Green Energy 19:72–83
Teles ASC, Chávez DWH, Oliveira RA et al (2019) Use of grape pomace for the production of hydrolytic enzymes by solid-state fermentation and recovery of its bioactive compounds. Food Res Int 120:441–448
Tits M, Elsen A, Bries J, Vandendriessche H (2014) Short-term and long-term effects of vegetable, fruit and garden waste compost applications in an arable crop rotation in Flanders. Plant Soil 376:43–59
Varzakas T, Zakynthinos G, Verpoort F (2016) Plant food residues as a source of nutraceuticals and functional foods. Foods 5:88
Velmurugan B (2011) Anaerobic digestion of vegetable wastes for biogas production in a fed-batch reactor. Int J Emerg Sci 1:478
Velusamy S, Subbaiyan A, Murugesan SR et al (2022) Comparative analysis of agro waste material solid biomass briquette for environmental sustainability. Adv Mater Sci Eng 2022:1. https://doi.org/10.1155/2022/3906256
Weimer ALA (2019) Anaerobic degradation of volatile fatty acids in continuous biogas reactors at very short hydraulic retention times, Thesis, Lund University. https://www.lunduniversity.lu.se/lup/publication/8992750
Zena Y, Periyasamy S, Tesfaye M et al (2023) Essential characteristics improvement of metallic nanoparticles loaded carbohydrate polymeric films-a review. Int J Biol Macromol 242:124803
Zhou H, Wen Z (2019) Solid-state anaerobic digestion for waste management and biogas production. Adv Biochem Eng Biotechnol 169:147–168

Chapter 9
Exotic Nutrients Content from Tamarind (*Tamarindus indica*) Seed is a Boon of Sustainable Healthy Diets

S. Parameshwari and C. Hemalatha

Abstract The objective of the book chapter research is to demonstrate the value of tamarind seeds as an alternative source of dietary staples for a sustainable diet. Foods that are good for our bodies and the environment are those that are sustainable to eat. Food products that contain tamarind seeds not only contribute to the conservation of the earth but also provide a healthy diet for the body. The tamarind tree (*Tamarindus indica*) is popularly used for its consumable pulp of fruit, used in cookery globally. The tamarind seeds were considered waste earlier but are now used to produce tamarind kernel powder (TKP), a high-value product. A seed is composed of approximately 60–65 of the kernel. In addition to being rich in proteins, carbohydrates, fiber, oils, polysaccharides, and sugars, TKP also contains calcium, magnesium, potassium, phosphorus, and vitamin C. Tamarind kernel powder has been identified as an alternate and inexpensive source of proteins as well as essential nutrients. It has considerable commercial feasibility and is readily available. TKP is nutritionally equal to the major aspects of pulses, legumes, and cereal crops. For the creation of unique food items, it can be used such as thickening, dissolving, emulsifying, gelling, binding, and solubilizing agent. The primary applications of TKP are in the preparation of baked goods, as well as ready-to-eat foods such as noodles, ice cream, sauces, and flavoring. As a result, tamarind seed is used to produce food items with minimal adverse effects on the environment, helping to ensure the security of food and nutrition as well as the health of the current and future generations.

Keywords Sustainable · Tamarind kernel powder · Thickening · Disintegrating · Nutritional value

---

S. Parameshwari (✉) · C. Hemalatha
Department of Nutrition and Dietetics, Periyar University, Salem, India

E. Cherian, B. Gurunathan (eds.), *Value Added Products From Food Waste*,
https://doi.org/10.1007/978-3-031-48143-7_9

9.1 Introduction

A major global focus is on raising diet quality while also lowering environmental effect. Food waste has not been accounted the relationship between the quality of diet and its sustainability implications, which have been traditionally focused on a limited range of factors. We investigate the connection between food waste and a sustainable, healthy diet to fill this crucial research gap (Conrad et al. 2018).

Seeds from fruit processing are wasted and unusable. Recent studies reveal that these discarded waste products may include nutrients and bioactive substances in significantly higher concentrations than those found in the fruit's meat. Studying the physicochemical characteristics and bioactive compound content of fruit seeds may help determine whether it would be beneficial and possible to use them more widely in the creation of functional foods (Siol et al. 2022).

Investigating plant-derived proteins as an affordable and environmentally friendly protein source for a healthy eating food could help solve the malnutrition issues (Sonawane and Arya 2018). Grains and seeds are examples of plant items that can be categorized as nutraceuticals since they contain proteins and peptides. Any functional food extract having well-being and medicinal uses, especially for humans, is referred to as a nutraceutical (Moldes et al. 2017). Proteins derived from cereals and seeds are essential components of dietary systems that fight protein-calorie malnutrition in underdeveloped nations (Hackler 1985). Nutritionally balanced protein foods from the legume family are known as "Poor Man's Meat" for people who do not have money for purchasing dairy, meat, and fish (Tharanathan and Mahadevamma 2003). Grain and seed proteins open doors of opportunity for rural communities with limited access to natural resources by lowering poverty levels, enhancing nutrition and health, conserving the ecosystem and enhancing food safety. Grain and seed protein are typical sources of B vitamins, calories, carbs, minerals, and protein.

A crucial tree for the economy is *Tamarindus indica*, also known as the tamarind. Every aspect of the tamarind tree has a use, making it a versatile tree species. The food, biochemical, pharmaceutical, textile, timber, fodder, and fuel sectors all use various parts of the tamarind tree (Sheikh and Shivanna 2022). Throughout Asia's southern and southeast regions, people frequently eat tamarind, which can be utilized for preparing chutneys, marinades, beverages, sauces, and desserts. However, the majority of people are unaware that tamarind seeds and leaves can also be eaten (Srivastava 2021).

The production of tamarind pulp yields seeds as a by-product. After soaking and boiling the seed with water, the tannins and other coloring agents present in the testa are eliminated, making the entire seed suitable for consumption. Tannin, glue, and polysaccharide (jellose) are all products made from tamarind seed as a raw material. The seed kernels are rich in protein, while the seed coat contains a significant amount of fiber and tannins (considered antinutritional factors). According to a study by Emmy De Caluwé et al. (2010), there has been a rise in the utilization of seeds as an alternative protein source. In addition, they contain plenty of essential minerals including calcium, magnesium, potassium, and phosphorus as well as being an exclusive source high in fat acids. The textile, paper, and jute industries utilize a small quantity of the seed, commonly referred to as tamarind kernel powder, as a sizing ingredient (Pugalenthi et al. 2007).

9.2 Scope of the Study

- The nutritional composition of tamarind seed is highly advantageous, boasting elevated levels of protein, antioxidant, and a range of essential minerals.
- Tamarind seed kernel powder can be seamlessly incorporated into cereal flour, enabling the creation of a wide array of food products.
- Utilizing tamarind seed kernel powder as a protein ingredient in food presents an opportunity to reduce dependence on conventional protein supplements derived from commonly utilized legumes.

9.3 Objectives of the Study

- To demonstrate the potential of tamarind seeds as an alternative source of dietary staples, promoting a sustainable diet.
- To give the physicochemical and nutritional properties of tamarind seed kernel powder.
- To show the health benefits of tamarind seed kernel powder.

9.4 Tamarind Plant

The Indian subcontinent is home to several of this native African tree. It is a drought-resistant plant, making it a vital resource for many industries (Altrafine.com).

9.4.1 Origin

According to different sources, various places are the origins of tamarind tree. It was initially thought that the tamarind fruit was originating from an Indian plant because in Persian the name "tamar-i-hind" meant "date of India." Its Sanskrit name, "amlika," indicates that it has long been a part of the country (Mishra 1997). It was described between 1200 and 200 BC in the Indian Tamarind Brahmasamhita literature, according to El-Siddig et al. (2006). According to many academics, the arid savannahs in tropical Africa spanning from Ethiopia, Tanzania, Sudan, and

Kenya to Senegal in sub-Sahelian Africa are believed to be the origin of the tamarind tree, although Morton (1987) attributed its origin to India. There are claims that the tamarind tree first appeared in Madagascar (Hocking 1993; Von Maydell 1986).

9.4.2 The Various Applications

Tamarind plant has multiple uses. For a veritably long time, the fruit pulp, particularly in the southern region of India, was utilized for its culinary use as a spice in Asian dishes. The culinary, pharmaceutical, chemical, or cloth sectors as well as those in the fodder, energy diligence, and timber used nearly every corridor of the tamarind tree (Dagar et al. 1995; George and Rao 1997; Rao and Mary Mathew 2001; Pugalenthi et al. 2004). Tamarind leaves, fruits, and seeds, among other tamarind products, have been extensively used in traditional Indian and African remedy (De Caluwé et al. 2009).

9.4.3 Phytochemistry

It has been shown how minerals, fatty acids, and amino acids are present in various tamarind plant components. It is possible that variations in genetic strains, harvesting stages for plant parts, and growing conditions are what caused the variances in values discovered in the literature (Glew et al. 2005).

The fruit is the most valued and frequently utilized component (Saideswara Rao and Mary Mathew 1999). Due to the fact that almost every part of a tamarind tree has some form of value, it can be considered a highly versatile tree. The fruit has roughly a 55% pulp content, a 34% seed content, an 11% shell content, and a pod's worth of fiber (Kumar and Bhattacharya 2008).

9.5 Tamarind Seeds

Tamarind seeds are readily available in India, where 11 million tons are produced each year. The dicotyledonous, flat, and glossy tamarind seeds are solid and vary in hue from red to purplish-brown. Hard cotyledons line the seed are coated with a firm or thin skin (Sheikh and Shivanna 2022). The eight- to ten-seed pods that hold its seeds range in length from three to six inches (Altrafine.com). Tamarind seeds have between 57% and 80% white kernels and 30 –40% crimson hulls as shows in Fig. 9.1.

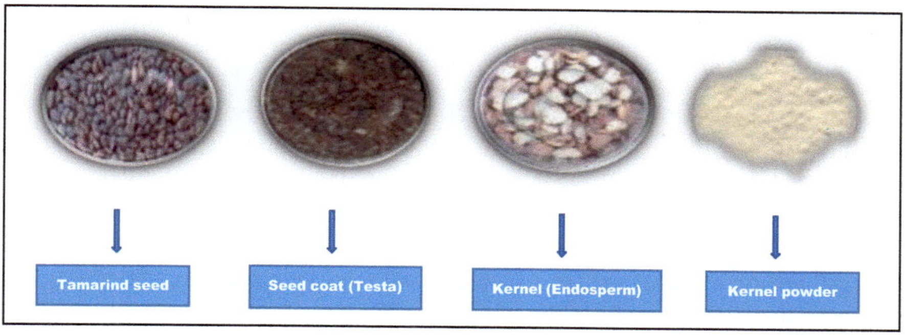

Fig. 9.1 Tamarind seed, kernel, and kernel powder

Table 9.1 Antinutritional factors in dehulled tamarind seed

Antinutritional factors	Values (soaked seed)	Values (roasted seed)
Phytate	2.15%	2.53%
Tannins	4.84%	8.34%
Alkaloid	3.34%	3.40%
Trypsin inhibitor	3.553 TIU/100 g	2.854 TIU/100 g
Saponin	0.8%	1.00%
Hydrogen cyanide	3.00 mg/mL	1.00 mg/mL

9.5.1 Antinutrient Factors

The existence of tannins, phytate, and oxalate in food can be described as a hindrance to nutrition and is commonly known as an antinutrient factor. These substances interact with the intestinal tract to affect the nutritional quality and also lessen amino acid absorption and protein digestibility (Nowacki 1980). Unless these elements are destroyed through heat or other treatments, when used by animals and people, they may have negative physiological effects (Liener 1994). Table 9.1 shows the antinutritional factors present in dehulled tamarind seed (Akajiaku et al. 2014). The antinutrients are reduced after the use of different processing methods such as soaking, boiling, roasting, and dehulling the tamarind seeds.

9.6 Benefits of Whole Tamarind Seed

Rural people and members of some ethnic groups, including Malayali and Dravidian tribes, roast the seeds, mechanically remove the seed coverings, and dehull the tamarind seeds before soaking them in water for an entire night before eating them. The flowers and leaves of the tamarind tree can be consumed as vegetables. Additionally, the gum derived from tamarind seeds is used to create tamarind gum, which has been utilized as a thickening agent in various Japanese dishes, as mentioned by

Akajiaku et al. (2014) and Khairunnuur et al. (2009). Furthermore, cheap tamarind seed kernel powder possibly will be an upright alternative to costly pectin for preparing jelly (Pugalenthi and Vadivel 2007). When preparing bread or cakes, tamarind powder is used as a raw ingredient. Nowadays, people use tamarind powder to cure ulcers, substitute starch in the maintenance of textiles, and coagulate fruit juice. The decorticated seed powder of tamarind exhibits impressive gelling and adhesion properties, which find applications in the cosmetics, pharmaceutical, and food industries, as highlighted by Pugalenthi et al. (2007).

A thickening agent or gum in the textile and food industries is provided by the polysaccharide found in seeds. Infections of the teeth, diarrhea, arthritis, indigestion, and eye health are all treated with seeds. Additionally, it has uses as an antioxidant, antidiabetic, antibacterial, and many others (Sheikh and Shivanna 2022).

9.7 Antioxidant Activities

The structural elucidations of the components of tamarind seed husk extract (TSE), as well as its antioxidant effects, have all been the subject of several investigations. According to a study by Wandee et al. (2022), 100 g of tamarind seeds yielded 38.51 g of seed coat, which provided 0.87 g of tamarind seed coat (testa) extract (2.0% w/w). The extract predominantly contained phenolic compounds (106.40 mg/g) rather than flavonoids (0.45 mg/g). The tamarind seed extract exhibited superior antioxidant activity (IC_{50} 3.0 μg/mL) compared to standard antioxidants such as ascorbic acid (IC_{50} 6.30 μg/mL) and catechin (IC_{50} 10.9 μg/mL). This information is represented in Table 9.2 and is determined based on the 5% inhibitory concentration (IC_{50}) derived from concentration response curves obtained from the DPPH assay.

The tamarind seed coat was extracted using a solution of HPLC chromatogram with a concentration of 2.5 mg/mL, and A1, blue chromatogram contained a noticeable peak at retention time of 7.50 min with a distinctive ultraviolet spectrum pattern and a maximum absorption wavelength of 209/239/279 nm, which was related to the position of the average catechin's peak at RT for 7.36 min as A1, red chromatogram with UV spectrum having a maximal absorption. The chromatograms of the catechin-infused tamarid seed extract mixture showed that catechin and the TS

Table 9.2 The total antioxidant content in whole tamarind seeds

Antioxidant activity	IC_{50} 3.0 μg/mL
Total flavonoid content	0.45 mg Q E/g of seed extract
Total phenolic content	106.40 mg GA E/g of seed extract

GA gallic acid, *Q* quercetin, *E* equivalence

extract's standout peak were eluted simultaneously. Catechin was therefore shown to be a substantial component of TS extract, with levels reaching 429 ±22.29 mg/g of extract, generalizing from the catechin calibration curve (Roongrawee Wandee et al. 2022).

9.8 Tamarind Seed Kernel Powder

After that the protective layer is removed, the tamarind kernel can be obtained. A type of trash that is still not being used to its full potential is the tamarind kernel. The kernel can be converted into powder to boost its usefulness as a normal hydrocolloid basis because it contains gum and starch. Kernel shell separation is necessary for powder processing. The process of making powder includes the separation of the testa from the tamarind kernel, because tamarind kernel testa causes desolation and other mental health issues (Havinga et al. 2010). The kernel of the tamarind contains a high concentration of essential amino acids, making it a valuable source of protein and various other nutrients.

9.8.1 Miscellaneous Applications of Tamarind Kernel Powder

The jute and textile industries use as a sizing material. Because it is simpler to remove from a spun fabric, it is preferable to starch for sizing spun viscose. When it comes to the processes of dying, printing, and finishing fabrics, it's thickening ability opens up a lot of possibilities. A tanning substance used in the leather industry. In the process of creating bricks, it can also be used to stabilize and condition the soil, as well as to bind the sawdust while preparing briquettes (Jiayu Ji 2005). It contributes to being a great paper adhesive when it is boiled in water. It can be used as food for cattle, pigs, and pets, because it contains nutrients and fiber. It has been shown to treat pet indigestion (Patel 2013). Because of its thickening properties, the it can be used to make tinned pet food for cats and dogs.

9.9 Tamarind Seed Kernel Powder (TKP) Preparation

The tamarind seed is often overlooked as a by-product of tamarind pulp industrial processing. However, when the seeds and pulp are separated, tamarind kernel can be produced. According to Pulungan et al. (2001), cleaned seeds must be roasted at 140 °C for 10 min before being de-hulled. The white endosperm component of the seed is mechanically processed to the required sieve size in order to yield a powder (Thombare Nandkishore et al. 2014).

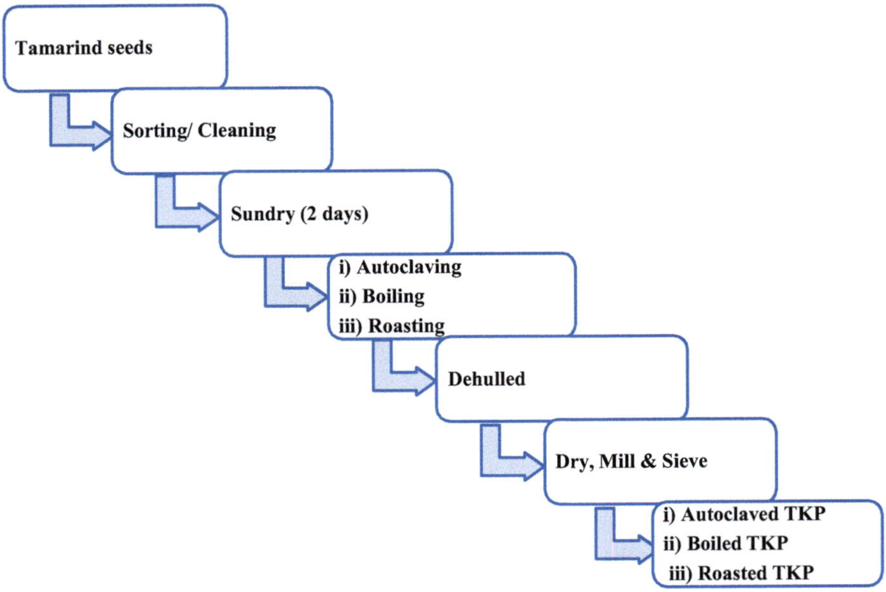

Fig. 9.2 The process of preparing tamarind seed kernel powder

There is also a soaking method for making this powder. After 14 days of water sowing, the seed coat should be removed. Seeds are dried and then ground into flour. A 72-h sun-drying period is required for the pulverized flour. The flour is transferred to a 1-mm mesh sieve and then kept well-sealed (Akajiaku et al. 2014). The process of preparing tamarind seed kernel powder is showed in Fig. 9.2.

9.10 Physicochemical and Proximate Composition

9.10.1 Physicochemical Properties of Tamarind Seed Kernel Powder

The suitability of tamarind seed kernel powder and its properties, such as moisture content, compressibility, density, water retention, pH, melting point, flow, surface tension, and swelling index, have been demonstrated for various applications, notably in the development of pharmaceutical formulations (Phani Kumar et al. 2011; Kumar et al. 2001; Lang et al. 1992; Deveswaran et al. 2010). Table 9.3 displays the aforementioned properties.

Table 9.3 Physical properties of tamarind seed kernel powder

Physical parameters	Values
True density (g/mL)	1.015–1.9
Moisture content (%)	8.10
Tapped density (g/mL)	0.363–0.781
Melting point (°C)	240–260
Angle of repose (°)	13–29.5
Swelling index (in water) (%)	12–17
Compressibility index (%)	15.33–16.64
Water retention (%)	20.00
Particle size in range (μm)	60–90
Hausner ratio (1% w/v TSP)	0.14–1.23 pH, 6–6.81
Average molecular weight (g/mol)	0.9–2.1 × 106
Bulk density (g/mL)	0.24–0.651
Surface tension (dynes/cm)	61.3–83.26

Table 9.4 Major nutrient composition of tamarind kernel flour

Nutrients	Amount (%)
Moisture (g)	7.65 ± 0.01
Ash (g)	2.74 ± 0.06
Carbohydrate (g)	62.13 ± 0.11
Protein (g)	19.46 ± 0.06
Fat (g)	5.7 ± 0.11
Crude fiber (g)	2.32 ± 0.06

All the values are mean ± SD of three observations

9.10.2 *Proximate Composition of Tamarind Seed Kernel Powder*

Tamarind seed kernel powder has major nutrient content such as moisture, ash, carbohydrates, protein, fat, and fiber. This information is denoted in Table 9.4.

9.10.3 *Micronutrients Composition and Amino Acids Profile*

TKP contains micronutrients composition, and this data is presented in Table 9.5. It has good amino acids profile (Gitanjali et al. 2020). Table 9.6 shows the amino acids content in tamarind seed kernel powder (Heuze et al. 2015).

Table 9.5 The mineral composition of tamarind seed kernel flour

Minerals	Amount (mg/100 g)
Calcium (mg)	109.25 ± 0.11
Magnesium (mg)	247.51 ± 0.07
Iron (mg)	11.37 ± 0.01
Potassium (mg)	1245.11 ± 0.05

All the values are mean ± SD of three observations

Table 9.6 Amino acids profile of tamarind seed kernel powder

Essential Amino acids	Amount (%)	Nonessential amino acids	Amount (%)
Histidine	3.3	Alanine	4.3
Isoleucine	4.5	Arginine	6.6
Leucine	7.5	Aspartic acid	11.3
Lysine	6.5	Tyrosine	5.7
Methionine	1.2	Glutamic acid	18.4
Phenylalanine	4.8	Glycine	8.8
Threonine	3.2	Proline	5.3
Valine	4.4	Serine	5.9

9.10.4 Polyphenolic Compounds

Tamarind seeds are known to be a rich source of polyphenolic compounds, which exhibit strong antioxidant and antibacterial activities (Sandesh et al. 2014; Sarkar et al. 2016). These seeds also contain flavonoids such as eriodictyol, taxifolin, luteolin, apigenin, and naringenin in relatively lower levels. The polyphenolic compounds present in tamarind seeds include (+)-catechin, (−)-epicatechin, procyanidin pentamer, procyanidin tetramer, procyanidin hexa, and procyanidin trimer (Razali et al. 2015; Sudjaroen et al. 2005).

The bioactivities found in tamarind seed extracts are thought to be a result of these chemical components, making tamarind seeds a valuable agro-industrial by-product with potential for usage in cosmetics, food, and pharmaceutical applications. However, for sensible and long-term use, it is necessary to gather supporting scientific data that addressed issues related to physicochemical composition, safety, analytical method, and positive health benefits.

9.11 Health Benefits

The metabolic processes of the human body, as well as its overall health and well-being, are significantly influenced by proteins and peptides derived from grains and seeds. These proteins and peptides can be classified based on their mode of action such as antibacterial, antihypertensive, immunomodulatory, and antioxidative properties (Sánchez et al. 2017). Bioactive peptides, in particular, are emerging as a novel class of physiologically active regulators that are being utilized to enhance health and address various diseases (Lemes et al. 2016).

Tamarind seeds have demonstrated in vitro antioxidant capacities, as evidenced by their total antioxidant capacity, total phenol content, and DPPH radical scavenging activity. Additionally, they have been found to impact blood sugar levels, indicating both antioxidant and antidiabetic properties (Nahar et al. 2014). Mucoadhesive microspheres, formulated using a combination of tamarind seed alginate and polysaccharide with varying ratios and calcium chloride concentrations as a cross-linker, have shown effectiveness in maintaining blood glucose levels and promoting patient compliance (Pal and Nayak 2012). Furthermore, a trypsin inhibitor derived from tamarind seeds has demonstrated satiety effects in rats by reducing leptin levels in obesity and increasing cholecystokinin levels in normal weight individuals. Tamarind kernel powder holds potential as an anti-obesity agent for the treatment of obesity (Medeiros et al. 2018; Ribeiro et al. 2015).

9.12 Tamarind Seed Kernel Powder in Food Processing Industry

Tamarind seed kernel has traditionally been eaten in rural regions after being dehulled and then roasted; the flavor and look are similar to groundnuts. It's powder has been utilized as a substitute for wheat flour and incorporated with it to make baked goods as a result of this information (Chakraborty et al. 2016).

It is widely utilized in the food processing industry due to its high water absorption capacity. It serves as an adhesive and is used as a stabilizer and emulsifier in food products due to its odorless, colorless nature and the ability to maintain high viscosity over an extended period. It is used as a flavoring component in baked goods and various ready-to-eat foods such as instant noodles, ketchup, soup, and ice cream. It is commercially available as a food additive for enhancing the consistency and viscosity of processed foods (El-Siddig et al. 2006). The uses of TKP in various food processing industries are given in Fig. 9.3.

Tamarind kernel serves as a source of xyloglucan, which functions as a gelling agent and enhances viscosity in food products, even under varying temperature and pH conditions. According to a study, It's powder can be used in place of xanthan gum to improve the texture of mango sauce (Betlach et al. 1987). The naturally occurring polysaccharide, commonly known as "jellose" or "polyose," is used in the production of jellies as well as a fruit preserver. TKP is a heat-resistant, low-acid emulsifier used in meals, salad dressings, and the synthesis of essential oils. TKP, as a food additive, has the ability to form gels in a wide range of food products. It can be utilized as a thickening and gelling ingredient, as well as a bulking agent, in various preparations such as candy, sour milk gel, puddings, desserts, yoghurt, low water release gel, jelly, jams, sauces, vegetable pancakes, pie fillings, protein-free

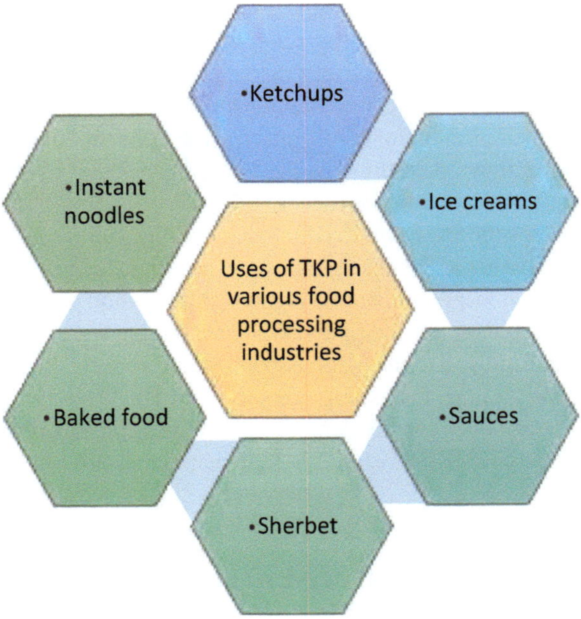

Fig. 9.3 Uses of TKP in various food processing industries

food, and acidic or neutral to basic pH food preparations. Addition of it to biscuits has also been shown to enhance their nutritional and functional value to varying degrees (Asma et al. 2015).

Tamarind kernel powder is used as an active ingredient in processed food products, particularly in the form of gum. It functions as an emulsifier in freezing foods, including freeze-dried gel, frozen food, and jelly-coated ice cream. It also serves as a stuffing agent, odor modifier, and coating agent in various food applications. Additionally, it is used as a deformation-preventive agent during the cooking process of food (Bhavini et al. 2018).

9.13 Innovative Food Products

The application of *Tamarindus indica* seed as a protein constituent in nourishment will support reduce the overdependency on conventional protein enhancements, specifically soybean and other common pulses. TKP possesses nutritional value comparable to that of pulses, legumes, and major cereal crops. It can be effectively utilized as a thickening agent, suspending agent, gelling agent, binding agent, solubilizing agent, disintegrating agent, and emulsifying agent in the development of innovative food products. It is mostly used for preparing food products such as baked food, sauces, ice creams, flavoring, and ready-to-eat edibles.

9.13.1 Antioxidant Properties of the TKP-Incorporated Food Products

In a study conducted by Priyadarshini Chakraborty et al. (2016), it was observed that the incorporation of TKP in cake samples resulted in enhanced antioxidant properties. The cakes demonstrated significantly higher scavenging activity, with values ranging from approximately 4.2 to 15.4 times greater than the control product.

9.14 The Future Potential of Tamarind Seed Kernel Powder

The tamarind seed utilization, although relatively small within the global agrochemical industry, has been experiencing notable growth. Despite being underutilized, tamarind seed holds significant potential for various applications. Exploring the utilization of tamarind seed as a composite ingredient in food and industrial products can be an intriguing avenue to explore. Key manufacturers are actively investing in research and development for the advancement of food products incorporated with tamarind seed kernel powder. This strategic approach is anticipated to propel the expansion of the tamarind seed utilization food industry in the coming years.

9.15 Conclusion

Boosting tamarind seed production and consumption could aid in food fortification, dietary diversity, and food demand alleviation of issues related to malnourishment in the nation. In accordance with the idea of "Sustainable Development Goals" (SDGs), increasing the utilization of agricultural wastes of tamarind seed can be a wonderful strategy for creating a more sustainable world. As a result, it is important to boost the numerous of antioxidants consumed by humans, and one way to do this is by including seeds that are high in phytochemicals into food products. Tamarind kernel powder enables us to create novel food items such as quick noodles, bread, cakes, cookies, and soup powder. Tamarind seed-derived food products not only help to preserve the environment but also provide the body a balanced diet. The exotic nutrient richness of tamarind (*Tamarindus indica*) seed is therefore a blessing for long-term healthy diets.

References

Akajiaku LO, Nwosu JN, Onuegbu NC, Njoku NE, Egbeneke CO (2014) Proximate, mineral and anti-nutrient composition of processed (soaked and roasted) tamarind (*Tamarindus Indica*) seed nut. Curr Res Nutr Food Sci 2(3):136–145

Betlach MR, Capage MA, Doberty DH, Hassler RA (1987) In: Yalpani M (ed) Industrial polysaccharides: genetic engineering, structure/property relations and application. Elsevier Science Publishers, Amsterdam, pp 35–50

Chakraborty P, Chakraborty N, Bhattacharyya DK, Ghosh M (2016) Effect of tamarind kernel powder incorporation in property and quality aspects of biscuit, bread and cake making. Sch Res Libr Arch Appl Sci Res 8(1):30–39. http://scholarsresearchlibrary.com/archive.html

Conrad Z, Niles MT, Neher DA, Jahns L (2018) Relationship between diet quality, food waste, and environmental sustainability. PLoS One 13(4):e0195405. https://doi.org/10.1371/journal.pone.0195405

Dagar JC, Singh G, Singh NT (1995) Evolution of crops in agroforestry with teak (Tectoma grandis), maharukh (Ailanthus excelsa) and tamarind (Tamarindus indica) on reclaimed salt-affected soils. J Trop For Sci 7(4):623–634

De Caluwé E, Halamová K, Van Damme P (2009) Tamarind (Tamarindus indica L.): a review of traditional uses, phytochemistry and pharmacology. ACS Symp Ser 1021:85–110. https://doi.org/10.1021/bk-2009-1021.ch005

De Caluwé E, Halamouá K, Van Damme P (2010) Tamarindus indica L. – a review of traditional uses, phytochemistry and pharmacology. Afr Focus 23(1):53–83

Deveswaran R, Bharath S, Furtado S, Abraham S, Basavaraj BV, Madhavan V (2010) Isolation and evaluation of tamarind seed polysaccharide as a natural suspending agent. Int J Pharm Biol Arch 1:360–363

El-Gindy AA, Youssif ME, Youssif MRG (2015) Chemical studies and utilization of *Tamarindus indica* and its seeds in some technological application. Egypt J Nutr Health 10(1):93–106

El-siddig K, Gunasena HPM, Prasad BA, Pushpakumara DKNG, Ramana KVR, Vijayanand P, Williams JT (2006) Tamarind, *Tamarindus indica L*, vol 1. International Centre for Underutilised Crops University of Southampton, p 188

George CK, Rao YS (1997) Export of Tamarind from India. In: Proc. Nat. Sym. on Tamarindus indica L., pp 156–161

Gitanjali VS, Shashi J (2020) Nutritional properties of tamarind (Tamarindus indica) kernel flour. Int J Curr Microbiol Appl Sci 9(05):1359–1364. https://doi.org/10.20546/ijcmas.2020.905.153

Glew RS, Vanderjagt DJ, Chuang LT, Huang YS, Millson M, Glew RH (2005) Nutrient content of four edible wild plants from West Africa. Plant Foods Hum Nutr 60:187–193

Hackler LR (1983, 1985) Lásztity R, Hidvégi M, editors (2012) Cereal proteins in human nutrition. In: Amino acid composition and biological value of cereal proteins: proceedings of the International Association for Cereal Chemistry symposium on amino acid composition and biological value of cereal proteins, vol 1. Springer Science & Business Media, Netherlands, pp 81–104

Havinga RM, Hartl A, Putscher J, Prehsler S, Buchmann C, Vogl CR (2010) Tamarindus indica L. (Fabaceae): patterns of use in traditional African medicine. J Ethnopharmacol 127(3):573–588

Heuzé V, Tran G (2015) Tamarind (Tamarindus indica), vol 11. Feedipedia, a programme by INRAE CIRAD AFZ and FAO, p 48. https://www.feedipedia.org/node/249

Hocking D (1993) Trees for drylands. Tamarindus indica L., Family Leguminosae: Caesalpinoideae. Oxford and IBH Publishing Co, New Delhi, pp 305–357. http://www.ind

https://www.altrafine.com, Applications & advantages of tamarind kernel powder. https://www.altrafine.com/blog/category/tamarind-kernel-powder/

Jiayu Ji Z (2005) Biological polysaccharide high molecular microbial fixing beta-galactosidase and its preparing method. CN patent 1587404

Khairunnuur FA, Zulkhairi A, Azrina A, Moklas MAM, Khairullizam S, Zamree MS (2009) Nutritional composition, in vitro antioxidant activity and Artemia salina L. lethality of pulp and seed of Tamarindus indica L. extracts. Malays J Nutr 15(1):65–75

Kumar CS, Bhattacharya S (2008) Tamarind seed: properties, processing and utilization. Crit Rev Food Sci Nutr 48:1–20

Kumar R, Patil SR, Patil MB, Paschapur MS, Mahalaxmi R (2001) Isolation and evaluation of the emulsifying properties of tamarind seed polysaccharide on castor oil emulsion. Sch Res Libr Der Pharm Lett 2:518–527

Lang P, Masci G, Dentini M, Crescenzi D, Cooke D, Gidley MJ, Fanutti C, Reid JSG (1992) Tamarind seed polysaccharide: preparation, characterisation and solution properties of carboxylated, sulphated and alkylaminated derivatives. Carbohydr Polym 17:185–198

Lemes AC, Sala L, Ores JDC, Braga ARC, Egea MB, Fernandes KF (2016) A review of the latest advances in encrypted bioactive peptides from protein-rich waste. Int J Mol Sci 17:950

Liener IE (1994) Implications of anti-nutritional components in soybean foods. Crit Rev Food Sci Nutr 34(1):31–67

Medeiros AF, Costa ID, Carvalho FM, Kiyota S, Souza BB, Sifuentes DN, Serquiz RP, Maciel BL, Uchôa AF, Santos EA, Morais AH (2018) Biochemical characterisation of a Kunitz-type inhibitor from Tamarindus indica L. seeds and its efficacy in reducing plasma leptin in an experimental model of obesity. J Enzyme Inhib Med Chem 33(1):334–348

Mishra RN (1997) Tamarindus Indica L: an overview of tree improvement. In: Proceedings of National Symposium on Tamarindus indica L., 27–28 June. Tirupathi (A.P.) organized by Forest Department of Andhra Pradesh, pp 51–58

Moldes AB, Vecino X, Cruz JM (2017) Nutraceuticals and food additives. In: Current developments in biotechnology and bioengineering. Elsevier, pp 143–164

Morton J (1987) Tamarind. In: Morton JF (ed) Fruits of warm climates. Florida Flair Books, Florida, pp 115–121

Nahar L, Nasrin F, Zahan R, Haque A, Haque E, Mosaddik A (2014) Comparative study of antidiabetic activity of *Cajanus cajan* and *Tamarindus indica* in alloxan-induced diabetic mice with a reference to in vitro antioxidant activity. Pharm Res 6(2):180–187

Nandkishore T, Sanjay S, Arnab RC (2014) Multipurpose applications of tamarind seeds and kernel powder. Sci Report 51:32–33

Nowacki E (1980) Heat-stable and anti-nutritional factors in leguminous plants. In: Advances in Legume Science. Royal Botanic Gardens Kew Richmond Survey, UK, pp 171–177

Pal D, Nayak AK (2012) Novel tamarind seed polysaccharide-alginate mucoadhesive microspheres for oral gliclazide delivery: in-vitro–in-vivo evaluation. Drug Deliv 19(3):123–131

Patel A (2013) Physical, biological and rheological properties of tamarind gum. Altrafine Blog.html, pp 11–12

Phani Kumar GK, Battu G, Kotha NS, Raju L (2011) Isolation and evaluation of tamarind seed polysaccharide being used as a polymer in pharmaceutical dosage forms. Res J Pharm Biol Chem Sci 2:274–290

Pugalenthi M, Vadivel V (2007) L- Dopa (l-3,4 Dihydrooxyphenylalanine): a non-protein toxic amino acid in Mucuna pruriens seeds. FoodReview 13:22–27

Pugalenthi M, Vadivel V, Gurumoorthi P, Janardhanan K (2004) Compositive nutritional evaluation of little known legumes, Tamarindus indica, Erythirna indica and Sesbania bispinosa. Trop Subtrop Agroecosystems 4:107–123

Pugalenthi M, Vadivel V, Janki P (2007) Comparative evaluation of protein quality of raw and differentially processed seed of underutilised food legumes, Abrus precatorius L. Livest Res Rural Dev 19:168–172

Pulungan MH, Sukardi, Sri Maryani EF, Atifah N, Sucipto (2001) Tamarind Kernel Powder (*Tamarindus indica*) processing (reviews on roasting temperature and roasting duration). J Teknol Pertanian 2:150–158

Rao YS, Mary Mathew K (2001) Tamarind. In: Peter KV (ed) Handbook of herbs and spices, vol 1. Woodhead, Cambridge, pp 287–296

Razali N, Mat Junit S, Ariffin A, Ramli NS, Abdul Aziz A (2015) Polyphenols from the extract and fraction of *T. indica* seeds protected HepG2 cells against oxidative stress. BMC Complement Altern Med 15:438. https://doi.org/10.1186/s12906-015-0963-2

Ribeiro JA, Serquiz AC, Silva PF, Barbosa PB, Sampaio TB, Araújo Junior RF, Oliveira AS, Machado RJ, Maciel BL, Uchôa AF, Santos EA (2015) Trypsin inhibitor from Tamarindus indica L. seeds reduces weight gain and food consumption and increases plasmatic cholecystokinin levels. Clinics 70:136–143

Saideswara Rao Y, Mary Mathew K (1999) Tamarind (Tamarindus indica L.) research – a review. Spices Board of India

Sánchez A, Vázquez A (2017) Bioactive peptides: a review. Food Qual Saf 1(1):29–46

Sandesh P, Velu V, Singh R (2014) Antioxidant activities of tamarind (*Tamarindus indica*) seed coat extracts using in vitro and in vivo models. J Food Sci Technol 51:1965–1973. https://doi.org/10.1007/s13197-013-1210-9

Sarkar A, Ghosh U (2016) Classical single factor optimisation of parameters for phenolic antioxidant extraction from tamarind seed (*Tamarindus indica*). Plant Sci Today 3:258–266. https://doi.org/10.14719/pst.2016.3.3.242

Sheikh H, Shivanna GB (2022) Tamarindus indica seeds and their nutraceutical applications. J Food Process Preserv. https://doi.org/10.1111/jfpp.17208

Siol M, Sadowska A, Król K, Najman K (2022) Bioactive and physicochemical properties of exotic fruit seed powders: mango (*Mangefiera indica* L.) and rambutan (*Nephelium lappaceum* L.) obtained by various drying methods. Appl Sci 12(10):4995. https://doi.org/10.3390/app12104995

Sonawane SK, Arya SS (2018) Plant seed proteins: chemistry, technology and applications. Curr Res Nutr Food Sci J 6(2):461–469. https://doi.org/10.12944/CRNFSJ.6.2.20

Srivastava S (2021) Don't throw tamarind seeds, instead use them to boost your health. Diet Nutr. https://www.herzindagi.com/diet-nutrition/use-tamarind-seeds-to-boost-your-health-article-162701

Sudjaroen Y, Haubner R, Würtele G, Hull WE, Erben G, Spiegelhalder B, Changbumrung S, Bartsch H, Owen RW (2005) Isolation and structure elucidation of phenolic antioxidants from Tamarind (Tamarindus indica L.) seeds and pericarp. Food Chem Toxicol 43:1673–1682. https://doi.org/10.1016/j.fct.2005.05.013

Surati BI, Minocheherhomji FP (2018) Benefits of tamarind kernel powder – a natural polymer. Int J Adv Res 6(3):54–57

Tharanathan RN, Mahadevamma S (2003) Grain legumes a boon to human nutrition. Trends Food Sci Technol 14:507–518

Von Maydell MJ (1986) Trees and shrubs of Sahel; their characteristics and uses. Deutsche. Geseuschft Fuer Techhische Zusammenarboit, Eschborn, p 111

Wandee R, Sutthanut K, Songsri J, Sonsena S, Krongyut O, Tippayawat P, Tukummee W, Rittirod T (2022) Tamarind seed coat: a catechin-rich source with anti-oxidation, anti-melanogenesis, anti-adipogenesis and anti-microbial activities. Molecules 27(16):5319. https://doi.org/10.3390/molecules27165319

Chapter 10
Valorization of Wastes and By-products of Cane-Based Sugar Industry

Tatek Temesgen, Selvakumar Periyasamy, Dinsefa Mensur, Belay Berhane, Sunaina, and Mani Jayakumar

Abstract Cane-based sugar industries have different wastes and by-products that can be further processed into useful products or applied as an energy source. Many wastes and by-products from cane-based sugar industries, such as bagasse, molasses, ashes, filler mud, trash tops, and muds, have been used for valorization. Valorization of these by-products contributed to the concept of resource efficiency and waste load reduction. Applying these wastes as a source of energy or input for other production units has proved to be of great importance for the improved echo-industrial nature of sugar industries. The ample production of these wastes from sugar industries also attracted the interest of other industries to form a symbiosis with sugar industries. Recently, the calorific value, fiber type, starch content, and ash properties of these waste types are attracting researchers to work on these wastes for different applications.

T. Temesgen
Department of Chemical Engineering, School of Mechanical, Chemical and Materials Engineering, Adama Science and Technology University, Adama, Ethiopia

Department of Chemistry, University of Calgary, Calgary, Canada

S. Periyasamy (✉)
Department of Chemical Engineering, School of Mechanical, Chemical and Materials Engineering, Adama Science and Technology University, Adama, Ethiopia

Department of Biomaterials, Saveetha Dental College and Hospitals, SIMATS, Saveetha University, Chennai, India

D. Mensur · B. Berhane
Department of Materials Science and Engineering, School of Mechanical, Chemical and Materials Engineering, Adama Science and Technology University, Adama, Ethiopia

Sunaina
Department of Chemistry, University of Calgary, Calgary, Canada

M. Jayakumar
Department of Chemical Engineering, Haramaya Institute of Technology, Haramaya University, Dire Dawa, Ethiopia

Department of Biotechnology, Faculty of Engineering, Karpagam Academy of Higher Education, Coimbatore, India

E. Cherian, B. Gurunathan (eds.), *Value Added Products From Food Waste*,
https://doi.org/10.1007/978-3-031-48143-7_10

Keywords Sugarcane industry · Bagasse · Biorefinery · Value-added products · Bioenergy

10.1 Introduction

Global sugarcane production is about 21% of the total primary crop production on average for over the past two decades. This reaches around 1.91 billion tons of sugarcane per annum. Brazil accounted for 40% of world sugarcane production, followed by India (20%) and China (6%) in 2020. In 2016 there were over 100 countries producing sugarcane now (Bhatnagar et al. 2015). Sugarcane is a cash crop in many countries around the world. It accounts for around 80% of the sugar produced and the rest is covered by sugar beets (Dotaniya et al. 2016). It makes income and contributes significantly to each nation's economy (Singh et al. 2021). This economic potential attracts the massive production of sugarcane for industrial applications. Moreover, the new and accelerated innovations in the industry as well as the mechanization sector led to a tremendous leap in sugarcane productivity (Periyasamy et al. 2022a). This massive sugarcane production entails a proportional increase in sugar production and sugar industry waste. The increasing trend in sugarcane production in the past two decades resulted in an environmental burden on the sugar industries to face severe environmental harm because of the limited availability of sustainable options in the waste management (Bhatnagar et al. 2015; Razia et al. 2020).

Environmental pollution due to industrial waste is a burning issue (Selvakumar et al. 2022). Legal restrictions are in place to prevent or limit emissions from different industries (Rathod et al. 2022). To this end, the possible wastes from a given industry have to be either prevented from generation by using different cleaner production techniques or have to be properly managed following the proper waste management hierarchies. One of the management techniques in the sugar industry waste management hierarchy is waste valorization. The common methods applied to recover the generated wastes and convert them to more useful products include reusing, recycling, and composting (Abdelbasir 2021). These valorization processes convert wastes into chemicals, materials, fuels, or other energy sources (Periyasamy et al. 2022b; Mohamed et al. 2022).

Waste valorization puts waste as a resource. It highly encourages the concept of type III industrialization proposed by Graedel (1996), where this type of industrialization mimics the global biological biosystem. The current global biosystem achieved complete cyclicity by evolving over the long term to the point where the boundary between resources and waste is not defined. This is because the waste in one part of a system represents a resource to another (Graedel 1996). Therefore, in this chapter, the wastes generated by cane-based sugar industries that have the potential to be used as a resource are considered as by-products of the sugar processing. The tendency to achieve a circular economy through applying by-product/waste valorization approaches is a way for sustainable manufacturing. The source and characteristics of waste are the major factors for valorization. The type of waste/by-product for valorization has to be classified under nonhazardous waste group. Valorization of the by-products in sugar industry have been in practice. This chapter discusses the major types of waste/by-product types, their sources, and some important resource recovery options. The main valorization approaches discussed include waste-to-energy production, pulp and paper, composites, biorefinery, organic fertilizer, and processed animal food.

10.2 Types of Sugar Industry Wastes and By-products

The type of waste in the sugar industry is of two types, solid and liquid. The major types of solid by-products are (1) sugarcane trash, (2) bagasse, (3) bagasse fly ash, and (4) press mud. Other suspended solids such as fat molecules, grease, and oil found in the effluent are also a part of solid wastes but are removed in the treatment process. The liquid waste/by-products from the sugar industry include washing water waste and molasses (Bhatnagar et al. 2015; Razia et al. 2020; Singh et al. 2021).

10.2.1 Solid Wastes of the Sugarcane Industry

The major solid wastes/by-products of sugarcane industries discussed here are mainly sugarcane trash, bagasse, bagasse fly ash, and press mud because of their potential for valorization.

10.2.1.1 Sugarcane Trash

Sugarcane trash is a by-product of sugarcane farming. It is produced during the harvesting period of the ripened sugarcane for the production of sugar. It constitutes the cane leaves, the remaining roots, and spoiled canes. This mass could contain soil and mud, and care has to be taken during valorization before mixing or processing it with other wastes.

10.2.1.2 Bagasse

The mill house is the leading process operation unit in cane sugar production. It is used for the extraction of juice from sugarcane. The by-product produced during the juice extraction is bagasse. Generally, sugarcane produces 3% of bagasse on a wet basis (Hofsetz and Silva 2012). It is highly fibrous (40–45%) in nature and contains 45–70% moisture content after the extraction process (Sahu 2018; Meghana and

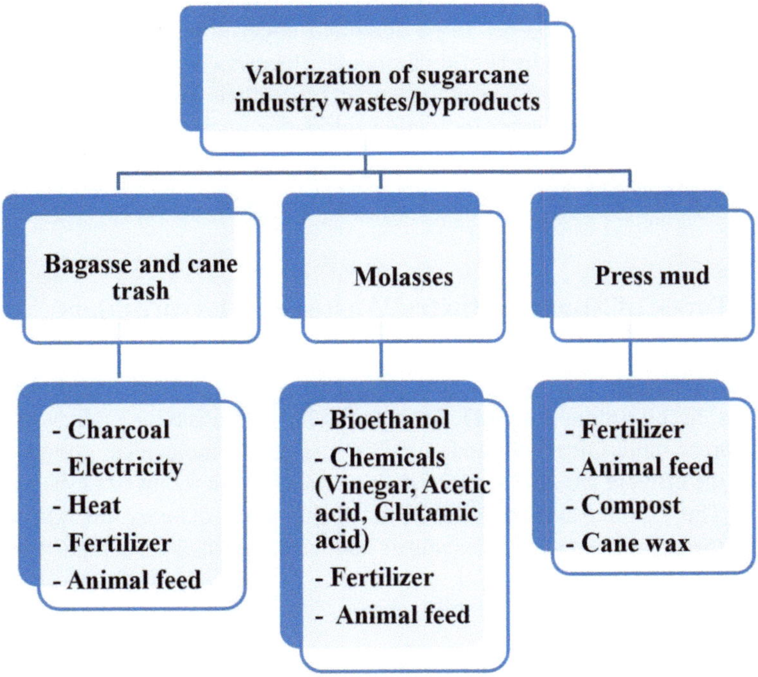

Fig. 10.1 Valorization of cane-based sugar factory wastes/by-products

Shastri 2020). It is mainly composed of lignin (20–30%), cellulose (40–45%), and hemicelluloses (30–35%) and a small amount of sugar (2–5%) (Cardona et al. 2010; Sahu 2018). The sugars in the bagasse act as a nutrient source for the microbial degradation of the bagasse. The microbial degradation rate depends on environmental factors such as temperature and moisture content. The bagasse as a by-product is used for energy generation, paper and board making, composting, and animal feed production (Sahu 2018). The valorization options in the cane-based sugar industry are presented in Fig. 10.1.

10.2.1.3 Bagasse Fly Ash

Bagasse fly ash is a solid waste/by-product produced by sugar mill boilers by burning bagasse and cane trash at 600–800 °C to produce energy and steam for power. It is composed of silica with excellent pozzolanic behavior. Bagasse fly ash was considered a source of particulate emission in environmental pollution. Recently, it has become a value-added product that is immensely used for absorption, partial input for cement industries, landfill stabilization, and manufacturing paper and wood boards (Bhatnagar et al. 2015; Tesfamariam et al. 2022).

10.2.1.4 Press Mud

Press mud/press cake/filter mud/filter cake is the solid waste produced after clarifying the cane juice in the sugar industries. Around 3% of press mud cake is expected for every 100 tons of sugarcanes crushed. The proximate analysis of crude press mud showed an approximate presence of moisture (68–77%), ash (19–31%), sugar (12–13%), fiber (5–10%), and crude wax (7–11%) (Bhatnagar et al. 2015). Elemental analysis shows macro- and micronutrients with an approximate proportion of calcium (21–30%), silicon (9.5–10%), phosphorus (8.–10%), sulfur (4–7%), and potassium (3–4%) in crude press mud; calcium (20–32%), potassium (9–14%), sulfur (8–168%), and magnesium (48–8%) in water-soluble portion press mud; and calcium (22–31%), potassium (2–3%), sulfur (4–7%), phosphorus (9–11%), aluminum (2.–3), and silicon (11%) in water in-soluble portion press mud (Saleh-e-In et al. 2012). Press mud can be produced from the alcohol distillation of molasses using chemicals and solvents. If it is disposed of untreated, it can pose environmental pollution. Press mud is appliable as a fertilizer, compost, and in fats and wax production (Sahu 2018).

10.2.2 Liquid Wastes from Sugar Industry

There are two major types of liquid waste in the cane-based sugar industry. These are the wastewater containing dissolved chemicals produced in the sugar manufacturing units and the molasses by-product.

10.2.2.1 Wastewater in a Sugar Industry

Sugar industries are under agro-industries that require a large quantity of fresh water for processing and discharge half of the ratio as effluent (Sahu 2018). Water is an essential input in the sugar industry. Almost all the units in sugar manufacturing industries require a huge amount of water as an input. Wastewater is mainly produced in mill houses, sulfur furnaces, lime hydrator units, and boiler houses. The wastewater from these units contains a heavy load of organic matter. Besides these units, excess water is required for cooling and condensation processes. Since water reuse in sugar industries is not practiced, fresh water is required for all operations (Bhatnagar et al. 2015).

10.2.2.2 Molasses

Molasses is a by-product of sugar industries. It has many applications in alcohol distilleries, foodstuff, and animal feed processing industries (Dotaniya et al. 2016). It is produced from the crystallization process of the sugar industry in three stages boiling for sugar extraction from the juice. Light molasses with a mellow flavor is produced in the first boiling stage. In the second boiling stage, full-flavored robust molasses, also called dark molasses, is produced. It is not sweet compared to light molasses. The black strap is the by-product of the last stage, and most of its sugar is already extracted. It is dark, viscous, and usually used for nutritional supplements (Bhatnagar et al. 2015). The composition of molasses varies based on different factors such as agro-climatic conditions, variety of cane, sugar manufacturing processes, and transportation and storage, but presented a sample composition with 30–35% sucrose, 10–25% glucose, and 23–23% moisture content, 16–16.5% ash content, 4.8–5% calcium and potassium, 2–3% non-sugar compounds, and 1–2% other mineral contents (Dotaniya et al. 2016).

10.3 Energy Production

The world energy demand has grown significantly since the twentieth century with a drastic increase in world population, urbanization, and industrialization. This caused a high consumption of fossil fuels resulting in the depletion of the limited resources (Alao et al. 2022). Therefore, study on alternative sources became an area of focus in energy studies. One of these alternatives is the application of waste materials as a source of energy production. Promoting biomass by-products for energy has become one of the key energy strategies in many countries. Sugarcane is considered an excellent storage of solar energy as biomass with a high energy-to-volume ratio (Arshad and Ahmed 2016). This stored energy can be extracted as heat using cogeneration, biogas, biochar, and hydrogen gas (Arshad and Ahmed 2016; Agarwal et al. 2022).

10.3.1 Cogeneration as an Energy Recovery

Cogeneration has been the main focus of waste-to-energy utilization in most cane-based sugar production industries from bagasse. Cogeneration is the production of electrical energy with valuable heat energy from a single fuel source. The process is referred to as the combined heat and power cycle. In sugar industries, cogeneration boilers are used to generate electricity and useful heat. The electricity generated by the cogeneration boilers is used internally by the sugar industry as a source of energy to produce sugar. The surplus amount is exported to the national grids for external consumer use. During the production of the electrical energy, heat energy as condensed steam is generated from the cogeneration boilers. This is a useful heat for processing cane juice to produce sugar.

Experiences of countries show the potential of cogeneration as a surplus energy source for a national grid supply. Around 50% of the bagasse, which is a 3% yield of the produced cane having a 48% moisture content with a net calorific value of 8 MJ/kg, fulfills the energy and steam usage of sugar mills and distilleries. In Brazil, a ton of bagasse generates 12 kWh of electricity, 330 kWh of heat energy, and 16 kWh of mechanical energy. The Brazilian cogeneration sugar industry produced more than 3000 MW of power. Considering the internal energy requirements of about 15–20 kWh/ton in Thailand, a minimum of 104 kWh/ton of surplus energy can be supplied to the national grid. In India, the potential of about 7000 MW of electricity from 650 sugar mills was estimated, of which 2250 MW has been achieved from 200 sugar mills (Gopinath et al. 2018). The other technique used to utilize the chemical energy from bagasse is the application of thermochemical conversion processes such as pyrolysis. The outputs of biomass pyrolysis can vary to a wide range of gaseous, liquid, or solid hydrocarbon products. This part will be covered in the next section of the chapter.

Cogeneration in sugar industries has a value addition in revenue and environmental load reduction. The revenue is gained by making the sugar industry energy independent from the grid system and getting profit through the export of electricity to the grid. In the environmental aspect, a sugar mill with 2500 TCD can save the release of 0.166 million tons of CO_2 by exporting 22 MW (Arshad and Ahmed 2016). Furthermore, it reduces the landfill cost or methane emission due to excess piling up of the bagasse during sugar production. The process flow diagram for the cogeneration plant is summarized in Fig. 10.2.

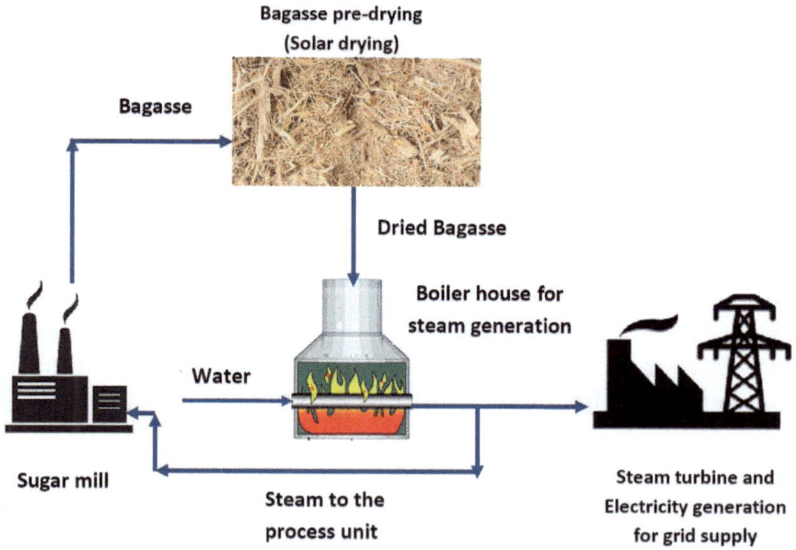

Fig. 10.2 Cogeneration from sugarcane bagasse in sugar industry

10.3.2 Chemical Energy Conversion Technologies for Sugar Industry By-products

In the sugar industry, two broad energy conversion technologies are used to change chemically bound energy in biomass/biomaterial to useful energy sources. These two technologies are thermochemical and biological processes. Comparing the two technologies, the biological conversion processes are time-consuming, while the thermochemical methods are energy intensive but have a better ability to break down organic compounds. The biological conversions are very slow because it is limited by the slow enzymatic or metabolic catalysis processes of microorganisms. The biochemical conversion technologies for sugar industry by-products include alcoholic fermentation, photobiological hydrogen production, and aerobic digestion. Applicable thermochemical processes mainly include pyrolysis, gasification, liquefaction, and biochar production (Li et al. 2022).

10.3.2.1 Thermochemical Conversion Processes

Pyrolysis is one of the thermochemical processes used in the conversion of sugar industry by-products into syngas (comprising of methane, hydrogen, carbon dioxide, carbon monoxide, and other hydrocarbons), solid (biochar), and liquid (bio-oil) products using a temperature range of 300–900 °C at atmospheric pressure in the absence of oxygen. It is a more efficient method to produce gas, char, and bio-oil due to the interaction of the decomposed bagasse and its intermediate products (Arshad and Ahmed 2016). The pyrolysis yield for solid, gas, and liquid (bio-oils) products depends on the applied pyrolysis conditions. It is possible to control the product outcome of pyrolysis. Figure 10.3 shows the block diagram for the pyrolysis process from the sugar industry.

10.3.2.2 Biological Energy Conversion Processes

Biological process for converting chemical energy in the sugar industry by-product to valuable products is represented in Fig. 10.4. Sugarcane bagasse and molasses are used for ethanol production through this process. Bioethanol can be blended with fuel and used as an energy source worldwide. Two types of ethanol, anhydrous and hydrous, are produced from sugar industry by-products. The hydrous ethanol (94% ethanol, 6% water) is used directly in E100 vehicles and also in flex-fuel vehicles (Nogueira et al. 2020), whereas the anhydrous ethanol (99.5% ethanol by volume at 15.6 °C) is blended with gasoline up to 20–27% to make the E(20–27) used in common gasoline engines (Gopinath et al. 2018; Nogueira et al. 2020). There are recent researches that show the possibility of hydrogen production from molasses. This can be a future source for applying hydrogen as an energy source (Keskin and Hallenbeck 2012).

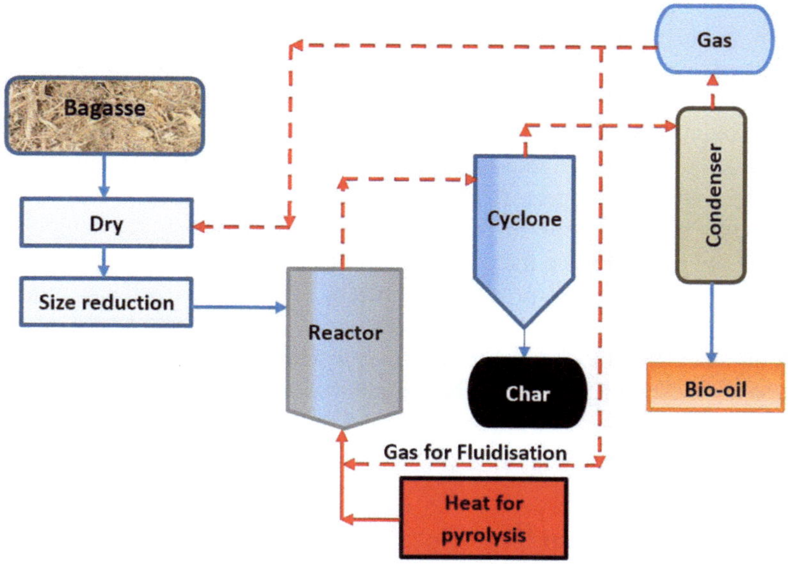

Fig. 10.3 Schematic representation of pyrolysis of bagasse in the sugar industry

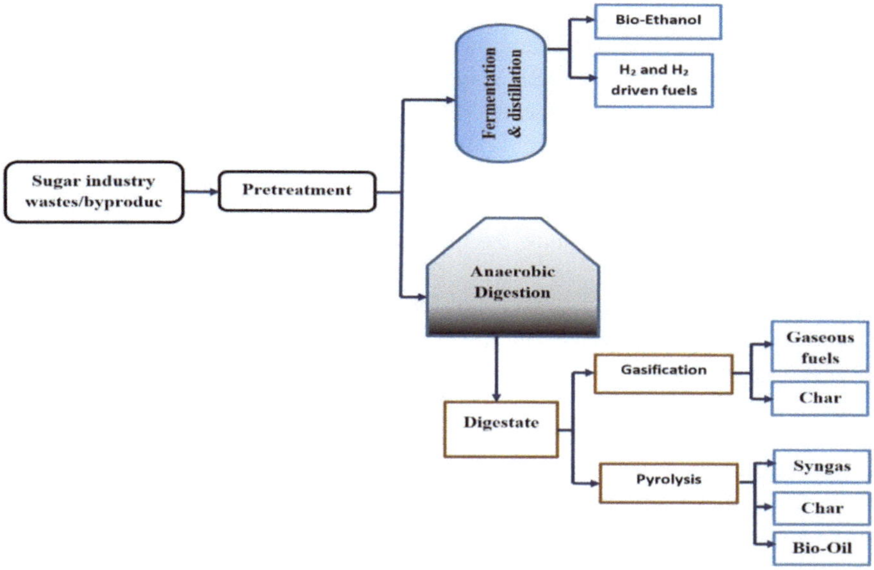

Fig. 10.4 The schematic representation of biological energy conversion processes from sugar industry wastes

Anaerobic digestion is the other option for biochemical energy conversion from sugar by-products. It is a low-cost technique for recovering bioenergy from lignocellulose or biowaste. It is activated by a community of anaerobic microorganisms to transform biodegradable by-product components into energy-rich biogas (consisting mainly of about 50–75% CH_4, 25–50% CO_2 and H_2O) and digestate residue. The process of aerobic digestion occurs through a series of well-known stages, namely hydrolysis, acidogenesis, acetogenesis, and methanogenesis consecutively. The biomass composition (carbon-to-nitrogen ratio), temperature, liquid-to-solid ratio, and pH can influence the product yield in the aerobic digestion process and therefore optimizing these process parameters for efficient methane production (Agarwal et al. 2022). The aerobic digestion of bagasse is reported in mono-digestion when pretreated bagasse is used as a substrate alone and in co-digestion when bagasse is co-applied with other organic wastes such as food waste, manure, or press mud (Agarwal et al. 2022).

10.4 Pulp and Paper

There is an increase in pulp and paper production worldwide due to urbanization and dependence on pulp and paper products. However, these industries are highly dependent on forest sources, which directly affects our climate due to deforestation. Therefore, alternative input sources have to come into the picture. In this aspect, sugarcane bagasse is a potential alternative to substitute forest products as an input for pulp and paper industries (Bhatnagar et al. 2015). In most cane-based sugar production mills, the surplus bagasse is considered to environmental threat and risk point for fire. Despite this fact, very limited research is reported. Currently, around 30 countries are believed to use bagasse to produce pulp and paper. According to estimates, around 2–5% of the global pulp and paper production uses bagasse as input. Bagasse is used to produce the main paper categories: packaging, printing, writing, and photocopier paper, tissues, and newsprint (Rainey and Covey 2016).

The application of bagasse, instead of forest-based raw materials, has the added advantages of reducing deforestation and less chemical application for bleaching because bagasse requires fewer bleaching chemicals than wood pulp to achieve a bright, white sheet of paper. This contributes to the sustainable production of paper by reducing the impacts of materials and reducing costs due to the application of limited bleaching chemicals such as chlorine (Sotoodehnia and Amiri Roodan 2012). Study shows that sugarcane bagasse showed good characteristics for producing paper. It contains a good amount of cellulose (32–44%), equivalent to other potential sources such as bamboo, preferably used by paper mills (Bhardwaj et al. 2019). Bagasse storage is a critical step to be managed properly due to its vulnerability to microbial degradation. The microbial activity on the bagasse leads to a yield reduction. Both chemical and mechanical pulping processes are used in the processing of bagasse paper making. Compared to wood lignin, bagasse lignin is much more reactive and the pulping conditions are very mild (Rainey and Covey

2016). Bagasse pulp is producing in a rotary digester at high pressure and temperature following soda pulping, kraft process, and sulfite process (Singh et al. 2021). The soda pulping process could be an alternative for converting sugarcane trash to a fibrous pulp with a yield of 46% (Bhardwaj et al. 2019).

The high pith content of the stalk, which is 30–35% by weight, is the greatest obstacle to bagasse pulping. Its pith cells consume chemicals while being pulped and are unsuitable for creating paper. Therefore, it is mandatory to depithing. It must be done to remove the pith, dirt, and soil materials in the pulp mill. It removes the shortest bagasse fibers, which is important for producing high-grade pulp. Depithed bagasse can be used in soda pulping or other techniques, such as the kraft and sulfite process, for pulp-making from the bagasse. For soda pulping, the chemical charge is 12–16% of sodium hydroxide (as NaOH on dry fibers) at 170–175 °C; for Kraft pulping, it is 11–13% active alkali with 15–20% sulfidity. Pulp produced from bagasse has a similar property to hardwood eucalypt pulp than softwood pulp with the fiber diameter of ~1.0–1.2 mm. The major features of bagasse pulp include reduced initial freeness hindering drainage triggered by the remaining pith, lower opacity, high stiffness, low tear, and tensile strength caused by the oversmoothness and the moderately thick cell walls (Rainey and Covey 2016).

10.5 Composites

There are continuous efforts by sugarcane industries to the conversion of their industrial waste to value-added products to enhance the benefits such as maximum revenue generation, improved product yield, decreased waste generation, and to produce goods according to the commercial demand for certain products such as organic fertilizers, biochemical, and ceramics. This section describes the utilization of sugarcane industry waste for the composite design. Bagasse is one of the important wastes produced, which can be converted to useful products individually or in combination with other sugarcane wastes with some simple modification. The composition of bagasse, as described in Table 10.1, is the key factor for its versatile use in different composite products.

Because of the high fibrous content in bagasse, it is widely used in polymer and epoxy-based composites to increase their strength. It is also used as an additive in thermal insulating boards to decrease the air conditioning load. Bagasse fiber and

Table 10.1 Chemical composition of sugarcane bagasse

Nutrients	Percentages
Cellulose	45–55
Hemicellulose	20–25
Lignin	18–24
Pectin	0.6–0.8
Ash	1–4
Extractives	1.5–9

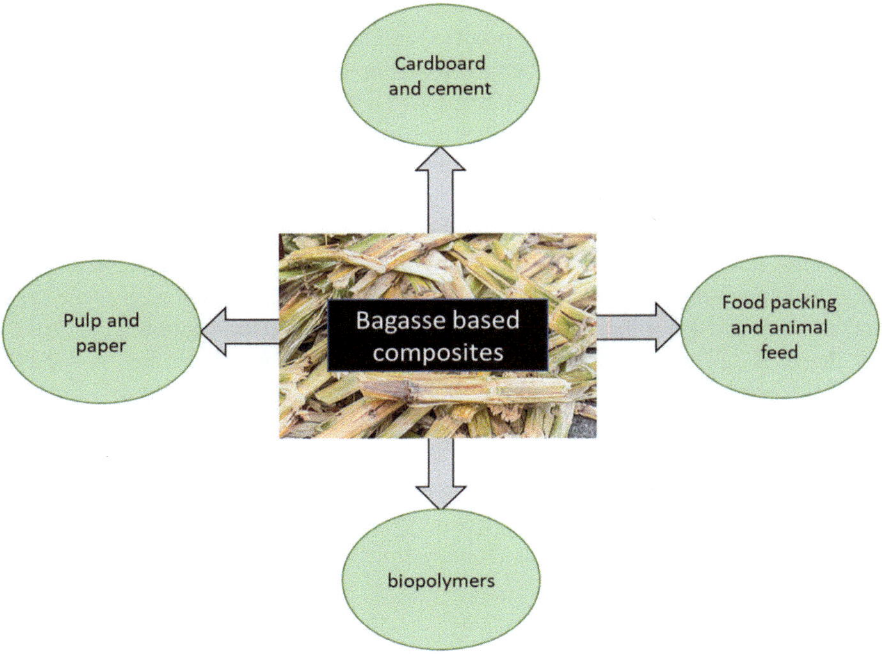

Fig. 10.5 Application of bagasse-derived composites in different industries

sugarcane ash are generally used as additives for clay, glass, and ceramics products. It can be added to cement composites, ceramic tiles, and other building products as a reinforcement material. In automobile industries, polypropylene is incorporated with natural fibers of bagasse instead of talc and it has also replaced glass fiber-reinforced polypropylene (Zabihzadeh 2010). It also serves as one of the components of acoustic protection. Bagasse can be easily modified into useful products. The application of bagasse-derived composites in different industries is indicated in Fig. 10.5. Researchers have found that there is not much change in the burned and unburned bagasse in terms of their composition; besides, the unburned bagasse has a larger number of polar groups. This favors the enhanced adhesion property within the phenolic matrix with the unburned bagasse. This composite has shown better storage modulus and impact strength. Therefore, this bagasse fiber/phenolic matric-based composite can be utilized in building materials, stiff packaging, and automobile applications (da Silva and Frollini 2020). The researchers are recently been working on developing eco-friendly polymers derived from natural resources. For example, bagasse is used to prepare bagasse/pectin composite-based films with excellent physicochemical and photodegradation properties. These films can be used for food packaging applications. Bagasse fibers are also used for enhancing the mechanical and water uptake properties of the polylactic acid-based composites. These polymers are very light and low-cost. Hence, they can be utilized for the interior parts of automobiles and consumer goods. In food industries, bagasse-derived nanocellulose is applied in fillings, chips, wafers, soups, puddings, and gravies (Mohit and Selvan 2020; Mendes et al. 2019).

There are many advantages associated with the utilization of bagasse for composites. The major points are as follows:

1. It has compatibility with thermoplastic as well as thermoset matrices.
2. It is biodegradable and eco-friendly.
3. It is a sustainable approach due to the renewable nature of bagasse.
4. It can replace the traditional wood-based products used in household.

Despite many advantages of mixing the bagasse in composites, a large amount is still unutilized and disposed of in open fields or incinerated to get rid. This results in environmental pollution. So, these practices can be minimized if we can maximize the use of bagasse as a binder in different composite materials.

10.6 Biorefinery

This section describes the use of sugarcane industries' waste as a raw material for biorefineries. Figure 10.6 shows the sugar industry waste and biorefinery potential product chain. Since sugarcane industries generate a large amount of biomass-based

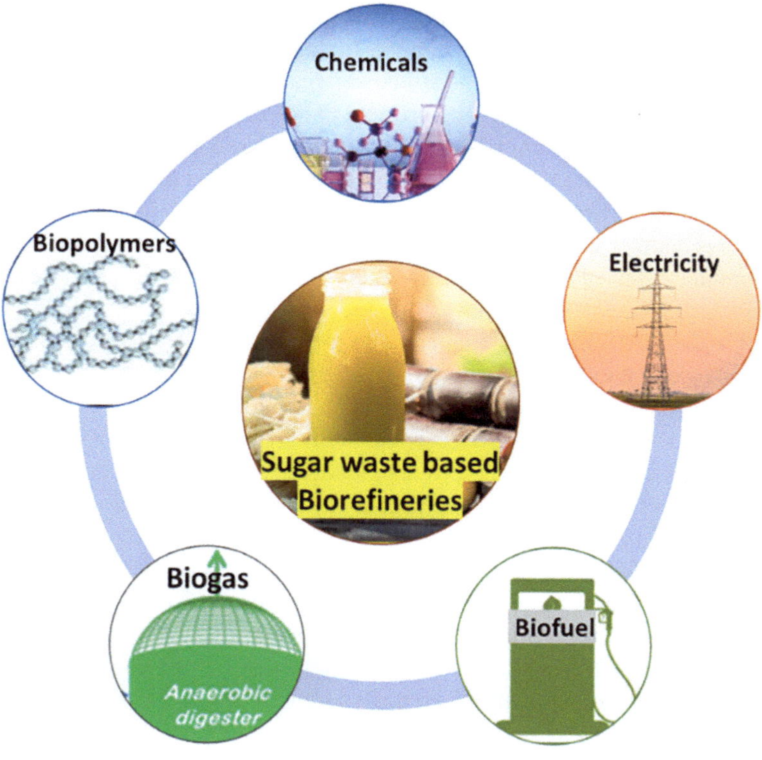

Fig. 10.6 Sugar industry waste and biorefinery potential product chain (Longati et al. 2020)

waste, there is a huge scope for developing the concept of biorefineries for producing chemicals and fuels that favors strategic, environmental, and economic growth. It is advantageous for biorefineries to convert their various biowaste and other intermediate waste to different products and enhance the value derived from the biomass feedstock as per the market demand and availability of biomass. This approach is sustainable in nature and renders additional revenue to these industries. With the increasing population and decreasing natural resources, there is a need for a circular bioeconomy for a sustainable future. For this, the efforts are being directed toward designing sustainable future biorefineries annexed to existing sugar industries to convert their biowaste to commercially important biofuels/biochemicals. Currently, biogas and bioethanol are the main side products from sugar-based biorefineries. Recently, efforts have been directed at producing other commercially important high-demand chemicals such as succinic acid, lactic acid, and carbon capture from biorefineries (Farzad et al. 2017; Oliveira et al. 2020).

One of the major products derived from sugarcane industry waste is bioethanol. The production of bioethanol is generally consists of four major steps. First, it is pretreated either by lime or hydrogen peroxide to render the accessibility of cellulose, followed by hydrolysis (either by adding enzymes or acid treatment) to release the monomeric sugars. Further, fermentation is done to convert the sugars into ethanol and distillation for product recovery. Biogas production is another alternative for utilizing this waste. This is also an environmentally friendly method for eliminating organic matter from the effluent. The generation of biogas is dependent on the composition of bagasse. The variability in the composition of bagasse is due to the soil quality of the land of cultivation, sugarcane species, and harvesting time. The cellulosic and hemicellulosic component of bagasse is majorly used for biogas generation, whereas lignin mostly remains unaffected and can be used for heat and energy production. Biogas and power are generated from the process residues of ethanol production (Ungureanu et al. 2022).

Xylitol is another important compound that can be derived from sugar wastes. It has potential in food and pharmaceutical applications. Its synthesis via the chemical route is energy extensive and gives a low yield, making it expensive. So, the biosynthesis of xylitol from different biomass is in progress. The hemicellulose hydrolysate from bagasse is converted to xylitol. It is generally coproduced with bioethanol. It can be produced either by using a single organism *Kluyveromyces marxianus*, for two products or using the two different yeasts, *S. cerevisiae* and *C. tropicalis*, which are best known for their ability to ferment glucose and xylose to ethanol and xylitol, respectively. Lactic acid can also be produced from bagasse (Muhammad Zohaib 2021). Lactic acid is an important chemical that is applied in the food, pharmaceutical, and chemical industries. It is also a monomer unit for polylactic acid, receiving great attention for being biodegradable. Bagasse can be used as raw material for the preparation of lactic acid, making it economically attractive and competitive with lactic acid production from sugar. It can be achieved by enzymatic liquefication performed at a high solid loading of pretreated lignocellulosic biomass with uncompromised and concentrated sugar yields. Furfural is another value-added compound that can be extracted from lignocellulose waste of sugar industries. Furfural has a high market value for plastics, resins, and fuel additives. Currently, the methods applied for converting biomass to furfural are not eco-friendly. So, there is a need for a better process design for sustainable conversion. The current process of furfural synthesis from sugar waste involves the high-temperature hydrolysis of lignocellulose under acidic conditions. The degradation of cellulose liberates xylose, mannose, glucose, and acetic acid. The xylose at high temperatures and pressure is further converted to furfural (Modelska et al. 2017).

In addition, instead of discarding the carbon dioxide into the atmosphere from sugarcane biorefineries, it can be captured at different stages (burning of bagasse or sugarcane juice fermentation). The captured CO_2 can be utilized as a valuable resource for other industries, such as the production of artificial synthetic carbohydrates in combination with hydrogen (Formann et al. 2020). Some simple biorefinery, ideas have been implemented in some sugar industries. They extended their product range from sugar to bioethanol, biogas, electricity generation, and biopolymers production. Although researchers and governments are making many efforts to design and implement such a complex system, still it is very challenging. It needs further improvements to be employed for future sugar biorefineries.

10.7 Organic Fertilizer

Lately, the scientific community has moved its interest toward using by-products from sugar industries mainly due to the energy and raw material crisis. In sugar industries, sugarcane is processed to form sugar and biomass. The huge generation of organic-based by-products from different processing stages, as well as their inadequate disposal in the environment have made the sugar industries less eco-friendly. The major waste products generated from the sugar industries are bagasse, press mud, molasses, and sugarcane bagasse ash. Bagasse consists of fibers used in boilers for steam production in sugar industries. Molasses is used for alcohol production in distilleries. Sugarcane bagasse ash is utilized for cement and ceramic industries because of the high content of silica and alumina. Press mud majorly consists of lignin and fiber pith. Therefore, it has a large application in the biochemical and microbial fields (Raza et al. 2021). There is a massive demand for organic fertilizers because of the high agricultural dependency. Composting is referred to as a conventional treatment to recycle the organic waste from the industries, since it can provide the essential plant nutrient and may affect the soil's physical, chemical, and biological properties. Recently, with the increased cost of fertilizers, these industries are releasing the potential of their by-products to substitute synthetic fertilizers. Therefore, scientists are focusing their research on finding quick and novel ways to decompose the press mud to form fertilizers that will be organic and free from any additive artificial chemicals. Sugarcanes consist of about 12–15% sugar and each ton of sugarcane can be crushed to generate about 70–90 kg of press mud (Dotaniya et al. 2016). The representative composition of press mud is given in Table 10.2.

Table 10.2 Composition of press mud generated from the sugar industries

Nutrients	Percentages
Moisture	50–65
Fiber	20–30
Crude wax	7–15
Sugar	5–12
Crude protein	5–10
Nitrogen	2–2.5

Apart from the listed components, a small quantity of silicon, iron, manganese, calcium, magnesium oxide, and phosphorus pentoxide are present in press mud. Press mud contains a variety of essential nutrients for plant growth. Hence, it is worth processing the press mud to organic fertilizers to enhance the soil quality for better agricultural crop production. In this regard, there are some reports wherein researchers have utilized the press mud in combination with urea (1:1 ratio) to enhance sugarcane yield. Press mud increases the accessibility of macro- and micronutrients in the soil. In addition, it provides carbon-to-soil microorganisms, favoring the breakdown and nutrient transformation processes.

Some reports are on using press mud with inorganic fertilizers to increase the fertility of red soil. The regular addition of press mud to the agricultural land for crop production results in the enhancement of organic carbon content in the soil and in 5–6 years, soil health is expected to improve. It also enhances the bacterial diversity as well as the bacterial population of the soil by enhancing the carbon content, which is the food for their growth (Prado et al. 2013). Press mud also increases the availability of phosphorous by enhancing the organic content of the soil. The carboxylate group of press mud and bagasse directly and indirectly affects phosphorous fixation. In direct impact, phosphorous is directly released from the organic content to the soil. In indirect impact, the organic acid released from the decomposition of organic content mobilizes the phosphorous and makes it available to the plants.

Bagasse is also used to enhance the fertility of the soil. Before adding the bagasse to the agricultural land, it is recommended to chop it properly. It should be added at least 1 month prior to seed sowing in the field for proper decomposition. Application of these by-products also increases the cation exchange capacity of the soil for a long time, generating cations such as Mg^{2+}, Ca^{2+}, and K^{+} during the organic content degradation. In addition, they also minimize the loss of essential nutrients by leaching caused by water and wind erosions. The use of vinasse, another residue of sugarcane industries, is common to improve the soil quantity. Before utilizing it as fertilizer, it can be concentrated via evaporation or microfiltration (Dotaniya et al. 2016). Overall, sugarcane industries improve soil health by increasing the essential micro- and macronutrients, resulting in better root growth and consequently enhancing crop production. The positive effects of applying organic waste to agricultural land proved to be a good, sustainable, and eco-friendly approach to tackle the waste storage problem and shortage of plant nutrients.

10.8 Processed Animal Food

Sugarcane and its waste are widely used as a fibrous source for cattle feed production. However, it is not suitable for human consumption but has been utilized as an additive to mice, cats, goats, and sheep's diets to improve their digestion and, therefore, general health. There are some limitations of using sugarcane and its biowaste for animal feed, such as the presence of high sugar levels in sugarcane favors the rapid growth of microbial in the rumen, relatively low nutrition value in terms of proteins and minerals in comparison to other bioproducts (corn meal and soybean meal) and poor polysaccharides digestibility. Despite these limitations, the global animal production from sugarcane cannot be overlooked (Harrison 2016; Costa et al. 2015).

Sugarcane industries-based wastes are often an unutilized source of roughage for the ruminant's diet. Bagasse primarily consists of insoluble polysaccharides (cellulose and hemicellulose) and lignin (aromatic polymer). Hence, bagasse provides protein, lipids, and minerals to animal feed. Some research in this direction proves that incorporating bagasse without any further processing into animal feed can enhance digestibility. It is also good for the diet of sheep and some nonruminant (Harrison 2016). Sugarcane molasses is another significant component in animal feedstock, which is also a low-cost energy source. It is regarded as an energetic animal feed because of the presence of high fermentable sugars in it. Additionally, it is also highly rich in salts and minerals. Its relatively low cost makes it more popular for feedstuff to replace cereals in formulations.

Molasses has around 35–35% water and can provide approximately 8 MJ of metabolizable energy per kg to cattle. However, molasses has low nitrogen levels, which restrict the capability of rumen microbial sources to use the energy sources present in molasses completely. Therefore, to enhance the nutritional benefits of the molasses, the addition of nitrogen sources is required in the form of protein or urea for the diet of ruminants. The ratio of molasses in the feedstock of animals can be different as liquid feeds utilize molasses exclusively, whereas solid feeds use less than 5% of molasses. Using liquid feed of molasses with a source of nitrogen is an excellent choice for drought-hit areas where dry feed is very low. However, special care should be taken while adding the nitrogen source to prevent molasses toxicity (also called drunkenness, a state raised from vitamin B deficiency and bloating). Another point that should be kept in mind is that molasses is rich in sulfur. So, the sulfur sources should not be added as a supplement. The high sulfur level in molasses can decrease copper absorption because of the ruminal thiomolybdates formation. Despite having high nutritional value, the molasses-based animal feed can be harmful, whereas molasses-urea-containing animal feed makes them more prone to cause diseases (Mordenti et al. 2021).

Another constituent of great value for animal feed is sugarcane spent juice, which is utilized as an additive to animal feed. For pig feeding, this is the main energy source. The main components of sugarcane juice are sugars (glucose, sucrose, and fructose), chlorophyll, wax, and fibers. Because of the high level of soluble sugar, it is more suitable to access the entire available energy through the intake of pigs (Harrison 2016). Filter cake from the sugar industries waste consists of dry matter with a high level of organic matter (~80%) and inorganic ash (~20%). The moisture content can be variable depending on the location, climate conditions, processing method, and cultivation practices. The organic part of the filter cake is made up of fats, proteins, and carbohydrates (fiber and sugars). This shows that it can be used as a raw material for animal feed production. Generally, the filter cake is dried in sunlight to reduce its moisture content to enhance its shelf life. The low moisture content reduces the bacterial, fungal, and other degradation of filter cake-derived animal feed (Abera et al. 2020). The biowaste from the sugar industries can potentially enhance the nutritional value of animal feed. However, there is a need for more research in this area to find out which feed is good for which animals as per their physiological conditions to provide better protein, minerals, vitamins, and energy for their better growth and development.

10.9 Conclusion

This chapter discussed waste and by-product valorization of cane-based sugar industries. The types of sugar industry wastes and by-products, such as sugarcane trash, bagasse, bagasse fly ash, press mud, and molasses, are covered. Furthermore, the major valorization opportunities, such as energy production, pulp and paper production, application for composites, utilization in biorefinery, and opportunities as an organic fertilizer, are covered. The valorization of these wastes/by-products has a contribution to the sustainable management of sugar industries. The valorization of the wastes/by-products improves the industries' environmentally friendly nature, making it economically feasible and socially acceptable. Moreover, it tried to summarize and show the application of these wastes/by-products as a source of energy or input for other production. This is believed to have great importance for the improved echo-industrial nature of sugar industries.

References

Abdelbasir SM (2021) Recycling, management, and valorization of industrial solid wastes. In: Makhlouf ASH, Ali GAM (eds) Waste recycling technologies for nanomaterials manufacturing. Springer International Publishing, Cham, pp 25–63

Abera AA, Duraisamy R, Seda TB (2020) Characterization of sugar industry waste (filter cake) and agro-waste crop residue as potential source of livestock feed raw materials. Research Square

Agarwal NK, Kumar M, Ghosh P, Kumar SS, Singh L, Vijay VK, Kumar V (2022) Anaerobic digestion of sugarcane bagasse for biogas production and digestate valorization. Chemosphere 295:133893

Alao MA, Popoola OM, Ayodele TR (2022) Waste-to-energy nexus: an overview of technologies and implementation for sustainable development. Cleaner Energy Syst 3:100034

Arshad M, Ahmed S (2016) Cogeneration through bagasse: a renewable strategy to meet the future energy needs. Renew Sust Energ Rev 54:732–737

Bhardwaj NK, Kaur D, Chaudhry S, Sharma M, Arya S (2019) Approaches for converting sugarcane trash, a promising agro residue, into pulp and paper using soda pulping and elemental chlorine-free bleaching. J Clean Prod 217:225–233

Bhatnagar A, Kesari KK, Shurpali N (2015) Multidisciplinary approaches to handling wastes in sugar industries. Water Air Soil Pollut 227(1):11

Cardona CA, Quintero JA, Paz IC (2010) Production of bioethanol from sugarcane bagasse: status and perspectives. Bioresour Technol 101(13):4754–4766

Costa DA d, Souza CL d, Saliba E d OS, Carneiro J d C (2015) By-products of sugar cane industry in ruminant nutrition. Int J Adv Agric Res 3(1):1–9

da Silva CG, Frollini E (2020) Unburned sugarcane bagasse: bio-based phenolic thermoset composites as an alternative for the management of this agrowaste. J Polym Environ 28(12):3201–3210

Dotaniya ML, Datta SC, Biswas DR, Dotaniya CK, Meena BL, Rajendiran S, Regar KL, Lata M (2016) Use of sugarcane industrial by-products for improving sugarcane productivity and soil health. Int J Recycl Org Waste Agric 5(3):185–194

Farzad S, Mandegari MA, Guo M, Haigh KF, Shah N, Görgens JF (2017) Multi-product biorefineries from lignocelluloses: a pathway to revitalisation of the sugar industry? Biotechnol Biofuels 10:87

Formann S, Hahn A, Janke L, Stinner W, Sträuber H, Logroño W, Nikolausz M (2020) Beyond sugar and ethanol production: value generation opportunities through sugarcane residues. Front Energy Res 8. https://doi.org/10.3389/fenrg.2020.579577

Gopinath A, Bahurudeen A, Appari S, Nanthagopalan P (2018) A circular framework for the valorisation of sugar industry wastes: review on the industrial symbiosis between sugar, construction and energy industries. J Clean Prod 203:89–108

Graedel TE (1996) On the concept of industrial ecology. Annu Rev Energy Environ 21:69–98

Harrison MD (2016) Sugarcane-derived animal feed. In: Sugarcane-based biofuels and bioproducts. Wiley & Sons, United States of America, pp 281–310

Hofsetz K, Silva MA (2012) Brazilian sugarcane bagasse: energy and non-energy consumption. Biomass Bioenergy 46:564–573

Keskin T, Hallenbeck PC (2012) Hydrogen production from sugar industry wastes using single-stage photofermentation. Bioresour Technol 112:131–136

Li J, Li L, Suvarna M, Pan L, Tabatabaei M, Ok YS, Wang X (2022) Wet wastes to bioenergy and biochar: a critical review with future perspectives. Sci Total Environ 817:152921

Longati AA, Batista G, Cruz AJG (2020) Brazilian integrated sugarcane-soybean biorefinery: trends and opportunities. Curr Opin Green Sustain Chem 26:100400

Meghana M, Shastri Y (2020) Sustainable valorization of sugar industry waste: status, opportunities, and challenges. Bioresour Technol 303:122929

Mendes JF, Martins JT, Manrich A, Sena Neto AR, Pinheiro ACM, Mattoso LHC, Martins MA (2019) Development and physical-chemical properties of pectin film reinforced with spent coffee grounds by continuous casting. Carbohydr Polym 210:92–99

Modelska M, Berlowska J, Kregiel D, Cieciura W, Antolak H, Tomaszewska J, Binczarski M, Szubiakiewicz E, Witonska IA (2017) Concept for recycling waste biomass from the sugar industry for chemical and biotechnological purposes. Molecules 22. https://doi.org/10.3390/molecules22091544

Mohamed BA, Bilal M, Salama ES, Periyasamy S, Fattah IR, Ruan R, Awasthi MK, Leng L (2022) Phenolic-rich bio-oil production by microwave catalytic pyrolysis of switchgrass: experimental study, life cycle assessment, and economic analysis. J Clean Prod 366:132668

Mohit H, Selvan VAM (2020) Effect of a novel chemical treatment on the physico-thermal properties of sugarcane nanocellulose fiber reinforced epoxy nanocomposites. Int Polym Process 35(2):211–220

Mordenti AL, Giaretta E, Campidonico L, Parazza P, Formigoni A (2021) A review regarding the use of molasses in animal nutrition. Animals 11. https://doi.org/10.3390/ani11010115

Muhammad Zohaib N (2021) Bio-purification of sugar industry wastewater and production of high-value industrial products with a zero-waste concept. Crit Rev Food Sci Nutr 61(21):3537–3554. v. 3561 no. 3521

Nogueira LAH, Souza GM, Cortez LAB, Brito Cruz CH d (2020) 9 – biofuels for transport. In: Letcher TM (ed) Future energy, 3rd edn. Elsevier, pp 173–197. https://doi.org/10.1016/B978-0-08-102886-5.00009-8

Oliveira RA, Schneider R, Lunelli BH, Rossell CEV, Filho RM, Venus J (2020) A simple biorefinery concept to produce 2G-lactic acid from sugar beet pulp (SBP): a high-value target approach to valorize a waste stream. Molecules 25(9):2113

Periyasamy S, Karthik V, Senthil Kumar P, Isabel JB, Temesgen T, Hunegnaw BM, Melese BB, Mohamed BA, Vo DVN (2022a) Chemical, physical and biological methods to convert lignocellulosic waste into value-added products. A review. Environ Chem Lett 20:1129–1152

Periyasamy S, Isabel JB, Kavitha S, Karthik V, Mohamed BA, Gizaw DG, Sivashanmugam P, Aminabhavi TM (2022b) Recent advances in consolidated bioprocessing for conversion of lignocellulosic biomass into bioethanol-a review. Chem Eng J 453:139783

Prado R d M, Caione G, Campos CNS (2013) Filter cake and vinasse as fertilizers contributing to conservation agriculture. Appl Environ Soil Sci 2013:581984

Rainey TJ, Covey G (2016) Pulp and paper production from sugarcane bagasse. In: Sugarcane-based biofuels and bioproducts. John Wiley & Sons, New Jersey, pp 259–280

Rathod ML, Shivaputra A, Umadevi H, Nagamani K, Periyasamy S (2022) Cloud computing and networking for SmartFarm AgriTech. J Nanomater 2022:6491747

Raza Q-U-A, Bashir MA, Rehim A, Sial MU, Ali Raza HM, Atif HM, Brito AF, Geng Y (2021) Sugarcane industrial byproducts as challenges to environmental safety and their remedies: a review. Water 13. https://doi.org/10.3390/w13243495

Razia M, Maheshwari Nallal VU, Sivaramakrishnan S (2020) Chapter 16 – Agro-based sugarcane industry wastes for production of high-value bioproducts. In: Krishnaraj Rathinam N, Sani RK (eds) Biovalorisation of wastes to renewable chemicals and biofuels. Elsevier, Amsterdam, pp 303–316

Sahu O (2018) Assessment of sugarcane industry: suitability for production, consumption, and utilization. Ann Agrar Sci 16(4):389–395

Saleh-e-In MM, Yeasmin S, Paul BK, Ahsan M, Rahman MZ, Roy SK (2012) Chemical studies on press mud: a sugar industries waste in Bangladesh. Sugar Tech 14(2):109–118

Selvakumar P, Adane AA, Zelalem T, Hunegnaw BM, Karthik V, Kavitha S, Jayakumar M, Karmegam N, Govarthanan M, Kim W (2022) Optimization of binary acids pretreatment of corncob biomass for enhanced recovery of cellulose to produce bioethanol. Fuel 321:124060

Singh SP, Jawaid M, Chandrasekar M, Senthilkumar K, Yadav B, Saba N, Siengchin S (2021) Sugarcane wastes into commercial products: processing methods, production optimization and challenges. J Clean Prod 328:129453

Sotoodehnia P, Amiri Roodan R (2012) Environmental benefit of using bagasse in paper production – a case study of LCA in Iran. In: Bharat Raj S (ed) Global warming. IntechOpen, Rijeka, Ch. 8

Tesfamariam BB, Seyoum R, Andoshe DM, Terfasa TT, Ahmed GMS, Badruddin IA, Khaleed HMT (2022) Investigation of self-healing mortars with and without bagasse ash at pre- and post-crack times. Materials 15(5):1650

Ungureanu N, Vlăduţ V, Biriş S-Ş (2022) Sustainable valorization of waste and by-products from sugarcane processing. Sustainability 14. https://doi.org/10.3390/su141711089

Zabihzadeh S (2010) Flexural properties and orthotropic swelling behavior of bagasse/thermoplastic composites. Bioresources 5(2):650–660

Part V
Waste Utilization from Meat, Poultry and Fish

Chapter 11
Keratinase: A Futuristic Green Catalyst and Potential Applications

Mani Jayakumar, S. Venkatesa Prabhu, C. Nirmala, M. Sridevi, and Magesh Rangaraju

Abstract Chicken feathers (CF) are a significant type of residual waste produced by the meat processing industry. Prolonged accumulation of such a leftover CF could be a serious problem for solid waste management. Consequently, CF can be exploited for widespread production of keratinaceous around the world that calls for their justifiable use. CF mostly consist of keratins, which are widely applied in a variety of industries. Since CF has potent resistance to protease breakdown, untreated feathers could create environmental pollution. Studies revealed that instated of treating the resistant pollutant (keratin), microbial digestion of keratin waste can provide the appreciable opportunity to obtain a commercially significant enzyme, keratinase. This review study provides a broad overview of the potential applications of several bacterial and fungi species for the keratinase synthesis employing keratinaceous wastes as substrates. Additionally, a briefly outlook has been provided on the exploitation of microbial keratinases with its biochemical and

M. Jayakumar (✉)
Department of Chemical Engineering, Haramaya Institute of Technology, Haramaya University, Dire Dawa, Ethiopia

Department of Biotechnology, Faculty of Engineering, Karpagam Academy of Higher Education, Coimbatore, Tamil Nadu, India
e-mail: drjayakumarmani@haramaya.edu.et

S. V. Prabhu
Centre for Food Nanotechnology, Department of Food Technology, Faculty of Engineering, Karpagam Academy of Higher Education, Coimbatore, Tamil Nadu, India

C. Nirmala
Department of Biotechnology, Paavai Engineering College, Paavai Institutions, Namakkal, Tamil Nadu, India

M. Sridevi
Department of Biotechnology, Vinayaka Missions Kirupananda Variyar Engineering College, Vinayaka Missions Research Foundation (Deemed to be University), Salem, Tamil Nadu, India

M. Rangaraju
Department of Chemical Engineering, Wachemo University, Hossana, Ethiopia

E. Cherian, B. Gurunathan (eds.), *Value Added Products From Food Waste*,
https://doi.org/10.1007/978-3-031-48143-7_11

functional properties in the synthesis of cattle feed, food, nutrition, detergents, organic fertilizers as well as leather, cosmetics, and medicinal products.

Keywords Keratinase · Bacteria · Chicken feather · Keratin · Fungi · Green catalyst

11.1 Background

Keratin is a prominent structural protein that commonly presents in pervasive fauna's keratinous tissues, i.e., claws, hair, wool, horns, feathers, hooves, and nails (Rodziewicz and Łaba 2006). Although the disulfide bond-breaking occurs during keratin degradation, keratinase facilitates the peptide cleavage that ensures as an integral part in feather industries. Due to their complex structure to cleavage disulfide bond, keratin resists to degrade by essential proteases, which is trypsin and pepsin (Perţa-Crişan et al. 2021; Wang et al. 2016a, b). Keratin, a structural protein present in human and other animal hard tissues, is generally broken down by proteolytic enzymes known as keratinases. They achieve this by rupturing the disulfide and peptide bonds in the polypeptide chain (Babbar et al. 2022).

Keratinase is an important instrument for turning keratin-rich wastes, such as feathers from the meat processing industry, into useful goods with additional value (Gupta et al. 2023). Keratin-based wastes, such as feathers from the chicken industries, can be a desirable source of nitrogen, carbon, and sulfur, which can be used to create novel goods (Zhang et al. 2023; Bhari et al. 2021). Keratinases are endo-acting proteases that belong to the protease family M36 (MEROPS database) (Qiu et al. 2022). It was observed that with keratinase 140 only 1 day was needed for complete keratin degradation into peptide with molecular weight ranging from 0.48 to 10 kDa. Similarly, the purified recombinant keratinolytic protease, namely RecGEOker, was expressed in *E. coli* BL21 and was cloned from the thermophilic *Geobacillus stearothermophilus* AD-11, which stimulates green development (DE3) (Gegeckas et al. 2015).

Additionally, it has been noted that a number of bacterial strains release extracellular keratinases (Brandelli et al. 2010). A good alternative to transforming substrates at a cheap energy cost is the synthesis of keratinases by psychrophilic and psychrotrophic strains (Joshi and Satyanarayana 2013). Similarly, dermatophytic species such as *Trichophyton rubrum* and *Microsporum canis* and nonpathogenic fungus such as *Onygena equina* are also some among the keratinase-secreting fungi (Huang et al. 2015; Qiu et al. 2020).

Recently, keratinase manufacturing is receiving global interest for its utility in sustainable improvement and purifier manufacturing. The exploitation of keratin by microbial keratinases for the creation of amino acid and peptides products for use in functional animal feed, fertilizers, value-added goods, skincare products, or as an ingredient to laundry and dishwashing detergents has recently come to light (Gupta et al. 2013b; Srivastava et al. 2020; Jin et al. 2018). Some reports have mentioned that keratinase is an optimistic protease that is considered to bring green development due to its perspective in the biocatalysis fields of waste recycling, textile, and leather industries (Gupta and Ramnani 2006).

Due to their unique properties, keratinases can be used for a variety of purposes, including the manufacture of animal feed and protein supplements, other personal care products, and hair removal creams in the pharmaceutical and cosmetic industries, animal feed and protein supplements in the poultry processing industry, bioremediation and wastewater treatment, prion degradation in medicine, the synthesis of biofuels, manufacture of nanoparticles, and the improved drug delivery (Adelere and Lateef 2016; Lateef et al. 2015). Upon the need of vigorous interest in keratinase enzyme, this chapter has been articulated about the keratinase biochemical and functional properties, method adopted for keratinase production, route to improve the keratinous productivity, keratinase applications, recommendation, and futuristic focus.

11.2 Biochemical and Functional Properties

Keratinases are produced by wide range of bacterial, actinomycetes, and fungal organisms that include *Bacillus*, *Clostridium*, *Chryseobacterium*, *Fervidobacterium*, *Kocuria*, *Kytococcus*, *Lysobacter*, *Microbacterium*, *Nesternkonia*, *Nocardiopsis*, *Stenotrophomonas*, *Streptomyces*, *Xanthomonas*, *Aspergillus*, *Doratomyces*, *Myrothecium*, *Paecilomyces*, *Scopulariopsis*, *Trichoderma*, *Trichophyton* specie (Brandelli et al. 2010; Hassan et al. 2020a). These microorganisms are isolated from poultry feather waste, sheepfold soil, hydrothermal hot spring, poultry farm, brown algae *Zonaria tournefortii*, goat skin, dump site, agro-wastes, cattle-yard utilizing donkey hairs, slaughterhouse soil, soil sample of a tea plantation, poultry dumpsite, poultry farm waste, and soil samples from various regions and characterized by biochemical and molecular methods. In general, keratinases are produced as extracellular enzyme (Kerouaz et al. 2021) yet intracellular production as reported elsewhere (Gupta and Ramnani 2006). They prevail preponderantly as serine, metallo, thiol, or subtilisin type of proteases in few yeast strains, also occurs as aspartic protease (Hassan et al. 2020a). The robust enzyme exhibits a diverse biological activity with a wide temperature and pH optima to unveil maximum activity based on the organisms and the sources from which they are isolated. In general, neutral to alkaline pH range of 7.5–9.0 was observed for the enzyme from different sources yet highest activity at acidic conditions (pH 1.5–3) are also reported (Habbeche et al. 2014; Sharma and Gupta 2016; Nnolim et al. 2020a, b; Vanaki et al. 2022; Sharma et al. 2022). Predominantly, the enzymes were active in mesophilic conditions of 40–60 °C (Babbar et al. 2022); however, a few thermophile keratinase found stable at extreme conditions of 80–100 °C are also identified (Jagadeesan et al. 2020; Pawar et al. 2018; Kerouaz et al. 2021).

Furthermore, a recombinant keratinase with enhanced activity and stability was successfully expressed in various host systems by random mutagenesis and molecular engineering strategies (Wang et al. 2016a, b; Zhang et al. 2020; Yong et al. 2020; Yahaya et al. 2021). Immobilization of the enzyme on substrates such as chitosan and chitosan-grafted β-cyclodextrin also improved the activity and thermostability over wide ranges of temperature conditions (Srivastava et al. 2020). Commonly, keratinases exist in monomeric form with the molecular weight ranging from 8 to 240 kDa; multimeric keratinases are also quantified by advanced chromatographic techniques (Laba et al. 2015; Elhoul et al. 2016; Hamiche et al. 2019; Sharma et al. 2022).

Additionally, the impact of inhibitors and metal ions on keratinolytic activity has been researched (Keshapaga et al. 2023). Usually, divalent metal ions such as Ca^{2+}, Cd^{2+}, Mg^{2+}, Mn^{2+}, Na^{+}, Mg^{2+}, Zn^{2+}, Ba^{2+}, Fe^{3+}, sodium sulfite, and β-mercaptoethanol increased the keratinases. These advantageous results may be attributed to the stability of the enzyme–substrate complex and the active enzyme configuration. Additionally, these metal ions may shield the enzyme from heat denaturation and increase activity (Habbeche et al. 2014; Pawar et al. 2018; Akram et al. 2020; Jana et al. 2022). A two- to eight-fold enhancement in the activity was noted about when there was H_2O_2, $NaHClO_3$, Triton X-100, DTT, and β-mercaptoethanol (Rajput et al. 2010; Babbar et al. 2022). Different keratinases respond differently to reducing substances, and the tendency to stimulate the activity of the enzymes are also identified. The effect of organic solvents, detergents, and heavy metals such as Cu^{2+}, Ag^{+}, Hg^{2+}, and Pb^{2+} on inhibiting the keratinolytic enzymes by reducing their cysteine bridges were also reported. Phenylmethanesulfonyl fluoride (PMSF), diiodopropyl fluorophosphates (DFP), ethylenediaminetetraacetic acid (EDTA), iodoacetamide, and 1,10-phenanthroline inhibited the activity of the enzyme (Nnolim et al. 2020a, b; Jana et al. 2022) (Table 11.1).

11.3 Production Strategies of Keratinase

In order to meet market demand, large-scale fermentation is much required to produce these enzymes (Jansen and van Gulik 2014). Owing to utilize the wastes, feathers are an industrial waste that can be used in culture media (Gafar et al. 2020). In this context, different wastes, such as poultry, meat processing, wheat and soybeans, can also be utilized as a substrate for the synthesis of the keratinase enzyme (Hassan et al. 2020b). Before they may be utilized for enzymatic characterization and other applications, keratinase enzymes need to be purified (Goda et al. 2021). One can produce keratinase enzyme with a high degree of purity using a number of techniques. Purification techniques include ammonium sulfate precipitation, gel filtration chromatography, and ion-exchange chromatography (Farag and Hassan 2004).

Because keratinases enzyme can be synthesized by a number of bacteria, it is challenging to compare the environments and purifying procedures for them (Ghaffar et al. 2018). Currently, there are over 30 different types of microorganisms

Table 11.1 Selective reports on biochemical properties of keratinase from different microorganisms

Catalytic type	Microorganisms	Molecular mass (kDa)	Optimal pH	Optimal temperature	References
Alkaline serine	*Micrococcus* sp. B1pz	62, 185, 139, 229	9.4	55	Laba et al. (2015)
Serine	*Bacillus licheniformis* ER-15	30 28	11	70	Tiwary and Gupta (2010)
Serine	*Actinomadura viridilutea* DZ50	19.53610	11	80	Elhoul et al. (2016)
	Brevibacillus parabrevis CGMCC 10798	28	8	60	Zhang et al. (2016)
Serine	*Caldicoprobacter algeriensis* strain TH7C1(T),	33.246	7	50	Bouacem et al. (2016)
Serine	*Bacillus subtilis* S1–4	45	8	65	Wang et al. (2016a, b)
Subtilisin-like serine	*Bacillus altitudinis RBDV1*	43	8	85	Pawar et al. (2018)
Serine	*B. Amyloliquefaciens* S13	28 47	6.5 8	50 60	Hamiche et al. (2019)
	Bacillus paralicheniformis MKU3		7		Kalaikumari et al. (2019)
Alkaline	*Arthrobacter* sp. KFS-1		8	60	Nnolim et al. (2020a, b)
Alkaline	*Bacillus* sp.		8	60	Nnolim et al. (2020a, b)
Metallo	*Bacillus thuringiensis* strain MT1	80	9	50	Hassan et al. (2020a)
Alkaline serine	*Bacillus pumilus* AR57		9	45	Jagadeesan et al. (2020)
Serine	*Bacillus subtilis* S1–4	36.1	9–10	50–55	Yong et al. (2020)
	Bacillus sp. NKSP-7	25	7.5	65	Akram et al. (2020)
Serine	*Actinomadura keratinilytica* strain CPT20	71 19	8 7	50 40	Kerouaz et al. (2021)
Aluminum-tolerant serine protease	*Bacillus thuringensis*		6.2	50	Wang et al. (2021)
Metallo	*Chryseobacterium aquifrigidense* FANN1		8	40–50	Bokveld et al. (2021)
Serine	*Bacillus cereus*		8–9	40–45	Almahasheer et al. (2022)

(continued)

Table 11.1 (continued)

Catalytic type	Microorganisms	Molecular mass (kDa)	Optimal pH	Optimal temperature	References
Thiol	3 strains of *Bacillus licheniformis* PVKR6, PVKR15, and PVKR41		2, 2.5, and 3		Vanaki et al. (2022)
Serine-metalloprotease	*B. Velezensis* ncim 5802	100, 62.5, 36.5 25	10	60	Sharma et al. (2022)
Metallo-serine	*Bacillus cereus* IIPK35	42	9	55	Jana et al. (2022)
Serine	*Bacillus tropicus* LS27		7.35	40	Liya et al. (2023)

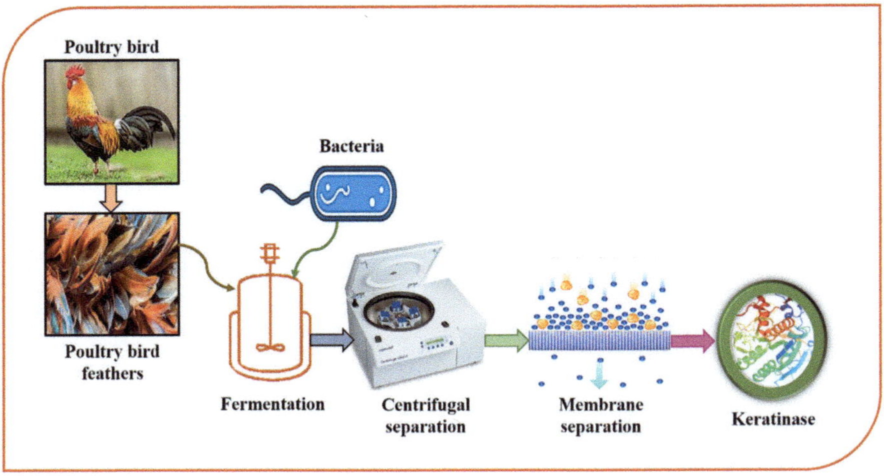

Fig. 11.1 Bacterial keratinase production practice

that can break down keratin (Lange et al. 2016). Because it needs a cheap growing medium containing salts and keratins as sources of carbon and other nutrients, respectively, keratinase synthesis is quite affordable (Ghaffar et al. 2018).

The keratinase enzyme production strategies include bacterial and fungal synthesis. The most widely used production practices are:

1. Bacterial keratinase production illustrated in Fig. 11.1.
2. Fungal keratinase production illustrated in Fig. 11.2.

Some insects and other microbes, including bacteria and fungi, produce keratinases (Khardenavis et al. 2009; Bohacz 2017). Numerous bacteria, particularly from areas where keratin-containing compounds are abundant, have been isolated from varied habitats (Li 2019). There have been reports of keratinase production by the bacterial

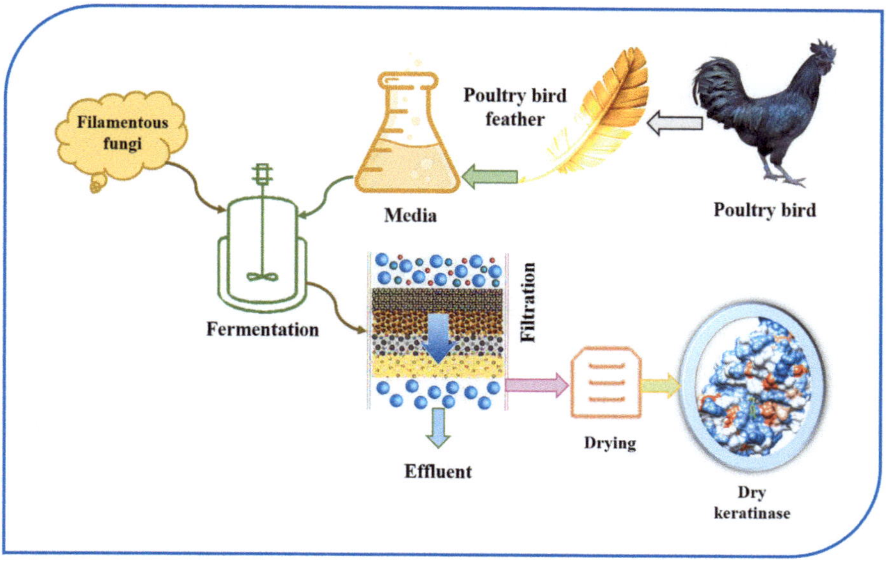

Fig. 11.2 Fungal keratinase synthesis practice

species *Bacillus*, *Vibrio*, *Chryseobacterium*, *Brevibacillus*, *Pseudomonas*, *Serratia*, *Fervidobacterium*, *Microbacterium*, *Aeromonas*, *Burkholderia*, *Stenotrophomonas*, *Rhodococcus*, *Geobacillus*, *Amycolatopsis*, *Meiothermus*, *Paenibacillus*, *Bacillus licheniformis*, and *Bacillus subtilis* (Nnolim et al. 2020a, b; Adibi et al. 2021).

It has been discovered that fungi can effectively break down the keratin-rich materials such as feathers. On the surface of human or animal skin, there occur some harmful fungus (Li 2019). So far, various microorganisms were reported for production of different amounts of keratinase, and these amounts differ in terms of the enzyme's existence as well as its structure, composition, stability, ideal reaction temperature, and pH value (Zhang et al. 2022). For instance, some types of keratinase are primarily found inside of cells, whereas others are primarily released outside of cells (Qiu et al. 2020). The fungus's keratinase can be found both inside and outside of cells (Filipello et al. 2000). In contrast to bacteria, fungus manufactures keratinases with distinct amino acid sequences (Li 2019). Because of the potential capacitance of keratinases production, the applications must be free of harmful fungus for safety reasons. The capacity of some nonpathogenic fungus to decompose feathers may one day lead to their approval for use as biofertilizer or animal feed (Li 2019).

As of now, there are different hydrolysis studies on keratinized tissues that are facilitated by keratinolytic fungi, which are known to naturally inhabit keratinous substrates. The reported keratinolytic fungi include *Aspergillus*, *Aphanoascus*, *Paecilomyces*, *Trichophyton*, *Doratomyces*, *Trichoderma*, *Fusarium*, *Acremonium*, *Onygena*, *Dermatophyte*, *Cladosporium*, *Microsporum*, *Chrysosporium*,

Lichtheimia, *Cnadina albicans*, and *Scopulariopsis* (Nnolim et al. 2020a, b; Adibi et al. 2021).

11.4 Microbial Degradation Mechanisms of Keratin

Studies revealed that the keratin is formed by a protein composed of forms of epidermal and epidermal appendages of vertebrates. Next to cellulose and chitin, keratin is identified as one of the toughest polymers that cannot be easily degraded naturally. Nevertheless, information on microbial degradation of keratin lack sufficient theory (Mazotto et al. 2013). So far, there are no appropriate proven mechanisms with regard to keratin microbial degradation. However, the possible mechanisms that underly for keratin degradation by microbial may provide possible strategies for improving the keratinase productivity by keratin degradation (Gupta et al. 2013b).

11.4.1 Breaking Disulfide Bonds by Denaturation

In parallel, during the growth of microorganism, keratin is getting progressively degraded to convert into smaller molecules such as peptides and amino acids. In this process, the three steps such as decomposition, denaturation, and transamination are involved (Mienda et al. 2011). However, keratin molecules are characterized by greater disulfide bonds, and the primary stage is to decompose the disulfide bonds,

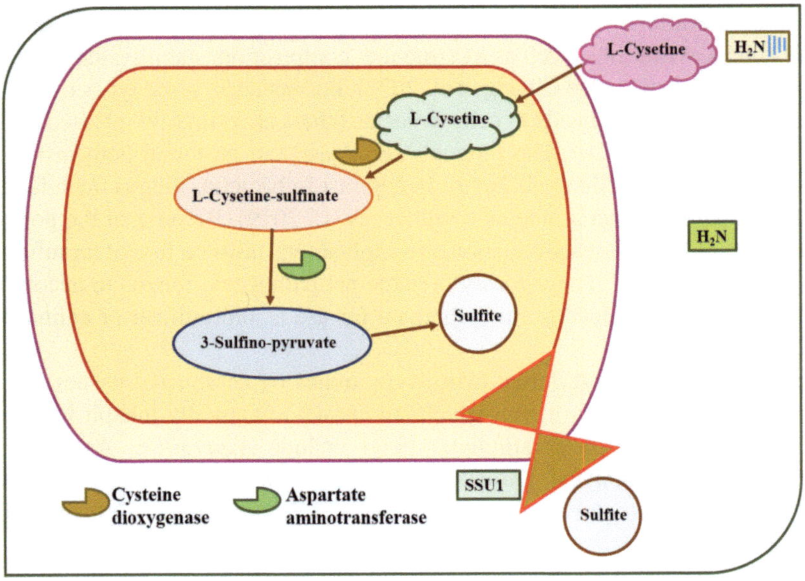

Fig. 11.3 Bacterial degradation of keratin by L-cysteine metabolism ("thiolysis" approach)

which loosens the structure. Recent studies reported that destruction mechanisms of disulfide bond can be occurred by four types such as mechanical pressure, thiolysis, biomembrane potential, and enzymatic hydrolysis. Another mechanism is based on cellulosic "thiolysis" approach (Fig. 11.3), while sulfite production process was reported by cysteine metabolism (Jeong et al. 2010).

11.4.2 Enzymatic Hydrolysis

Previous studies reveal that the keratinase obtained from different microbes may differ characteristics. Protein engineering strategies offer pure keratinase with the use of keratin hydrolysate. Nevertheless, pure keratinase could not hydrolyze the keratin without proper destruction of disulfide bond. Hence, disulfide bond reductase has a significant role on keratin degradation that has to be secreted continuously during the course of keratin decomposition (Khardenavis et al. 2009). There are various disulfide reductases involved for breaking the disulfide bond such as glutathione reductase, alkyl hydroperoxide reductase, ribonucleoside-diphosphate reductase, dihydrolipoyl dehydrogenase, phosphoadenosine phosphosulfate reductase, thioredoxin reductase, and peptide methionine sulfoxide reductase. In general, disulfide reductases preserve different reaction types as follows. Figure 11.4 illustrates the enzymatic mechanism of disulfide reductases:

1. thioredoxin disulfide +2′-deoxyribonucleoside diphosphate + H_2O → thioredoxin + ribonucleoside diphosphate,

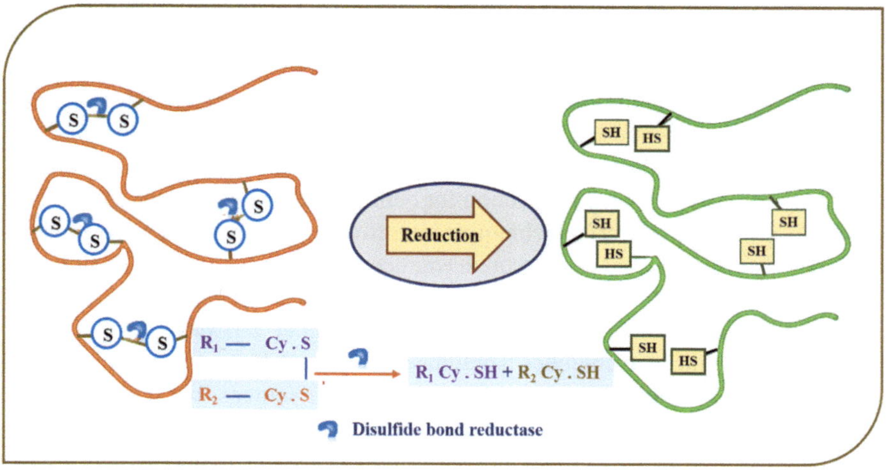

Fig. 11.4 Mechanism of disulfide reductases through various reactions

2. sulfite + adenosine 3′,5′-bisphosphate + thioredoxin disulfide → thioredoxin +3′-phosphoadenylyl sulfate
3. thioredoxin disulfide + peptide-L-methionine + H_2O → thioredoxin + peptide-L-methionine (S)-S-oxide
4. NAD^+ + protein N6-(dihydrolipoyl) lysine → NADH + protein N6-(lipoyl)lysine
5. glutathione + RX → R-S-glutathione + HX
6. $NADP^+$ + thioredoxin → NADPH + thioredoxin disulfide

11.4.3 Decomposition of Keratin

After disulfide bonds were subjected to break, microorganisms can secret different proteases such as trypsin, keratinase, pepsin, and papain to decompose these fibrous proteins into amino acids and peptides (Reddy and Yang 2007).

11.5 Novel Techniques for Improvement of Keratinases Production (Protein Engineering Approach)

It is proven that the biocatalysts can be successfully utilized in the chemical, food, and pharmaceutical industries over the past few decades. Due to their innate capacity to catalyze reactions with high velocities and great specificity under a variety of circumstances, they are frequently used for the synthesis of a variety of complicated drug intermediates, specialty compounds, and even commodity chemicals. The development of protein engineering strategies for inventive biocatalysts with new or superior capabilities has been pushed by the rising interest in using enzymes for industrial and home catalysis. Current developments in recombinant DNA technology, genomics, proteomics, and high throughput technologies have made it possible to create novel biocatalysts and biocatalytic processes. From the beginning of large-scale (recombinant) enzyme production for commercial usage, protein engineering has developed into a powerful technique to improve enzyme properties. By enhancing process conditions and using protein engineering, under industrial conditions, it is possible to produce enzymes with the required characteristics, such as increased activity, high thermo-stability, and specificity (Singh et al. 2013).

Using molecular biotechniques, keratinase genes have been successfully cloned and produced via heterologous way from a variety of sources, but the outcome is still unsatisfactory. In industrial applications and commercial markets, keratinases must be more productive, highly effective, and thermostable than they now are (Li et al. 2013; Su et al. 2020). In recent decades, protein engineering has grown astronomically, and methodologies are quickly maturing, offering tools to change protein structure, resulting in novel properties and increased productivity (Su et al. 2020; Bornscheuer et al. 2012). A promising method for creating extraordinary keratinase variations with improved thermo-stability and activity for the breakdown of feather debris is protein engineering (Akram et al. 2022)

11.6 Applications

Keratin is regarded as a good source of sulfur, nitrogen, and carbon, which can be changed in nature into a number of other chemicals. Keratinase enzymes offer a wide range of commercial and biotechnological applications because of their ability to degrade keratins. Numerous applications of keratinase enzyme are presented in Fig. 11.5.

11.6.1 Keratinase as Nutrition, Food Technology, and Livestock Feed

The minerals copper, calcium, phosphorus, potassium, zinc, magnesium, iron, manganese, and nitrogen as well as peptides and critical amino acids are thought to be abundant in chicken feathers (Kshetri et al. 2018). Due to the presence of enormous nutritional value, processed feather meal by microbial keratinase is more viable and simple to produce (Gurav et al. 2020).

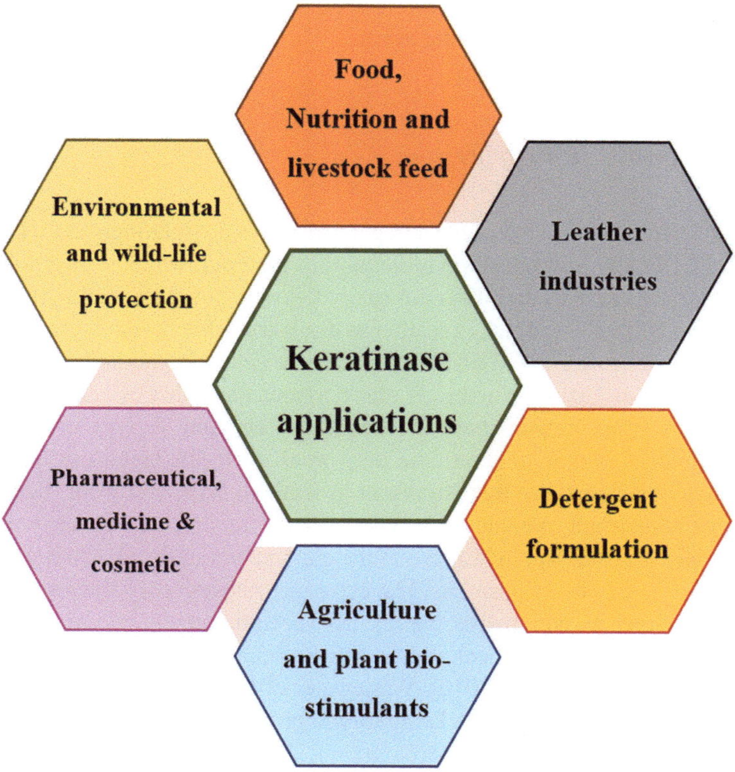

Fig. 11.5 Numerous applications of keratinase enzyme

Because there is no enzyme to breaking down the disulfide bonds in the protein in traditionally processed animal feeds of keratin origin, the metabolism is improper (Wang et al. 2006; Qiu et al. 2020). Therefore, the protein digestibility must be improved for the keratin-based feed-products so as to enhance the nutritive value and nutrient availability (Nnolim et al. 2020a, b). By using a submerged or solid state fermentative method, keratin solubilization by keratinase aids in enhancing the nutritional characteristics of keratin-based feed products (Łaba and Szczekala 2013). Keratinase-processed aquafeed can substitute cattle keratin for a portion of other protein sources (Jumini 2017). It is a wonderful substitute protein source that causes the body to experience less stress (Zhang et al. 2014) and increase the digestibility of feather keratin (Grazziotin et al. 2006).

Purified keratinase extracted from chicken feather waste was used in a Kshetri study that demonstrated its promise as a source of antioxidant and antityrosinase chemicals (Kshetri et al. 2020). Release of glucocorticoids, which the biological stress response is mediated by, has the potential to suppress immunity, growth, and reproduction when induced repeatedly. Corticosterone is a major glucocorticoid in birds and its amounts may be extracted from feathers and monitored (Alba et al. 2019). Study using chicken feathers as a substrate and biochemical and biophysical characterization of keratinase for greater comprehension and implications in industrial applications was published by Jana et al. (2022) to provide a new path for keratinase research. They reported that *B. cereus* IIPK35 keratinase displayed promising application potential as milk clotting agent.

11.6.2 Leather Industries

Chicken feather is a non-disposable and persistently accumulating waste (Brandelli et al. 2010). Due to its composition of keratinous material, which is primarily made up of keratin, chicken feather does not easily degrade in nature (Călin et al. 2017; Moran et al. 1966). Using chicken feather as a substrate, the keratinolytic bacterium (*Bacillus paralicheniformis* MKU3) is found to be generated keratinase, a protease (Kalaikumari et al. 2019). Numerous microorganisms such as Eucarya, Bacteria, and Archea domains manufacture the keratinase enzyme. A type of proteolytic enzyme known as microbial keratinase may break down the keratin structural protein, an insoluble substance that is present in feathers, hair, and wool. Keratin is a protein that is resistant to being broken down by normal proteolytic enzymes such as trypsin, pepsin, and papain because of the amount and molecular structure of the amino acids present in this protein. The ecosystem is seriously threatened by the enormous number of keratinase wastes that accumulate in nature (Gupta and Ramnani 2006). The biochemical and biophysical characteristics of keratinases are quite diverse (Brandelli et al. 2010). Wastewater effluent from the leather industry could be biologically treated with keratinolytic proteases to significantly reduce its volume and toxicity (Wang et al. 2016a, b).

The raw hide must first undergo a number of chemical processes before it can be turned into the lovely leather. This procedure involves bating, degreasing, deliming, dehairing, liming, soaking, and pickling (Kamini et al. 1999). Lime and sodium sulfide, which are both used to effectively dehair skin, are the two main leather industrial chemicals. Additionally, when skin is chemically treated, it has been claimed that dehairing is not completed and that this damages the skin and lowers the quality of the leather. The sheep skin is successfully dehaired by the extracellular keratinase (Briki et al. 2016; Kim et al. 2001). Study reported by Akhter et al. (2020) proved that the leather pieces were immersed in enzymes and chemicals (Na_2S, CaO) separately for a total of 16 h to demonstrate the superiority of keratinase enzyme over chemicals in the dehairing process. After 12 h, 100% enzyme dehairing proved to be more efficient than chemicals without damaging the leather. Na_2S damaged leather after 16 h, whereas CaO's dehairing was still not complete. Additionally, when chemical treatment was applied, the hair pulp had a harsh texture. Few investigations on the enzymatic leather process and quality evaluations have been published, despite the fact that many keratinases-producing strains have been documented in numerous research articles (Thanikaivelan et al. 2004).

The quality of the leather is greatly affected by the process of eliminating unfavorable proteins from animal hide or skin, which is known as "bating." Enzymatic treatment is necessary in the procedure, which has been carried out using a variety of techniques throughout history, to clean, smoothen, and fine-tune the grain surface of the final leather (Hameed et al. 1996). In several stages of leather processing, keratinases are used in many processes such as soaking, dehairing, and bating (Jaouadi et al. 2015).

11.6.3 Detergent Formulation

Drains and clothes stained with keratinous waste have been cleaned using keratinases as a detergent (Brandelli 2008). The use of *Aspergillus* sp. DHE7 keratinase in detergent formulations (as an additive) and in a number of biotechnological applications are proposed as potential options (El-Ghonemy and Ali 2021). Keratinases can clean quickly without compromising the fiber's strength and structure because of their substrate specificity (Paul et al. 2014). Keratinase when combined with detergents expel blood, turmeric strains, egg yolk, fruit juice, and chocolate stains from the cloth (Cavello et al. 2012; Manivasagan et al. 2014; Reddy et al. 2017). Detergent formulations can utilize keratinase in a variety of ways, such as adding it to clear keratinase waste-clogged drains and improving the wash performance of enzyme-based laundry detergents (Rai et al. 2009). The ideal detergent protease should be compatible with the constituents of the detergent, have good activity at the right pH and washing temperature, and be stable with oxidants and bleach (Rai et al. 2010; Manni et al. 2010). Except for a few scant examples, there are not many studies on detergent and oxidant-stable microbial alkaline keratinase that can function at room temperature (23 °C).

Presently, keratinase accounts for 90% of the total market sale in the detergent industry because of its stability toward a variety of surfactants, bleaches, and other additives in the detergent, thereby improving its washing performance and increased efficiency in stain removal (Paul et al. 2014). El-Ghonemy and Ali (2021) purified and characterized a thermo-surfactant stable extracellular keratinolytic enzyme produced by the recently discovered terrestrial fungus *Aspergillus* sp. DHE 7 (accession no. KX950801), which was grown in an ideal medium with submerged fermentation utilizing turkey feather as a keratinous substrate. Additionally, the sustainability and compatibility of enzymes with commercial laundry detergents had been researched. Drains and clothes stained with keratinous waste have been cleaned using keratinases as a detergent (Brandelli 2008). By removing blood stains from surgical cloths, the alkaline keratinase of *P. woosongensis* TKB2 demonstrated its use in the laundry industry. It also achieved excellent results in the removal of the composite stain of blood, egg yolk, and chocolate in a short period of time without altering texture, strength (27.18 g/tex), or fibers of the cloth. With the use of this detergent formula, there is no longer any impact on aquatic ecology from the effluents generated after material washing. Thus, by using this environmentally friendly detergent formulation, pollution issues related to the effluent generated after washing clothes can be avoided (Paul et al. 2014).

11.6.4 Agriculture and Plant Biostimulants

Due to its enormous nutritional value, processed feather meal by microbial keratinase is more viable and simple to produce (Gurav and Jadhav 2013). Additional research has revealed that it significantly improves plant development properties, thus it may be confidently marketed as an affordable and effective organic fertilizer (Adetunji and Adejumo 2018). By taking this activity, keratinous wastes will be converted into inorganic and sulfur compounds that plants can use right away (Nayaka and Vidyasagar 2013).

Huge amounts of keratinous wastes are produced annually, particularly by the poultry, textile, and leather industries. Additional research has revealed that it significantly improves plant growth properties, and it can be safely regarded as an affordable, effective, and promising organic fertilizer (Adetunji and Adejumo 2018). In order to biodegrade chicken feathers into a digestible and nutritionally beneficial feather lysate that is rich in free amino acids, peptides, and ammonium ions, keratinolytic bacteria are used.

In addition to providing plants with nitrogenous fertilizers, feather lysate can also be fed to animals as a protein-rich meal (Tamreihao et al. 2019). By doing this, keratinous wastes will be transformed into inorganic and sulfur compounds, which the plants may use right away (Nayaka and Vidyasagar 2013). By using a practical and environmentally beneficial way of recycling, microbial keratinases hasten the composting process of used chicken feathers into nitrogen-rich fertilizer. The comparative study between feather hydrolysate and keratinolytic bacteria shows that the feather hydrolysate, which is a rich source of nitrogen and carbon, can be used to encourage beneficial microbial activity in soil and accelerate plant growth. By producing indole-3-acetic acid, siderophore, solubilizing inorganic P, and other significant growth promotion features, keratinolytic bacteria exhibits biocontrol and plant growth promoting (PGP) capabilities that can shield plants from dangerous pathogens and boost their growth. In addition to promoting human health, environment, organic farming, agroecosystem, and soil biological activity, the employment of keratinolytic bacteria in agriculture increases crop productivity and raises awareness of sustainable agricultural practices.

11.6.5 Pharmaceuticals, Medicine, and Cosmetic Production

In the pharmaceutical sector, keratinases are seen as promising biocatalysts. In the biomedical industry, keratinases and keratinolytic organisms were crucial. The medication administration methods used in the pharmaceutical business are expected to change as a result of this technique (Gupta et al. 2013b). Studies have shown that keratinases have significant therapeutic results when used to treat hyperkeratosis such as in the case of corns, calluses, and psoriasis (Gupta et al. 2017). According to reports, some of these proteases are used in hair-improving products as they enhance hair quality. Similarly, keratinases are efficient in treating acne and pimples, which are the frequent skin conditions brought on by an excessive buildup of keratin in the sebaceous gland (Spyros 2003). It can be utilized as a supporting component to encourage the efficient adsorption of cosmetics and medications (Mohorčič et al. 2007). According to certain reports, keratinase helps medications penetrate the nail more deeply, enhancing their bioavailability (Tiwary and Gupta 2010). The effect of keratinase on skin and nails shows that keratins make it easier to remove dead skin and nail cells from surfaces. It has been applied topically to treat a number of skin conditions, including hyperkeratosis (a thickening of the skin brought on by a buildup of too much keratin) and acne, by removing extra keratin. There are now commercially available keratinase-infused face scrubs, anti-dandruff shampoos, and other personal care items for hair removal and glowing skin (Gupta et al. 2013b). It has many more applications in modern medicine. In the area of contemporary medicine, keratinases have a lot more applications. According to reports, keratinases have the ability to deactivate and disassemble misfolded prion proteins. This could be viewed as very encouragement for future treatments (Langeveld et al. 2003; Narayan and Dutta 2005). They can also exfoliate dead skin while substituting traditional treatments (Gupta and Ramnani 2006). These bum-forming cells in skin can also be unblocked using keratinase (Selvam and Vishnupriya 2012). When injected, keratinase can increase medicine delivery and break down viral proteins. It has been demonstrated to improve drug penetration through the nail plate, a very strong pharmacological barrier (Mohorčič et al. 2007). Due to their capacity to destroy prions, they are also helpful in sterilizing equipment and instruments (infectious protein) (Liang et al. 2010).

Keratinases have been used in topical hair removal products as an active ingredient in the cosmetic industry (Adelere and Lateef 2016). The enzyme has the power to significantly harden the hyperkeratosis (dead skin layer) present in appendages. Therefore, it might be a viable substitute for the salicylic acid that is now being used (Gupta and Ramnani 2006). Hydrolysates, which are used to manufacture hair and skincare products, have been made from a variety of protein sources, including wheat protein, wool keratin, and collagen (Kshetri et al. 2020). They typically provide improved sensation, hydration, and protect the skin's natural integrity (Barba et al. 2008; Villa et al. 2013). Acne is a common skin condition that develops when too much keratin blocks the sebaceous gland (Selvam and Vishnupriya 2012). The development of anaerobic bacteria in clogged sebaceous glands causes acne. The sebaceous gland opens and functions normally when the keratinase enzymes are applied. As long as the pores are treated and kept enzymatically open, the gland's internal environment will become aerobic, and the acne-causing bacteria would not be able to grow new colonies. In 2001, a keratinase-based product (enzymatic powder or tablet) that can be a successful adjuvant in the treatment of acne was patented (Spyros 2003). Both Keratopeel PB (Proteo Biotech) and Keratoclean Sensitive PB are commercially marketed products for delicate enzymatic peeling. Similarly, keratinases have uses in the treatment of scars and the regeneration of epithelium (Gupta et al. 2013b).

Less molecular weight peptides are produced during keratinase-mediated keratin biodegradation (Stiborova et al. 2016). These peptides have been used more frequently in the creation of cosmetic products such as moisturizers and conditioners since they tend to penetrate the hair or nail cuticles in comparison to hydrolysates from other sources (Villa et al. 2013). The biological formulation's hair-removal action was replaced by purified keratinase enzyme rather than thioglycolate (Sanghvi et al. 2016). Keratinase from *B. subtilis* has been shown to be effective at hair removing. It is intriguing that the cream containing keratinase removed hair more effectively than the chemical-based cream (Nnolim et al. 2020a, b).

11.6.6 Environmental and Wildlife Protection

Keratin biomass is generated in enormous quantities from the tannery, poultry, textile, wool industries, and so on, which are disposed into environment, causing pollution. Further, worldwide meat producing sectors grow vastly and they generate enormous offal, manure, and feather waste and pose a significant impact on environment, particularly on water, land, nitrogen oxide, carbon dioxide, soil erosion, eutrophication of water reservoirs, and so on (Godfray et al. 2018). Feathers from poultry accumulate more to the ecosystem and also it causes local disturbance such as enormous flies, rodents, bad odor, contamination, and eutrophication problems (Gerber et al. 2007; Sharma and Gupta 2016). The microbes *Vibrio* and *Salmonella* found in feather waste may cause the atmosphere's concentrations of ammonia, nitrous oxide, and hydrogen sulfide to increase (Tesfaye et al. 2017). Processing, such as burning, increases the greenhouse impact (Cheong et al. 2018), and controlled landfilling generates significant amounts of trash from organic materials (Tesfaye et al. 2017). The productivity, energy use, effluent generation, and waste management benefits of enzymatic hydrolysis of keratin may be greater. Keratinases increase the value of poultry and industrial waste, resulting in more sustainable economic and environmental practices (Srivastava et al. 2020). Peptides, vital amino acids, and minerals including calcium, potassium, phosphorus, and nitrogen are all present in feathers.

11.6.7 *Textile Industry*

Currently, the main techniques for limiting wool felting shrinkage are chlorine-Hercosett and enzyme treatments. Since decades, the chlorine-Hercosett technique has been utilized as a successful commercial shrink resist finishing. One of the main issues with this procedure is the issue of environmental degradation brought on by the disposal of absorbable organic chlorides. The creation of eco-friendly and regulated wool shrink-resist finishing techniques is urgently required. In the textile industry, keratinase from *Stenotrophomonas maltophilia*, a bacterial strain that produces keratinase, has been hailed as a potentially useful tool for removing wool scale layers and as a safe, efficient substitute for the traditional chemical technique. After being treated with enzymes, felting shrinkage was calculated and then the leather materials had been dried and dewatered. Fabric tensile strength was assessed using a material testing machine. The morphology of the treated and untreated wool fabrics was investigated using a scanning electron microscope. Using a scanning electron microscope, the morphology of the treated and untreated wool fabrics was examined (Cai et al. 2011; Zhang et al. 2016).

The elimination of bound fatty acids from the wool surface by keratinase treatment increased the hydrophilicity of wool fibers. The hydrophilicity of wool fibers is significantly recovered by this treatment. It was found that hydrophilicity decreases with increasing keratinase concentration (Gunes et al. 2018). In the textile industry's wool processing, the existence of wool scale was considered to be the primary factor, affecting directional friction, wool fiber shrinkage, and dyeing efficiency (Shen et al. 2007). It was anticipated that the scale will be removed to address these drawbacks. Dichlorodicyanuric acid is traditionally used to chlorinate wool fabrics, although this process has the potential to pollute the environment (Wang et al. 2009). Keratinase has drawn a lot of attention as a potential replacement for the standard chemical technique in the leather industry because of the growing concern about environmental protection (Gradisar et al. 2000) (Table 11.2).

Table 11.2 A brief summary of keratinase applications in various industries

S. no.	Source	Application	Field	References
1	*B. cereus* IIPK35	Milk clotting agent	Dairy industry	Almahasheer et al. (2022)
2	*Bacillus licheniformis*	Treatment of chicken feathers	Poultry industry	Babbar et al. (2022)
3	*Trichoderma harzianum* isolate HZN12	Detergent formulation	Leather and detergent industries	Bagewadi et al. (2018)
4	*Chryseobacterium* sp. L99	Biotreatment of wool fibers	Textile industry	Liang et al. (2010)
5	*Aspergillus* sp. DHE7	Detergent formulations	Detergent industry	El-Ghonemy and Ali (2021)
6	*Bacillus* sp. RCM-SSR-102	Antioxidant and antityrosinase compounds	Pharmaceutical industry	Kshetri et al. (2020)
7	*Bacillus* sp. Okoh-K1	Used as reducing agents, organic solvents, surfactants, and laundry detergents	Biotechnology and detergent industry	Nnolim et al. (2020a, b)

11.7 Recommendation and Futuristic Prospects

Keratin, one of the most prevalent hard substances in soil, is challenging to break down naturally. However, turning these chemicals into useful end products is simpler and less expensive when done by microbial and fungal breakdown (Gopinath et al. 2015). To increase keratinase synthesis, numerous microorganisms have been suggested and used in a range of industrial applications. Additionally, new keratinase research has made additional advancements (Daroit and Brandelli 2014; Gupta and Singh 2014). However, a very sensitive technique for keratinase testing is not yet available. Additionally, recombinant keratinase chimeras need to be enhanced in order to produce efficient keratinase. The manufacturing and detection of keratinase will be more quickly applied to industries and environmental waste management as more effective technologies are developed. Therefore, tremendous efforts are required to enhance the current production strategies and create a completely new, efficient way with few limitations. There is reason to anticipate that enhanced keratinase will soon be able to resolve the nutrition, food technology, livestock feed, leather industries, detergent formulation agriculture, plant biostimulants, pharmaceuticals, medicine, cosmetic, environmental, and wildlife protection applications, given the significant advancements already made and the increased interest and input in the field. Soon, further research and development in this area are needed to provide enough advancement in the valorization of food sector waste.

11.8 Conclusion

Keratinase enzyme is well-known industrial important enzyme that has a wide range of applications in different fields as green catalyst. This enzyme has countless opportunities to thank with numerous potential applications including nutrition, food technology, livestock feed, leather industries, detergent formulation agriculture, plant biostimulants, pharmaceuticals, medicine, cosmetic, environmental, and wildlife protection. It can be synthesized from microbial degradation practices using the waste chicken feathers via environmentally friendly method for managing poultry feather waste. In this context, a detailed purview has been provided in this chapter with respect to keratinase enzyme, derived from waste resources, and recent developments in keratinase enzyme production, with its biochemical and functional properties have been also covered.

References

Adelere IA, Lateef A (2016) Keratinases: emerging trends in production and applications as novel multifunctional biocatalysts. Kuwait J Sci 43(3):118–127

Adetunji CO, Adejumo IO (2018) Efficacy of crude and immobilizedenzymes from *Bacillus licheniformis* for production of biodegraded feather meal and their assessment on chickens. Environ Technol Innov 11:116–124

Adibi A et al (2021) Characterization and isolation of peptide metabolites of an antifungal bacterial isolate identified as *Bacillus amyloliquefaciens* subspecies plantarum strain FZB42. J Microbiol Biotechnol Food Sci 2021:1309–1313

Akhter M et al (2020) Microbial bioremediation of feather waste for keratinase production: an outstanding solution for leather dehairing in tanneries. Microbiol Insights 13:1178636120913280

Akram F, ul Haq I, Jabbar Z (2020) Production and characterization of a novel thermo-and detergent stable keratinase from *Bacillus* sp. NKSP-7 with perceptible applications in leather processing and laundry industries. Int J Biol Macromol 164:371–383

Akram F et al (2022) Multifarious revolutionary aspects of microbial keratinases: an efficient green technology for future generation with prospective applications. Environ Sci Pollut Res 29:1–20

Alba AC et al (2019) Using a keratinase to degrade chicken feathers for improved extraction of glucocorticoids. Gen Comp Endocrinol 270:35–40

Almahasheer AA et al (2022) Novel feather degrading keratinases from *Bacillus cereus* group: biochemical, genetic and bioinformatics analysis. Microorganisms 10(1):93

Babbar N, Sharma G, Arya SK (2022) Effective degradation of chicken feather waste by keratinase enzyme with triton X-100 additive. Biocatal Agric Biotechnol 44:102447

Bagewadi ZK, Mulla SI, Ninnekar HZ (2018) Response surface methodology based optimization of keratinase production from *Trichoderma harzianum* isolate HZN12 using chicken feather waste and its application in dehairing of hide. J Environ Chem Eng 6(4):4828–4839

Barba C et al (2008) Cosmetic effectiveness of topically applied hydrolysed keratin peptides and lipids derived from wool. Skin Res Technol 14(2):243–248

Bhari R, Kaur M, Sarup Singh R (2021) Chicken feather waste hydrolysate as a superior biofertilizer in agroindustry. Curr Microbiol 78(6):2212–2230

Bohacz J (2017) Biodegradation of feather waste keratin by a keratinolytic soil fungus of the genus Chrysosporium and statistical optimization of feather mass loss. World J Microbiol Biotechnol 33:1–16

Bokveld A, Nnolim NE, Nwodo UU (2021) *Chryseobacterium aquifrigidense* FANN1 produced detergent-stable metallokeratinase and amino acids through the abasement of chicken feathers. Front Bioeng Biotechnol 9:718

Bornscheuer UT et al (2012) Engineering the third wave of biocatalysis. Nature 485(7397):185–194

Bouacem K et al (2016) Novel serine keratinase from *Caldicoprobacter algeriensis* exhibiting outstanding hide dehairing abilities. Int J Biol Macromol 86:321–328

Brandelli A (2008) Bacterial keratinases: useful enzymes for bioprocessing agroindustrial wastes and beyond. Food Bioprocess Technol 1:105–116

Brandelli A, Daroit DJ, Riffel A (2010) Biochemical features of microbial keratinases and their production and applications. Appl Microbiol Biotechnol 85:1735–1750

Briki S, Hamdi O, Landoulsi A (2016) Enzymatic dehairing of goat skins using alkaline protease from *Bacillus* sp. SB12. Protein Expr Purif 121:9–16

Cai S-B et al (2011) Identification of a keratinase-producing bacterial strain and enzymatic study for its improvement on shrink resistance and tensile strength of wool-and polyester-blended fabric. Appl Biochem Biotechnol 163:112–126

Călin M et al (2017) Degradation of keratin substrates by keratinolytic fungi. Electron J Biotechnol 28:101–112

Cavello IA, Hours RA, Cavalitto SF (2012) Bioprocessing of "hair waste" by *Paecilomyces lilacinus* as a source of a bleach-stable, alkaline, and thermostable keratinase with potential application as a laundry detergent additive: characterization and wash performance analysis. Biotechnol Res Int 2012:369308

Cheong CW et al (2018) Chicken feather valorization by thermal alkaline pretreatment followed by enzymatic hydrolysis for protein-rich hydrolysate production. Waste Manag 79:658–666

Daroit DJ, Brandelli A (2014) A current assessment on the production of bacterial keratinases. Crit Rev Biotechnol 34(4):372–384

El-Ghonemy DH, Ali TH (2021) Effective bioconversion of feather-waste keratin by thermo-surfactant stable alkaline keratinase produced from *Aspergillus* sp. DHE7 with promising biotechnological application in detergent formulations. Biocatal Agric Biotechnol 35:102052

Elhoul MB et al (2016) Biochemical and molecular characterization of new keratinoytic protease from *Actinomadura viridilutea* DZ50. Int J Biol Macromol 92:299–315

Farag AM, Hassan MA (2004) Purification, characterization and immobilization of a keratinase from *Aspergillus oryzae*. Enzym Microb Technol 34(2):85–93

Filipello Marchisio V, Fusconi A, Querio FL (2000) Scopulariopsis brevicaulis: a keratinophilic or a keratinolytic fungus? Mycoses 43(7–8):281–292

Gafar A et al (2020) Response surface methodology for the optimization of keratinase production in culture medium containing feathers by *Bacillus* sp. UPM-AAG1. Catalysts (2073–4344) 10(8):848

Gegeckas A et al (2015) Keratinous waste decomposition and peptide production by keratinase from *Geobacillus stearothermophilus* AD-11. Int J Biol Macromol 75:158–165

Gerber P, Opio C, Steinfeld H (2007) Poultry production and the environment—a review. Animal Production and Health Division, Food and Agriculture Organization of the United Nations, Viale delle Terme di Caracalla 153:1–27

Ghaffar I et al (2018) Microbial production and industrial applications of keratinases: an overview. Int Microbiol 21:163–174

Goda DA et al (2021) Feather protein lysate optimization and feather meal formation using YNDH protease with keratinolytic activity afterward enzyme partial purification and characterization. Sci Rep 11(1):14543

Godfray HCJ et al (2018) Meat consumption, health, and the environment. Science 361(6399):eaam5324

Gopinath SCB et al (2015) Biotechnological aspects and perspective of microbial keratinase production. Biomed Res Int 2015:140726

Gradisar H, Kern S, Friedrich J (2000) Keratinase of Doratomyces microsporus. Appl Microbiol Biotechnol 53(2):196–200

Grazziotin A et al (2006) Nutritional improvement of feather protein by treatment with microbial keratinase. Anim Feed Sci Technol 126(1–2):135–144

Gunes GB et al (2018) Microbial keratinase production and application to improve the properties of wool fabrics. Int J Text Sci 7(2):43–47

Gupta R, Ramnani P (2006) Microbial keratinases and their prospective applications: an overview. Appl Microbiol Biotechnol 70:21–33

Gupta S, Singh R (2014) Hydrolyzing proficiency of keratinases in feather degradation. Indian J Microbiol 54:466–470

Gupta R, Sharma R, Beg QK (2013a) Revisiting microbial keratinases: next generation proteases for sustainable biotechnology. Crit Rev Biotechnol 33(2):216–228

Gupta R et al (2013b) Biotechnological applications and prospective market of microbial keratinases. Appl Microbiol Biotechnol 97:9931–9940

Gupta S et al (2017) Molecular modeling of cloned *Bacillus subtilis* keratinase and its insinuation in psoriasis treatment using docking studies. Indian J Microbiol 57:485–491

Gupta S et al (2023) Chicken feather waste hydrolysate as a potential biofertilizer for environmental sustainability in organic agriculture management. Waste Biomass Valorization 14:1–17

Gurav RG, Jadhav JP (2013) A novel source of biofertilizer from feather biomass for banana cultivation. Environ Sci Pollut Res 20:4532–4539

Gurav R et al (2020) Microbial degradation of poultry feather biomass in a constructed bioreactor and application of hydrolysate as bioenhancer to vegetable crops. Environ Sci Pollut Res 27:2027–2035

Habbeche A et al (2014) Purification and biochemical characterization of a detergent-stable keratinase from a newly thermophilic actinomycete *Actinomadura keratinilytica* strain Cpt29 isolated from poultry compost. J Biosci Bioeng 117(4):413–421

Hameed A, Natt MA, Evans CS (1996) Production of alkaline protease by a new *Bacillus subtilis* isolate for use as a bating enzyme in leather treatment. World J Microbiol Biotechnol 12:289–291

Hamiche S et al (2019) Purification and biochemical characterization of two keratinases from *Bacillus amyloliquefaciens* S13 isolated from marine brown alga *Zonaria tournefortii* with potential keratin-biodegradation and hide-unhairing activities. Int J Biol Macromol 122:758–769

Hassan MA, Abol-Fotouh D et al (2020a) Comprehensive insights into microbial keratinases and their implication in various biotechnological and industrial sectors: a review. Int J Biol Macromol 154:567–583

Hassan MA, Taha TH et al (2020b) Biochemical characterisation and application of keratinase from *Bacillus thuringiensis* MT1 to enable valorisation of hair wastes through biosynthesis of vitamin B-complex. Int J Biol Macromol 153:561–572

Huang Y et al (2015) Genome and secretome analyses provide insights into keratin decomposition by novel proteases from the non-pathogenic fungus *Onygena corvina*. Appl Microbiol Biotechnol 99:9635–9649

Jagadeesan Y et al (2020) Sustainable production, biochemical and molecular characterization of thermo-and-solvent stable alkaline serine keratinase from novel *Bacillus pumilus* AR57 for promising poultry solid waste management. Int J Biol Macromol 163:135–146

Jana A et al (2022) Efficient valorization of feather waste by *Bacillus cereus* IIPK35 for concomitant production of antioxidant keratin hydrolysate and milk-clotting metallo-serine keratinase. J Environ Manag 324:116380

Jansen MLA, van Gulik WM (2014) Towards large scale fermentative production of succinic acid. Curr Opin Biotechnol 30:190–197

Jaouadi NZ et al (2015) A novel keratinase from *Bacillus tequilensis* strain Q7 with promising potential for the leather bating process. Int J Biol Macromol 79:952–964

Jeong J-H et al (2010) Keratinolytic enzyme-mediated biodegradation of recalcitrant feather by a newly isolated *Xanthomonas* sp. P5. Polym Degrad Stab 95(10):1969–1977

Jin H-S et al (2018) Identification of matrix metalloproteinase-1-suppressive peptides in feather keratin hydrolysate. J Agric Food Chem 66(48):12719–12729

Joshi S, Satyanarayana T (2013) Biotechnology of cold-active proteases. Biology 2(2):755–783

Jumini S (2017) Alternative fish feed production from waste chicken feathers. Int J Sci Appl Sci Conf Ser 1:144–152

Kalaikumari SS et al (2019) Bioutilization of poultry feather for keratinase production and its application in leather industry. J Clean Prod 208:44–53

Kamini NR et al (1999) Microbial enzyme technology as an alternative to conventional chemicals in leather industry. Curr Sci 77:80–86

Kerouaz B et al (2021) Purification and biochemical characterization of two novel extracellular keratinases with feather-degradation and hide-dehairing potential. Process Biochem 106:137–148

Keshapaga UR et al (2023) Characterization of high-yield *Bacillus subtilis* cysteine protease for diverse industrial applications. Braz J Microbiol 54(2):739–752

Khardenavis AA, Kapley A, Purohit HJ (2009) Processing of poultry feathers by alkaline keratin hydrolyzing enzyme from *Serratia* sp. HPC 1383. Waste Manag 29(4):1409–1415

Kim JM, Lim WJ, Suh HJ (2001) Feather-degrading *Bacillus* species from poultry waste. Process Biochem 37(3):287–291

Kshetri P et al (2018) Feather degrading, phytostimulating, and biocontrol potential of native actinobacteria from North Eastern Indian Himalayan Region. J Basic Microbiol 58(9):730–738

Kshetri P et al (2020) Valorization of chicken feather waste into bioactive keratin hydrolysate by a newly purified keratinase from *Bacillus* sp. RCM-SSR-102. J Environ Manag 273:111195

Łaba W, Szczekala KB (2013) Keratinolytic proteases in biodegradation of pretreated feathers. Pol J Environ Stud 22(4):1101–1109

Laba W et al (2015) Keratinolytic abilities of *Micrococcus luteus* from poultry waste. Braz J Microbiol 46:691–700

Lange L, Huang Y, Busk PK (2016) Microbial decomposition of keratin in nature—a new hypothesis of industrial relevance. Appl Microbiol Biotechnol 100(5):2083–2096

Langeveld JPM et al (2003) Enzymatic degradation of prion protein in brain stem from infected cattle and sheep. J Infect Dis 188(11):1782–1789

Lateef A, Adelere IA, Gueguim-Kana EB (2015) *Bacillus safensis* LAU 13: a new source of keratinase and its multi-functional biocatalytic applications. Biotechnol Biotechnol Equip 29(1):54–63

Li Q (2019) Progress in microbial degradation of feather waste. Front Microbiol 10:2717

Li Q et al (2013) Commercial proteases: present and future. FEBS Lett 587(8):1155–1163

Liang X et al (2010) Enhancement of keratinolytic activity of a thermophilic subtilase by improving its autolysis resistance and thermostability under reducing conditions. Appl Microbiol Biotechnol 87:999–1006

Liya SM et al (2023) Optimized production of keratinolytic proteases from *Bacillus tropicus* LS27 and its application as a sustainable alternative for dehairing, destaining and metal recovery. Environ Res 221:115283

Manivasagan P et al (2014) Production, biochemical characterization and detergents application of keratinase from the marine *Actinobacterium actinoalloteichus* sp. MA-32. J Surfactant Deterg 17:669–682

Manni L et al (2010) An oxidant-and solvent-stable protease produced by *Bacillus cereus* SV1: application in the deproteinization of shrimp wastes and as a laundry detergent additive. Appl Biochem Biotechnol 160:2308–2321

Mazotto AM, Couri S, Mônica CT, Vermelho AB (2013) Degradation of feather waste by *Aspergillus niger* keratinases: comparison of submerged and solid-state fermentation. Int Biodeterior Biodegrad 85:189–195

Mienda BS, Idi A, Umar A (2011) Microbiological features of solid state fermentation and its applications—an overview. Res Biotechnol 2(6):21–26

Mohorčič M et al (2007) An investigation into keratinolytic enzymes to enhance ungual drug delivery. Int J Pharm 332(1–2):196–201

Moran ET Jr, Summers JD, Slinger SJ (1966) Keratin as a source of protein for the growing chick: 1. Amino acid imbalance as the cause for inferior performance of feather meal and the implication of disulfide bonding in raw feathers as the reason for poor digestibility. Poult Sci 45(6):1257–1266

Narayan SK, Dutta JK (2005) Creutzfeldt–Jakob disease. JAPI 53:791–795

Nayaka S, Vidyasagar GM (2013) Development of eco-friendly bio-fertilizer using feather compost. Ann Plant Sci 2(7):238–244

Nnolim NE et al (2020a) Exoproduction and characterization of a detergent-stable alkaline keratinase from *Arthrobacter* sp. KFS-1. Biochimie 177:53–62

Nnolim NE et al (2020b) Microbial keratinase: next generation green catalyst and prospective applications. Front Microbiol 11(December). https://doi.org/10.3389/fmicb.2020.580164

Paul T et al (2014) An efficient cloth cleaning properties of a crude keratinase combined with detergent: towards industrial viewpoint. J Clean Prod 66:672–684

Pawar VA et al (2018) Molecular and biochemical characterization of a thermostable keratinase from *Bacillus altitudinis* RBDV1. 3 Biotech 8:1–7

Perţa-Crişan S et al (2021) Closing the loop with keratin-rich fibrous materials. Polymers 13(11):1896

Qiu J et al (2020) Microbial enzymes catalyzing keratin degradation: classification, structure, function. Biotechnol Adv 44:107607

Qiu J et al (2022) Bioinformatics based discovery of new keratinases in protease family M36. New Biotechnol 68:19–27

Rai SK, Konwarh R, Mukherjee AK (2009) Purification, characterization and biotechnological application of an alkaline β-keratinase produced by *Bacillus subtilis* RM-01 in solid-state fermentation using chicken-feather as substrate. Biochem Eng J 45(3):218–225

Rai SK, Roy JK, Mukherjee AK (2010) Characterisation of a detergent-stable alkaline protease from a novel thermophilic strain *Paenibacillus tezpurensis* sp. nov. AS-S24-II. Appl Microbiol Biotechnol 85:1437–1450

Rajput R, Sharma R, Gupta R (2010) Biochemical characterization of a thiol-activated, oxidation stable keratinase from *Bacillus pumilus* KS12. Enzyme Res 2010:132148

Reddy N, Yang Y (2007) Structure and properties of chicken feather barbs as natural protein fibers. J Polym Environ 15:81–87

Reddy MR et al (2017) Effective feather degradation and keratinase production by *Bacillus pumilus* GRK for its application as bio-detergent additive. Bioresour Technol 243:254–263

Rodziewicz A, Łaba W (2006) Keratyny i ich biodegradacja. Biotechnologia 2(73):130–147

Sanghvi G et al (2016) A novel alkaline keratinase from *Bacillus subtilis* DP1 with potential utility in cosmetic formulation. Int J Biol Macromol 87:256–262

Selvam K, Vishnupriya B (2012) Biochemical and molecular characterization of microbial keratinase and its remarkable applications. Int J Pharm Biol Arch 3(2):267–275

Sharma S, Gupta A (2016) Sustainable management of keratin waste biomass: applications and future perspectives. Braz Arch Biol Technol 59:1–14

Sharma I et al (2022) Parametrically optimized feather degradation by *Bacillus velezensis* NCIM 5802 and delineation of keratin hydrolysis by multi-scale analysis for poultry waste management. Sci Rep 12(1):17118

Shen J et al (2007) Development and industrialisation of enzymatic shrink-resist process based on modified proteases for wool machine washability. Enzym Microb Technol 40(7):1656–1661

Singh RK et al (2013) From protein engineering to immobilization: promising strategies for the upgrade of industrial enzymes. Int J Mol Sci 14(1):1232–1277

Spyros T (2003) Use of dual compartment mixing container for enzyme mixtures useful to treat acne. Patent: US6627192

Srivastava B et al (2020) Microbial keratinases: an overview of biochemical characterization and its eco-friendly approach for industrial applications. J Clean Prod 252:119847

Stiborova H et al (2016) Transformation of raw feather waste into digestible peptides and amino acids. J Chem Technol Biotechnol 91(6):1629–1637

Su C et al (2020) The tale of a versatile enzyme: molecular insights into keratinase for its industrial dissemination. Biotechnol Adv 45:107655. https://doi.org/10.1016/j.biotechadv.2020.107655

Tamreihao K et al (2019) Feather degradation by keratinolytic bacteria and biofertilizing potential for sustainable agricultural production. J Basic Microbiol 59(1):4–13

Tesfaye T, Sithole B, Ramjugernath D (2017) Valorisation of chicken feathers: a review on recycling and recovery route—current status and future prospects. Clean Techn Environ Policy 19:2363–2378

Thanikaivelan P et al (2004) Progress and recent trends in biotechnological methods for leather processing. Trends Biotechnol 22(4):181–188

Tiwary E, Gupta R (2010) Subtilisin-γ-glutamyl transpeptidase: a novel combination as ungual enhancer for prospective topical application. J Pharm Sci 99(12):4866–4873

Vanaki P et al (2022) Isolation and identification of keratinolytic probiotic Bucillus licheniformis bacteria from the soil below poultry slaughterhouse waste. Braz J Biol 84:e257473

Villa ALV et al (2013) Feather keratin hydrolysates obtained from microbial keratinases: effect on hair fiber. BMC Biotechnol 13:1–11

Wang JJ, Garlich JD, Shih JCH (2006) Beneficial effects of versazyme, a keratinase feed additive, on body weight, feed conversion, and breast yield of broiler chickens. J Appl Poult Res 15(4):544–550

Wang P et al (2009) Effects of cutinase on the enzymatic shrink-resist finishing of wool fabrics. Enzym Microb Technol 44(5):302–308

Wang B et al (2016a) Keratin: structure, mechanical properties, occurrence in biological organisms, and efforts at bioinspiration. Prog Mater Sci 76:229–318

Wang XC et al (2016b) Improving production of extracellular proteases by random mutagenesis and biochemical characterization of a serine protease in *Bacillus subtilis* S1-4. Genet Mol Res 15(2):1–11

Wang T et al (2021) A thermostable aluminum-tolerant protease produced by feather-degrading *Bacillus thuringiensis* isolated from tea plantation. Protein Pept Lett 28(5):563–572

Yahaya RSR et al (2021) Molecular strategies to increase keratinase production in heterologous expression systems for industrial applications. Appl Microbiol Biotechnol 105:3955–3969

Yong B et al (2020) Recombinant expression and biochemical characterization of a novel keratinase BsKER71 from feather degrading bacterium *Bacillus subtilis* S1-4. AMB Express 10:1–10

Zhang Z et al (2014) Effects of partially replacing dietary soybean meal or cottonseed meal with completely hydrolyzed feather meal (defatted rice bran as the carrier) on production, cytokines, adhesive gut bacteria, and disease resistance in hybrid tilapia (Oreochromis nilot). Fish Shellfish Immunol 41(2):517–525

Zhang R-X et al (2016) Biochemical characterization of a novel surfactant-stable serine keratinase with no collagenase activity from *Brevibacillus parabrevis* CGMCC 10798. Int J Biol Macromol 93:843–851

Zhang R-X et al (2020) Recombinant expression and molecular engineering of the keratinase from *Brevibacillus parabrevis* for dehairing performance. J Biotechnol 320:57–65

Zhang R-X et al (2022) Production of surfactant-stable keratinase from *Bacillus cereus* YQ15 and its application as detergent additive. BMC Biotechnol 22(1):1–13

Zhang Y et al (2023) Molecular design and experimental study of deep eutectic solvent extraction of keratin derived from feathers. Int J Biol Macromol 241:124512

Chapter 12
Valorization of Aquatic Waste Biomass

J. Suresh Kumar and Veerapandi Loganathan

Abstract The affluence of livestock in India is enormously conferred with the rate growing at 6% per annum. The livestock industry, especially the fisheries industries, contributes around 1.07% of GDP to the Indian economy. Environmental pollutions and the economy of the fish processing industry have a direct impact on the utilization of by-products. Besides pollution and hazard aspects, wastes from aquatic products have a potential role in converting into products of higher value. Treated aquatic waste has explored numerous applications among which the most important are animal feed, biodiesel/biogas, dietetic products (chitosan), natural pigments (after extraction), and cosmetics (collagen). Additives utilized in food industries are derived synthetically, which creates an adverse allergic condition and also a negative impact on some consumers. Such failings can be overcome by the additives derived from natural sources such as aquatic by-products and they paves way for the production of additives such as chitin and chitosan, hydroxyapatite, antifreeze proteins, astaxanthin, and enzymes. This chapter summarizes the potential utilization of aquatic wastes as food additives.

Keywords Aquatic processing waste · Biosorbents · Chitin · Chitosan · Fish silage · Squalene

12.1 Introduction

Since it contains a broad spectrum of phytochemical compounds, such as phenol, enzymes, and polymers, ocean debris constitutes a biological substrate with such a wide range of compositional and physiological characteristics. There is an increasing necessity to suggest alternative food sources and alternative treatments due to

J. S. Kumar
Department of Food Technology, Saintgits College of Engineering, Kottayam, India

V. Loganathan (✉)
Nehru Institute of Technology, Coimbatore, Tamil Nadu, India

E. Cherian, B. Gurunathan (eds.), *Value Added Products From Food Waste*,
https://doi.org/10.1007/978-3-031-48143-7_12

the scarcity of terrestrial sources and the fast-growing aging population. Increasing awareness of the correlation among both health and nutrition is growing, increasing demand for innovative fish products that have enhanced nutritional and physiological attributes (al Khawli et al. 2019). Besides, musculoskeletal disorders are still the major leading cause of disability, as reported by the World Health Organization (WHO) (2021), Consequently, specific treatments, surgeries, and components for cell therapy are highly investigated. Despite it is usually underestimated, marine debris could be a better alternative source to solve a variety of social and economic problems (Ahmed et al. 2021). This study concentrates on current knowledge of the value-added possibilities from marine wastes. It also addresses the utilization of marine biomass and the valorization of seafood by-products, whether directly or through the separation of biomaterials, as prospective alternatives that could result in a significantly economically beneficial and eco-friendly use of ocean sources.

It is essential to start processing coastal debris or marine by-products as soon as possible to reduce physiological, biochemical, and microbiological deterioration. Refrigeration, cooling, or hydrolysis with organic substances are employed to retain their natural characteristics (Remme & Austnes, n.d.). By using proper collection, purification, and bioprocess approaches, by-products from waste or other streams are frequently utilized in their present condition (Fig. 12.1).

In addition to being improved for separation and synthesis of concentrates marine oil, beneficial rich protein ingredients and products, pharmacological biomaterials, and fabrics, alternative processes such as digestion with enzymes, ensilaging, acidification, and gelatinization (surimi synthesis from the fish protein content) are also produced (Kim and Mendis 2006; Teixeira et al. 2014). *Bacillus subtilis*, *Lactobacillus bulgaricus*, and *Pichia pastoris* have all been shown synthesize hyaluronic acid (Badri et al. 2019; Oliveira et al. 2020). The complicated regulatory systems and in vivo crosslinking remain significant barriers to the efficient production of biodegradable polymers utilizing microbes. However, utilizing molecular

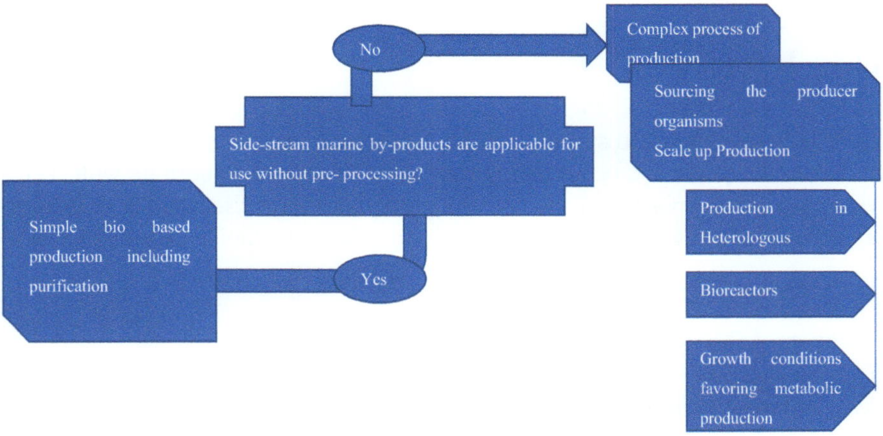

Fig. 12.1 Utilization of marine by-products

markers and transcriptional and/or controlled molecular switches markers, which including CRISPR-Cas tools, used to transcription of endogenous or exogenous genomes has already developed. Various by-products have been researched and developed from the marine wastes, which adds value in the market of waste utilization and is discussed further.

12.2 Biosorbents

Biosorbents are the removal of impurities using materials of biological source (biomass), and most of the aquatic biowastes are efficiently applied as biosorbents in residual water for the removal of radionuclides, inorganic and organic pollutants, and so on (Fig. 12.2). It is an operation generally independent of energy, exploiting dead or waste biomass of low cost. Based on the presence of functional group in biosorbents, the mechanisms of absorption of pollutants gets differ during the interaction with the target pollutants. Considering the aquatic environment, the aquatic pollutants such as metals, metal ions, and dyes are being removed extensively with microalgae, seagrasses, and macroalgae. The polysaccharide composition of algal cell wall has numerous functional groups that play as binding sites for pollutants and metals. Among aquatic biosorbents, brown algae are very efficient on removing the varied ions from the water because of the huge content of alginate and higher uptake capacities. It adds a valuable point that it is unlimited and spread over the vast region of the ocean. Davis et al. and Mazur et al. discussed that marine brown macroalgae remove toxic metals by acting as a natural cation exchanger. Such

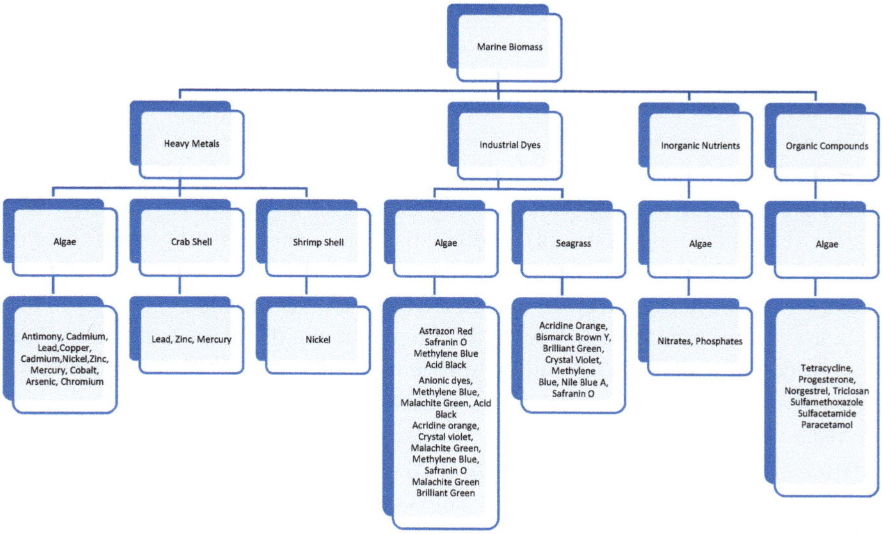

Fig. 12.2 Various applications as biosorbents

biomass from seagrasses and macroalgae can be obtained in larger quantities at a lower price without affecting their ecosystem in their aquatic environment. The new or emerging concept of magnetically responsive which have a significant application on eradicating various organic and inorganic pollutants from wastewater with magnetic modification to obtain smart biomaterials which exhibits numerous response types to an external magnetic field of marine-derived biosorbents. Usually, magnetic oxides of iron nano- and microparticles are used as magnetic tags for marine biomass changes (Vita Rudovica et al. 2021).

Several dyes such as methylene blue and safranin are being removed from the water by the magnetically modified algal biomass of *Sargassum swartzii*, *Cystoseir barabata*, and *Cymopolia barbata*, respectively. Figure 12.2 shows the various applications of aquatic biomass and their respective applications in removing the pollutants.

12.3 Biofuels

The world energy depend on the fossil fuel, which have significant effect on global warming with the emission of the greenhouse gases. The alternative for this emission is being targeted with the technologies behind the production of biofuel, and many studies are being conducted worldwide to overcome the global energy demands. Considering the above statements, biorefineries that process the waste are attracting significant interest globally, because sustainable waste management solutions (Vita Rudovica et al. 2021) meet the energy need and solution for the management of wastes. Production of the biofuels from corn, sugarcane, soybean, and palm oil creates ecology damage, shortage of water, and fuel and food debate. Considering such problems, the second- and the third-generations have alternative options for production of biofuel, which are respectively produced from microbes and waste materials (municipal sludge, plant, and agricultural waste). Microalgae are utilized to produce biofuel on incorporation with wastewater, which benefits both economically and environmentally. This process includes the microalgae as biosorbent for pretreatment and on applying conversion technologies, biofuels are produced, which includes the chemical conversion – extraction and transesterification (biodiesel), fermentation (bioethanol), and biochemical – anaerobic digestion (biogas) (Kumar et al. 2020).

Direct combustion, pyrolysis, and gasification are the alternative processes that include the non-fermentation process for the production of energy from microalgae (Vita Rudovica et al. 2021). A mild pyrolytic process called torrefaction, which is a destructive drying and slow pyrolysis, attracts recent attention that upgrades and pretreats the low quality fuels and also for the biochar production. This procedure may be ordered at scales ranging from extensive industrial facilities down to individual farms and even at the inland level, making it applicable to various socioeconomic situations. Figure 12.3 précises the most significant algal biofuels, their applications, and production mechanisms. Algae, especially microalgae, because of

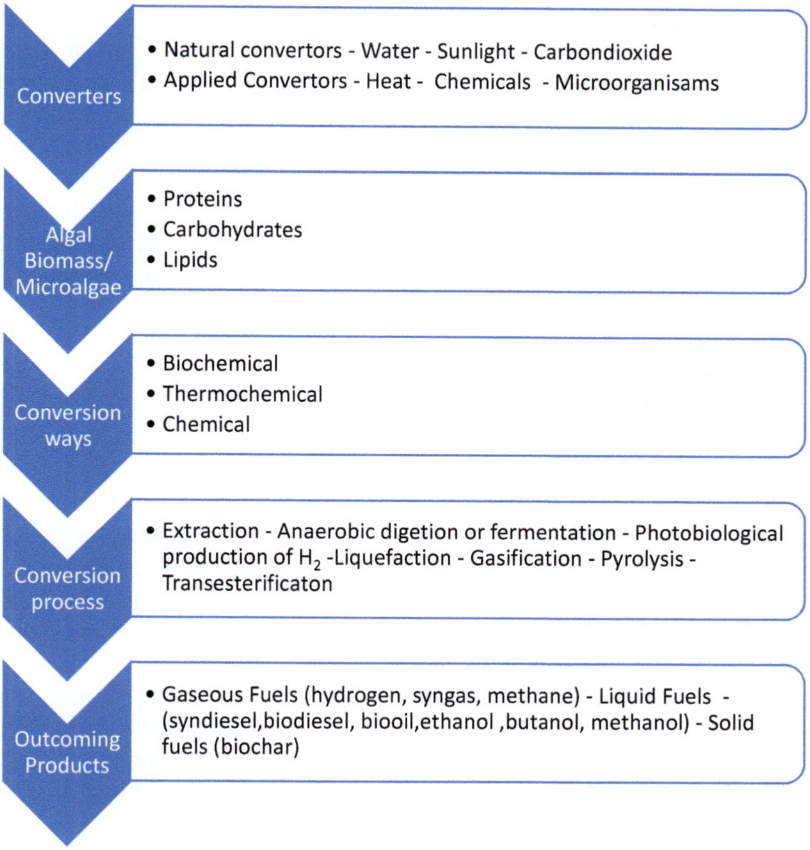

Fig. 12.3 Process of biofuel production

their huge availability and continuous supply, is considered as the most promising source of waste for biofuel production and the techniques of production differs on the algal type and end product. Thus, the use of algae can still be regarded as a feasible option for the next generation of biofuels.

12.4 Feed Supplements

Due to the high protein and lipid content of fish wastes, it is used extensively for the production of fishmeal decades before. However, recent studies and research showed that fishmeal is not the only choice, and products such as fish mince, gelatines, fish proteins hydrolysates, and oils can be manufactured from the discards of fish plants or hatches. Industrial processing of fish generates wastes such as offal and meats, which are rich in protein content that can be converted to peptones and are being used as growth medium for bacteria in laboratories. The rich contents of proteins, lipids, carbohydrates, minerals, and carotenoid pigments of mussel meat are having a potent application as feed supplements, somewhere as food supplement, preservatives, and enzymes. Seaweeds are the best source of aqua feed and cattle feed and being utilized traditionally. Seaweeds provide the rumens with great sums of rumen-degradable protein or can be used as a source of palatable and digestible bypass protein (Molina-Alcaide et al. 2017). *Asparagopsis taxiformis* is a seaweed that have anti-methanogenic activity over fermentation and able to cease methanogenesis at very low inclusion levels in the rumen (Machado et al. 2016).

12.5 Protein Hydrolysate

The breakdown of protein into peptides and free amino acids is referred to as protein hydrolysate. Because of the wide application of protein in food and pharma industries globally, industries are keen on recovery of protein from various sources and increase the biological and nutritional value of the final output. This wide application steps into aquaculture and aquaculture wastes and acts as an excellent source for the extraction and recovery of the protein by either enzymatic or chemical methods by forming peptides and functional products. The products from aqua wastes can be applied as stabilizers, protein supplements, and flavor enhancers. Proteins are extracted in the form of hydrolysates from the aqua wastes as shown in Fig. 12.4.

Fish protein hydrolysates (FPH) are amorphous powders and hygroscopic in nature, which are extracted entirely from the fish wastes. These powders contain higher protein of 81–93% when the fat is present at a lower level of 5%, 3–8% ash, and 1–8% moisture. Lean fish species or their processing wastes are the ideal raw material for FPH and their process is shown in Fig. 12.4. This FPH was being used in many applications as a gelling agent, nutritional supplement, and food binders. Besides, aquafeed and liquid fertilizers are utilizing this FPH as a cryoprotectant and nutritional additive (Vita Rudovica et al. 2021).

The sources of proteins from animals are more nutritious on comparing the plants due to the better stability of the dietary essential amino acids. Various products were continuously developed in recent days in protein hydrolysis resulted in various applications. Many studies of lipid profile and amino acid composition

Fig. 12.4 Process of production of protein hydrolysate

confirm the nutritional value and high quality of fish protein hydrolysate. The agricultural, food, cosmetic, pharmaceutical, and nutraceutical industries commonly use protein hydrolysate and could be employed as sources of protein in diets for farmed animals and aquatic organisms. Finally, the enzymatic hydrolysis results in the solid material, which can be used as a fertilizer (Ormanci et al. 2019).

12.6 Natural Pigments

Products from aquaculture and fish have a diverse range of colors, which impacts customers' choice to purchase them. Shrimps and prawns are one of the most important sources of natural carotenoids. The production of seafood, which is gradually increasing, generates a large volume of industrial wastes. The color of shells, skin, and exoskeleton of aquatic organisms, which is rich in yellow, orange, and red color, is caused by carotenoids, one of the most commonly known natural pigments obtained from industry waste after shrimp, crabs, trout, lobster, crayfish, salmon, snapper, and tuna processing, which can be found in other nonaquatic organisms as well (Paulina et al. 2017). Carotenoids can be extracted from body carapace, head, and rest of shrimp waste with different organic solvents such as isopropyl alcohol (IPA), methanol, ethyl acetate, hexane, ethylmethyl ketone, petroleum ether, ethanol, and solvent mixtures at diverse conditions of extraction (Paulina et al. 2017). Consumers start to avoid foods containing artificial colorants, which results in higher level of environmental awareness. Shrimps waste is a significant source of natural pigments, which could replace the addition of various synthetic colorants.

12.7 Products from Fish Wastes

The recent biotechnological improvements and the intervention of food processing made abundant changes in the field of utilization of wastes, especially of aquatic outputs. They serve as efficient alternatives to the chemical and mechanical process of scrubbing and produce various valuable products. The processing plants of fishes leaves scraps as waste and it was supposed to be dumbed and affects the environment as pollutants. An alternative to these pollutants emerges as the utilization of the scraps and converting into products as discussed earlier. Tables 12.1 and 12.2 explain various sources of aquatic wastes and their application. The products that are being utilized globally out of fish wastes are chitin and chitosan, shrimp extract, fish meal, fish body oil, pearl essence, isinglass, fish silage, and so on. Certain aquatic wastes recovered and their utilization are tabulated in Table 12.1.

Table 12.1 Sources of by-products and their active components and applications

By-product	Components of importance	Application
Heads	Proteins, lipids, peptides, collagen, gelatin, minerals including calcium	Food-grade hydrolysates, animal-grade hydrolysates, fish meal, fish oil, pet food, nutraceuticals, cosmetics
Frames (bones, flesh, fins)	Proteins, lipids, peptides, collagen, gelatin, minerals including calcium	Pet food, fish meal, food-grade hydrolysates, animal-grade hydrolysates, fish oil, nutraceuticals, cosmetics
Trimmings	Proteins, lipids, peptides	Fish meal, fish oil, food-grade hydrolysates, animal-grade hydrolysates, pet food
Viscera	Proteins, lipids, peptides, enzymes such as lipases	Food-grade hydrolysates, animal-grade hydrolysates, fish meal, fish oil, fuel, fertilizers
Skin (with belly flap)	Collagen, gelatin, proteins, peptides, lipids, minerals	Fish meal, fish oil, cosmetics, food, fish meal, nutraceuticals, leather, fuel, cosmetics, fertilizers
Blood	Proteins, peptides, lipids, thrombin and fibrin	Fuel, therapeutics, fertilizer

12.7.1 Chitin and Chitosan

The second most abundant biopolymer next to cellulose is chitin and chitosan of white, inelastic, hard polysaccharide and are added as ingredients in several products and processes including clarification, packaging films, and so on. The chitin and chitosan can be extracted from many sources but the abundant shrimp shell wastes act as great sources for the preparation of these substances. Deproteinization and demineralization are the two major operations that shells undergo and produce this chitin, and after deacetylation of chitin, chitosan is produced. Chitosan has several industrial and medicinal uses as a growth promoter in pets, feed ingredients, clarifying agents in juice, purification of water, wastewater treatment, textiles, cosmetics, and pharma industries.

12.7.2 Fish Meal

Fish meal is an excellent protein source mainly used as feed for aquaculture species and livestock. The major process aspects involved in fish meal product is cooking, pressing, drying and grinding from the scraps of the fishes at the moment of industrial processing. Typical species are small fatty species such as anchovy, sprat, herring, and krill. Next to the cooking and drying process is the fish that is turned into rough brown flour called as fish meal and packed for further logistics.

Table 12.2 Biowaste utilization of aquatic sources

Biowaste	Percentage of utilization	Sources		Valuable bioactive components
Head, shell	65–85	Crustacean	Shrimp/ prawn	Chitin, chitosan, *N*-acetyl-chitooligosaccharides, chitosan oligosaccharides, *N*-acetyl-d-glucosamine, pigment protein, d-glucosamine, protein hydrolysate, enzymes, flavor
Back shell, viscera, gills, claws shell	60–70		Crab	Chitin, pigment
Head, shell	<60		Lobster	Pigment, chitin, flavor
Head, shell	71–74		Krill	Chitin, protein, pigment, hydrolysate, oil
Head, shell	<85		Crayfish	Chitin, pigment, oil, flavor
Shell, nonedible body part	60–80	Mollusks including cephalopods	Scallop, clam, oyster, mussel, etc.	Enzyme, protein hydrolysate, flavor
Ink bag, gladius or pen, liver, other organs	25–32		Squid	Chitin, enzymes, bioactive peptides, collagen, gelatin
Intestine, mouth apparatus, eyes	10–20		Octopus	Collagen, gelatin
Shell, viscera		Coelenterate and echinoderm	Sea urchin	Gelatin, collagen,
			Sea cucumber	Protein hydrolysate, bioactive, collagen, gelatin
			Jelly fish	Protein hydrolysate, collagen, gelatin

12.7.3 *Pearl Essence*

Pearl essence is the product obtained by suspending the epidermal layer of guanine present in scales of pelagic fishes in the solvent. This guanine is crystalline in nature and largely present in the scales of fishes. The scales are soaked and allowed to separate the pearl essence from the water and protein mixture. The pearl essence is transferred into gasoline where it floats to the surface. The separation is then filtered into fine particles as pearl essence. These particles of pearl essence are used for the manufacture of artificial pearls, to give shiny coating to the trays, used for fishing rods, glazers in covers, textiles, and sometimes electrical switches.

12.7.4 Isinglass/Fish Maws

Fish maws, also called as isinglass, are produced out of fish air bladders (swim bladders) and by application of drying. The swim bladers of catfish, carps, eels, polynemids, sciaenids, sea bass, and so on are mostly preferred for the manufacture of fish maws. They are rich in collagen and used as a delicacy in Chinese soups. The bladders are first separated from the selected fish and blood and associated fat materials are scraped off and cleaned with running water after cut opening the swim bladder. Subsequently, the outer black membrane is removed and mined into pieces and dried under sun or mechanical dryers to less than 8% moisture. Wine, beer, and vinegar are being is used to clarify with the application of isinglass. Around 2–0.05% of the suspended solids in beer are removed and increase the filtration rate from 3000 to 11,000 L. It can also be used in confectionery products, Indian ink, and as an adhesive base for leather pottery and glass.

12.7.5 Fish Silage

Fish silage is a liquid product made from fish offal and trashed fish. The waster is enzymatically treated and liquified in the presence of an acidic environment. Bromelain was the most preferred enzyme for silage preparation with the presence of *Lactobacillus plantarum*. The enzymes break down the protein into minimum particle-sized soluble units and the process gets accelerated because of the acidic condition and prevents spoilage. Formic, phosphoric, and sulfuric acids are used for fish silage preparation at the level of 3–4%, and a pH of 4.0 or below is maintained. This fish silage is being used as fish feed and animal feed.

12.8 Importance of Value Addition

Value addition is one of the prospective ways to improve the sustainability of the seafood processing sector, and it is the most frequently used term in the industry, especially in that sector. Enhancing the value of aquatic species is known as value addition (Kim and Mendis 2006). An analysis of the strategic planning indicates that the development of value-added aquatic products contributes to overall value (Thorkelsson and Kristinsson 2009). Value addition enhances the product's worth or value after undergoing any of the basic or extensive processing operations listed in Table 12.3, as well as increases processing efficiency, maintains with customer demands, and provides a range of value-added aquatic products (Thorkelsson et al. 2009; Undeland et al. 2009).

An extremely perishable food product is fish meat because of high water activity (aw) and a neutral pH, and it provides microorganisms with favorable conditions for adequate nutrient availability (Adams and Moss 2000). A considerable surface

Table 12.3 Processing operations

Traditional methods	Modern methods
Drying	Chilling, freezing
Curing	Canning
Smoking	Retort processing
Fermentation/pickling	Baking
	Extrusion
	Fortification
	Minimal processing

Source: Thorkelsson et al. (2009), Undeland et al. (2009)

temperature in tropical locations enhances the problem caused in fish flesh's inherent compatibility as a substrate for bacterial growth. In these circumstances, fish was started spoiling after 12 h. The processing technologies of chilling, freezing, and canning can enhance the shelf life of fish (Onyuka and Ofulla 2013).

Due to poor postharvest facilities, postharvest fishery losses range around 10%, which is a significant problem in fish processing industries across the nation. The physicochemical and microbiological degradation that takes place after fish death is a common cause of postharvest fish wastage. When a fish is alive, its natural protection strives to prevent it from deteriorating. There are numerous things that can influence how quickly fish spoils, including time, temperature, handling practices and poor processing techniques (Ólafsdóttir et al. 2006).

In recent years, the amount of waste generated from the fisheries sector has increased, raising concerns worldwide, which are impacted by a variety of socio-economic, biological, and operational factors as well as technological and operational constraints (Nitta and Numata 2013). Low-value or underutilized fish species are referred to as fish wastes (Raveendran et al. 2013; Raveendran et al. 2013). Furthermore, more than 50% of the fish portions that are processed (such as fins, heads, skin, and other parts) are collected as "fish wastes" (Scotland 2015). Discards from global fisheries, which comprise "nontarget" species, by-products, and waste from fish processing, total to around 20 million tons per annum or 25% of the overall production of marine fishery catch (Buschmann et al. 2017). The fish's composition is influenced by its species, sex, age, nutritional state, weather, and health (Johansen et al. 2019).

12.8.1 Fish Pickle

The biochemical process of fermentation involves the conversion of sugars, carbohydrates, and organic acids by microorganisms and enzymes. Foods become much more nutritionally and easily digestible due to microorganisms such as bacteria and yeast. Other techniques of preserving, such as canning and freezing, can halt the fermentation process. Although having very slight similarity, both fermentation and pickling are processes of food preservation. Pickling is a fermentation process in which foods are preserved by the inclusion of acids.

Chop ginger, green chilli and garlic

↓

Grind to fine paste

↓

Heat oil and curry leaves and the paste to it (sauté for 30 min)

↓

In a separate pan roast mustard, asafoetida and cumin powder together and powder it

↓

Add this mixture to the sautéed paste

↓

Add chilli powder, garam masala, and salt to it

↓

Add required amount of vinegar and stir well

↓

Add citric acid and sodium benzoate to the sauce

↓

Fry the turmeric masala separately and add to it

↓

Leave it for overnight and pack it in required packaging material

Fig. 12.5 Fish pickle (Shikha et al. 2019)

Pickling is a technique that may be used to maintain or increase the shelf life of food. Because to its low pH, this can hold perishable foodstuffs fresher for months. Pickling really does have an influence on the food's aroma, mouthfeel, and taste. There are antibacterial characteristics in the materials utilized, such as garlic and mustard seed (Shikha et al. 2019). The process for making fish pickles is shown in Fig. 12.5, which is also applicable to manufacturing prawn pickle.

12.8.2 Dried Fish

Dried fish is one of the popular foods not only in India but also in other countries. The taste of dried fish is unique. It can be consumed either sprinkling with edible oil after roasting or by cooking or frying. Although the basis techniques of dried fish production are similar in principles, not all production methods are identically the same depending on the species and size of the fish. Dried fish is one of the main dishes for rural areas and fish-deficient areas such as upper Myanmar. One of the main quality concerns for dried fish is molding (Kent 2019; Siddhnath et al. 2022).

Dried fish can get molding easily if the moisture content exceeds 40%, and due to poor sanitation and packaging quality. Quality can be improved, and shelf life can be prolonged when vacuum packaging technique is used. By using vacuum packing, it is possible to increase quality and extend shelf life.

Dehydration is a component of drying. Fish that have been dried have already had their moisture removed, protecting it from pathogenic bacteria or enzymatic activity. Whenever fish are kept in humid temperatures and optimum moisture content drop by up to 10%, they would not become rotten. Both natural and artificial methods were employed to dry fish (Onyuka and Ofulla 2013; Singh et al. 2018).

Fish is quickly processed after being caught, skinned, dried whole or split, salted, and dried in an open place or sun-drying system. It has a pungent odor, is heavily salted, and tastes richly cheesy.

12.8.3 Smoked Fish

One technique of fish preservation is smoking, which operates as a kind of acidic coating on the fish's surface and inhibits the development of microbes. Dehydrating the fish during the smoking process assists in lowering the environment's capabilities for bacteria to exist. There are typically two forms of smoking performed. At high temperatures (150 °C), hot smoking has been done, while cold smoking has been done at temperatures of 100 °C or lower. Both the pre-smoking and the smoking processes can be followed.

There should be no remains of the kidney, gill, or intestine. Fish can be immersed in a strong brine (permitted colous may be added to the brine). The duration of contact time is influenced by the fish's size and saturated fat (Assogba et al. 2019; Degnon et al. 2013; Idah and Nwankwo 2013). The fish is then placed to dripping inside the kilns in which it dries as well and collects smoke on the surface. The brining and smoking durations of shellfish includes mussels, clams, oysters, and shrimp. The methodology for preparing smoked fish is depicted in Fig. 12.6.

12.8.4 Fish Noodles and Pasta

Pasta and noodles are both prepared by cooking fish in boiling water before consumption. These days, it is also eaten with ketchup or in soup, and occasionally it is fried. Noodles are lengthy pasta shapes that can be strained, extruded, flattened, and then trimmed into a wide range of shapes such as long waves, tubes, strings, thin strips, and others. There are many distinct forms and identities for short pasta. They can be refrigerated for a limited shelf life or dried and preserved. In order to enhance the product's nutritional value, fish is added. To make these products nutritious, different compositions might be utilized in their production (Jesmi et al. 2017).

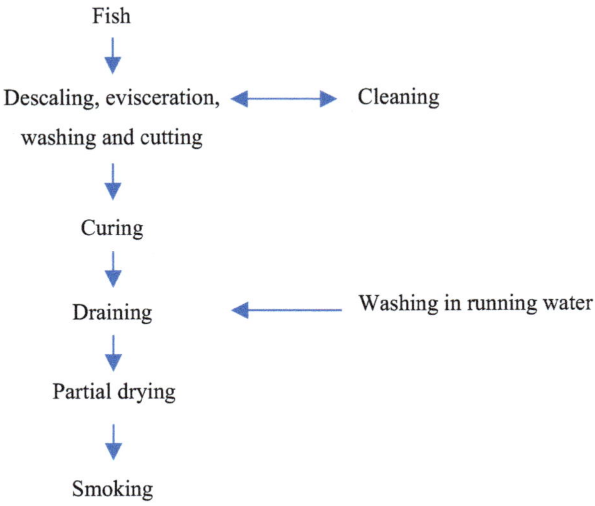

Fig. 12.6 Smoked fish (Assogba et al. 2019)

12.9 Future Aspects

The future of the fish waste utilization is moving toward a widespread zone of importance. The fish silage and fish bone mineral extraction processes are getting updated, which increases the profitability of the production unit. Squalene, an intriguing bioactive component, has been reported to be found in fish oil and their applications receive a wide scope in future for the treatment of cancer, diabetes, and tuberculosis. The agricultural field are moving toward the demand of the fertilizer and organic components produced with the fish wastes. Researchers are working toward the better utilization of fish wastes and bring new value-added products out of aquatic biowastes.

12.10 Conclusion

Products with additional value satisfy the needs of changing consumer lifestyles. Value-added products typically provide greater yields than raw fish and other fish. Also, there are certain drawbacks. Value addition elevates the price of a product and requires specialized technology, aesthetic packaging, and cold storage, among other things. There are other benefits, though. Value is a combination of quality, service, and price. Value-added fish items are typically considered to be those that have additional features such as a seasoning or flavor, are prepared, sliced, or in some other manner offer the consumer additional convenience. In terms of food, the functional and psychological advantages of excellent quality and nourishment, ease of preparation, and great food taste at an affordable price are the key benefits of value addition. Finding ways to use low-value fish species more effectively and producing protein-rich convenience products have been highlighted as the two main benefits of value addition.

References

Adams MR, Moss MO (2000) Microbiology of primary food commodities. Int J to Food Microbiology, pp 121–160

Ahmed N, Howlader N, Hoque MA-A, Pradhan B (2021) Coastal erosion vulnerability assessment along the eastern coast of Bangladesh using geospatial techniques. Ocean Coast Manag 199:105408

al Khawli F, Pateiro M, Domínguez R, Lorenzo JM, Gullón P, Kousoulaki K, Ferrer E, Berrada H, Barba FJ (2019) Innovative green technologies of intensification for valorization of seafood and their by-products. Mar Drugs 17(12):689

Assogba MF, Anihouvi DGH, Iko Afé OH, Kpoclou YE, Mahillon J, Scippo M-L, Hounhouigan DJ, Anihouvi VB (2019) Processing methods, preservation practices and quality attributes of smoked and smoked-dried fishes consumed in Benin. Cogent Food & Agriculture 5(1):1641255

Badri A, Raman K, Jayaraman G (2019) Uncovering novel pathways for enhancing hyaluronan synthesis in recombinant Lactococcus lactis: genome-scale metabolic modeling and experimental validation. PRO 7(6):343

Buschmann AH, Camus C, Infante J, Neori A, Israel Á, Hernández-González MC, Pereda SV, Gomez-Pinchetti JL, Golberg A, Tadmor-Shalev N (2017) Seaweed production: overview of the global state of exploitation, farming and emerging research activity. Eur J Phycol 52(4):391–406

Degnon RG, Agossou VE, Adjou ES, Dahouenon-Ahoussi E, Soumanou MM, Sohounhloue DCK (2013) Évaluation de la qualité microbiologique du chinchard (Trachurus trachurus) au cours du processus de fumage traditionnel. J Appl Biosci 67:5210–5218

Idah PA, Nwankwo I (2013) Effects of smoke-drying temperatures and time on physical and nutritional quality parameters of Tilapia (Oreochromis niloticus)

Jesmi D, Viji P, Rao BM, Prasad MM (2017) Nutritional and physical characteristics of noodles incorporated with green seaweed (Ulva reticulata) and fish (Pangasianodon hypophthalmus) mince

Johansen U, Bull-Berg H, Vik LH, Stokka AM, Richardsen R, Winther U (2019) The Norwegian seafood industry–importance for the national economy. Mar Policy 110:103561

Kent G (2019) Fish, food, and hunger: the potential of fisheries for alleviating malnutrition. Routledge

Kim S-K, Mendis E (2006) Bioactive compounds from marine processing byproducts–a review. Food Res Int 39(4):383–393

Kumar M, Sun Y, Rathour R, Pandey A, Thakur IS, Tsang DCW (2020) Algae as potential feedstock for the production of biofuels and value-added products: opportunities and challenges. Sci Total Environ 716:137116. https://doi.org/10.1016/j.scitotenv.2020.137116

Machado L, Magnusson M, Paul NA, Kinley R, de Nys R, Tomkins N (2016) Identification of bioactives from the red seaweed Asparagopsis taxiformis that promote antimethanogenic activity in vitro. J Appl Phycol 28:3117–3126. https://doi.org/10.1007/s10811-016-0830-7

Molina-Alcaide E, Carro MD, Roleda MY, Weisbjerg MR, Lind V, Novoa-Garrido M (2017) In vitro ruminal fermentation and methane production of different seaweed species. Anim Feed Sci Technol 228:1–12. https://doi.org/10.1016/j.anifeedsci.2017.03.012

Nitta SK, Numata K (2013) Biopolymer-based nanoparticles for drug/gene delivery and tissue engineering. Int J Mol Sci 14(1):1629–1654

Ólafsdóttir G, Lauzon HL, Martinsdottir E, Kristbergsson K (2006) Influence of storage temperature on microbial spoilage characteristics of haddock fillets (Melanogrammus aeglefinus) evaluated by multivariate quality prediction. Int J Food Microbiol 111(2):112–125

Oliveira J, Belchior A, da Silva VD, Rotter A, Petrovski Ž, Almeida PL, Lourenço ND, Gaudêncio SP (2020) Marine environmental plastic pollution: mitigation by microorganism degradation and recycling valorization. Front Mar Sci 7:567126

Onyuka JHO, Ofulla AVO (2013) Effect of salting on houseflies (Musca domestica) and beetles (Necrobia rufipes and Dermestes maculatus) infestation of fish from Lake Victoria, Kenya. J Pure Appl Microbiol 3(1):30–35

Ormanci HB, Künili İE, Colakoğlu FA (2019) Fish processing wastes: potential source of byproducts Çanakkale Onsekiz Mart University. Faculty of Marine Science and Technology, Turkey, p 2019

Pędziwiatr P, Zawadzki D, Michalska K (2017) Aquaculture waste management. Acta Innov. ISSN 2300-5599 22:20–29

Raveendran S, Yoshida Y, Maekawa T, Kumar DS (2013) Pharmaceutically versatile sulfated polysaccharide based bionano platforms. Nanomedicine 9(5):605–626

Remme JF, Austnes A (n.d.) Kvalitetsproteiner fra torskehoder—HEADS UP II; 2020: 00322; SINTEF Ocean AS. 2020. Rustad, T, Oslo, pp 2001–2014

Scotland ZW (2015) Sector study on beer, whisky and fish. Final Report. Zero Waste Scotland, Stirling

Shikha FH, Hossain MI, Shohan S (2019) Effect of low temperature on quality of Mola (Amblypharyngodon mola) fish pickle during storage and consumer preference towards. Bangladesh J Fish 31(1):125–135

Siddhnath, Ranjan A, Mohanty BP, Saklani P, Dora KC, Chowdhury S (2022) Dry fish and its contribution towards food and nutritional security. Food Rev Intl 38(4):508–536

Singh SM, Siddhnath RB, Aziz A, Pradhan S, Chhaba B, Narinder K (2018) Insect infestation in dried fishes. J Entomol Zool Stud 6(2):2720–2725

Teixeira JA, Vicente AA, Press CRC (2014) Engineering aspects of food biotechnology. CRC Press

Thorkelsson G, Kristinsson HG (2009) Bioactive peptides from marine sources. State of art. Report to the NORA fund. Food Res Inn Safe 1:1–20

Thorkelsson G, Slizyte R, Gildberg A, Kristinsson HG (2009) Fish proteins and peptide products: processing methods, quality and functional properties. Marine Funct Food:115–133

Undeland I, Lindqvist H, Chen-Yun Y, Falch E, Ramel A, Cooper M, Gildberg A, Luten J, Stenberg E, Nielsen HH (2009) Seafood and health: what is the full story. Marine Funct Food:17–87

Vita Rudovica V, Rotter A, Gaudêncio SP, Novoveská L, Akgül F, Akslen-Hoel LK et al (2021) Valorization of marine waste: use of industrial by-products and beach wrack towards the production of high added-value products. Front Mar Sci 8:723333

Part VI
Conversion of Food Waste into Biofuel and Electricity

Chapter 13
Valorization of Agro-Waste Biomass into Biofuel: A Step Towards Effective Agro-Waste Management

Pratyush Kumar Das, Bidyut Prava Das, Patitapaban Dash, Bikash Kumar Das, and Baskar Gurunathan

Abstract The increase in population and accompanying agricultural production generates tons of agro-wastes. Sustainable management of agro-waste is essential for efficient food production and minimization of environmental pollution. A million tons of agro-wastes produced per annum have enough renewable energy content. The energy-rich lignocellulosic biomass has the potential to generate biofuel that can bridge the gap between production and consumption in the available energy pool. The pretreatment and treatment of agro-wastes produce biofuels having multipurpose utility. Industrial development and change in lifestyle patterns increase the pressure on the consumption of conventional energy resources. Biofuel is a product of biomass valorization. It is a cheap source of nonconventional energy, having the ability to fulfill the energy demands of society, without environmental degradation. In view of the increased demand for agro-waste management, the transformation of these wastes to biofuel is a viable option. Further study on this aspect is required to improve the efficiency of this biofuel production system.

Keywords Agro-wastes · Biofuel · Biomass valorization · Energy · Environmental pollution · Renewable · Waste management

---

P. K. Das · P. Dash (✉) · B. K. Das
Centre for Biotechnology, Siksha 'O' Anusandhan (Deemed to be University), Bhubaneswar, Odisha, India
e-mail: patitapabandash@soa.ac.in

B. P. Das
Department of Botany, Sailabala Women's Autonomous College, Cuttack, Odisha, India

B. Gurunathan
Department of Biotechnology, St. Joseph's College of Engineering, Chennai, Tamil Nadu, India

School of Engineering, Lebanese American University, Byblos, Lebanon

E. Cherian, B. Gurunathan (eds.), *Value Added Products From Food Waste*,
https://doi.org/10.1007/978-3-031-48143-7_13

13.1 Introduction

The synchronization of economic growth at national and international levels with the generation and utilization of renewable energy resources can maintain an equation on a sustainable basis. Environmental health is becoming degraded with modern economic development, due to the overexploitation of nonrenewable natural resources (Sharma et al. 2020). It is adversely affecting our natural environment in the form of ozone depletion, acid rain, and greenhouse effects and is responsible for wide-scale pollution of environmental components. Economic growth is not at all acceptable at the cost of environmental health (Das et al. 2021). It is the time to broaden the horizon of the generation and consumption of renewable energy resources to meet the requirements of energy-demanding processes. The aim of this strategy is to relieve pressure on commonly used fossil-based energy resources and to promote the exploration of alternative renewable energy forms such as solar, wind, tidal, geothermal, and biomass-based energies.

The excessive consumption of nonrenewable energy resources is responsible for the depletion of the coal and petroleum reserves, increasing environmental toxicity, and is equally responsible for the disruption of biogeochemical cycles. The use of renewable energy is a possible alternative to restrict the present-day devastation, caused by the increase in carbon and heat budget of the environment. Under the prevailing situation, the harnessing of biofuel from potential biomass, using biodigester, looks promising (Barr et al. 2021). Plant biomass is a ready and rich source of chemical energy, formed by the transformation of radiant energy in solar light during photosynthesis (Das et al. 2009). The photosynthetic yield of the plant along with its catabolism, determines the quantity of its storable energy pool (Das et al. 2008). The storable energy of biomass acts as the source of the quantum of energy required for transformation into the heat/electrical energy of biofuel.

Biofuel, an eco-friendly, renewable, and alternative form of fuel, can be harnessed from agro-wastes, using biomass as feedstock (Prasad et al. 2020; Ajayi and Lateef 2023). The disposal of agro-waste biomass is a costly, labor-intensive, and time-consuming process during agronomical practices. These wastes are rich in carbohydrates, proteins, lipids, and their derivatives (Dixit and Yadav 2020). The feedstock, rich in polysaccharides and lipids, is a good source of organic carbon. Similarly, proteins and amino acids present in the feedstock serve as a good source of organic nitrogen. The chemical bonds of these polymeric molecules contain enough stored energy. This in turn makes biomass from agro-wastes to be a good source of energy having the potential for biofuel production (Das et al. 2022a, b; Kumar et al. 2023). Nonconventional energy sources, such as bioethanol, biodiesel, biohydrogen, and biogas, can reduce overexploitation and dependency on conventional sources of energy such as fossil-based fuels. To a certain extent, these biofuels can act as substitutes for fossil fuels to meet the demands of domestic, industrial, agricultural, and transport sectors.

The chapter highlights the current state of the nonconventional/renewable energy sector along with their possible sources of generation. It also lays emphasis on the conversion of agro-waste biomass into biofuels to cope with the current energy deficit, and environmental pollution scenario arising due to the rampant use of conventional fuel sources. Taking environmental sustainability into the equation (Jose and Ramakrishna 2021), the biochemical processes involved in the valorization of agro-waste biomass into biofuels in the form of bioethanol, biodiesel, biogas, and biohydrogen have been extensively discussed. Various factors impacting the biofuel generation process along with safety measures required to be taken during the valorization steps have been highlighted. Biofuel generation from agro-waste biomass unquestionably not only provides several opportunities in the renewable energy sector but also faces several challenges as far as its implementation is considered in developing nations like India. The chapter concludes by pointing out such challenges and emphasizing the need for improved government policies, infrastructural facilities, stringent emissions, and waste management standards. It invites further attention from researchers and policymakers in this regard to make it a cost-effective and widely acceptable technology.

13.2 Current Status of Nonconventional Energy Resources

The rise in global pollution coupled with dwindling fossil fuel reserves have initiated the need to look out for clean energy (nonconventional) sources. The energy sector is a vital part of a nation's economic growth and should be given more priority.

The utilization of renewable energy increased by 3% in 2020. It can be mostly attributed to a 7% additional generation of electricity from renewable sources. Figure 13.1 provides an account of available sources of renewable energy in India as of the year 2020. The contribution from nonconventional sources toward the generation of electricity on a global scale has increased from 27% in 2019 to 29% in the year 2020. China is set to be the highest renewable electricity-producing country followed by the countries such as United States of America (USA), European Union (EU), and India. China and USA are expected to contribute almost 50% of global energy output from the wind (Ji et al. 2020; Dey et al. 2022). This section provides the updated status of energy production from renewable resources on a global scale.

13.2.1 Solar Energy

It is one of the most widely available energies on the earth's surface. The quantity of this form of energy available on earth in an hour is enough to meet the per annum global energy requirements. The earth is absorbing approximately 385,000 exa joules of this energy. It includes the absorbance by air, ocean, and land components. The energy harvested from the sun is mostly used in the generation of electricity, water heating, and cooking (Elliott 2000).

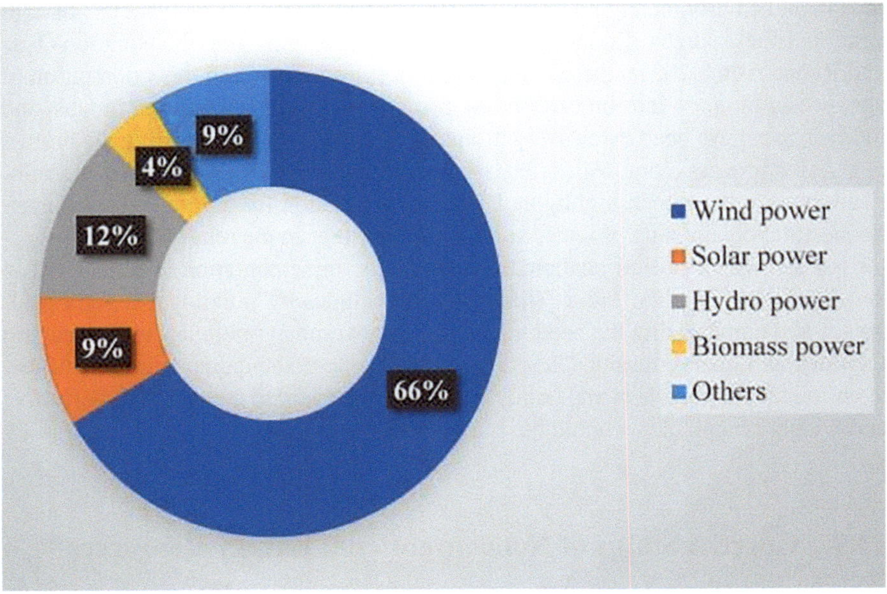

Fig. 13.1 Total installed capacity for generation of renewable energy in India. (Source: Ji et al. 2020; Dey et al. 2022)

13.2.2 Wind Energy

It is generally obtained through windmills that make use of wind flow to rotate a turbine. The mechanical energy obtained from the turbine is transformed into the required electrical form of energy. China boasts of the world's largest windmill with a production capacity of 6000 MW (megawatt). On a global scale, approximately 534.5 TWh (terawatt-hour) of wind power is produced (Pappas 2017). Energy generation from wind power in India increased from 1.58 GW (gigawatt) in the financial year 2018–19 to 2.07 GW in 2019–20, an almost 31% increase in comparison to the previous year (Ragaert et al. 2017).

13.2.3 Tidal Energy

It is one type of hydro-energy that makes use of tides that happens twice on a daily basis. This form of energy uses ocean tides to rotate turbines. India as a developing country has the potential for the generation of an 8000 MW tidal form of energy (Himri et al. 2009).

13.2.4 Geothermal Energy

Geothermal energy makes use of the gradient of heat energy present below the earth's surface to generate electricity. The 4760 MW capacity, Vindhyachal Thermal Power Station is the biggest of its kind in India. It is situated in the Singrauli district of Madhya Pradesh. As per the reports of the International Geothermal Association, around 24 countries in the world collectively produce 10,715 MW of geothermal power (Ruggero 2007).

13.2.5 Energy from Biomass

It involves the conversion of agricultural, industrial, and domestic solid biowastes into fuels of different types. The energy derived from these sources is economical and friendly to the environment. Plants or crops make use of sunlight during photosynthesis to produce biomass. These biomasses are further converted into usable fuel forms such as biogas, bioethanol, biodiesel, or biohydrogen. This implies plants utilize solar energy to convert it into biomass energy, which is further converted into desirable fuels. On a global scale, approximately billion tons of biomass are produced. In India, on an annual basis, biomass availability is projected to be around 750 MMT (million metric tons) (Wang et al. 2022). The power generation from biomass by the top five Indian states is provided in Fig. 13.2. Out of these states, Uttar Pradesh has the highest production capacity of 592.50 MW of power from biomass in India. It is followed by states such as Tamil Nadu, Maharashtra, Karnataka, and Andhra Pradesh with biomass power production capacities of 488.20, 403, 365.18, and 363.25 MW, respectively (Dey et al. 2022).

13.3 Types of Agro-wastes Produced and Their Sources

The rise in human population in the last few decades has brought about the necessity to increase agricultural production. The last half of the century has seen more than three times an increase in agricultural production. Technological advancements and the extension of land for farming are some of the other reasons for increased agricultural production (FAO 2017a; FAO and OECD 2019). The increased agricultural production involves the production of food for humans, industrial utilities, and feed for animals (FAO 2017b). Increased agricultural production implies the increased generation of solid agro-wastes at each and every step of production, processing, and consumption (Adejumo and Adebiyi 2020). This makes agriculture one of the leading sectors producing solid biowastes that not only accumulate but also pose a threat to environmental health (Bracco et al. 2018). Figure 13.3 shows the conventional use of postharvest rice residues in India. The residues are either used as cattle

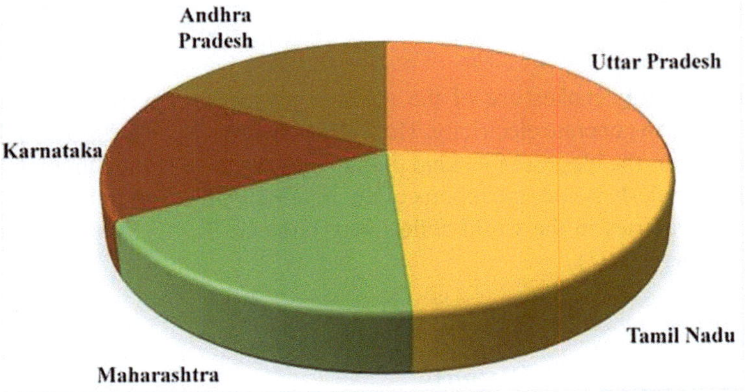

Fig. 13.2 Power generation from biomass (in MW) by the top five Indian states. (Source: Dey et al. 2022)

feed or dumped conventionally away from the sites or are burnt on the field, thereby causing pollution of the environment on a larger scale.

Agro-wastes can be classified into four different categories on the basis of sources of production like residues from crops, wastes generated from agro-industries, wastes from livestock, and wastes generated from fruits and vegetables (Fig. 13.4).

One of the commonly available and cost-effective forms of agro-waste is residues from crops. It mainly includes straws and husks of rice and wheat plant along with stover of corn and sugarcane bagasse. Besides, other residues include dried leaves, stems, and seed pods. Specifically, in India, by the year 2030, straw as waste from rice crops is estimated to reach 221.8 million tons per annum (Kumar et al. 2018). The agricultural residues mostly from rice and wheat cultivation are usually left on the fields. These solid wastes are to some extent used as animal fodder while a majority of these are burnt in the fields (Adhikari et al. 2018).

Wastes from agro-industries mainly include the by-products generated during food processing and after it. These include residues such as sugarcane bagasse, fruit and vegetable peels, starch, molasses, fruit pomace, seed cakes, animal meat, and fat residues. Sugarcane bagasse accounts for the largest waste generated from agro-industries (Saini et al. 2015). The global palm oil industry also accounts for 41% of waste generated from approximately 86 million tons of fresh palm fruit biomass (Sukiran et al. 2017). In a similar fashion, apple pomace as a waste account maximum up to 30% of the total apple biomass processed (Hijosa-Valsero et al. 2017).

The third major type of agro-wastes is generated from livestock, which mainly includes liquid manures in the form of urinary wastes and solid manures. Liquid manures are responsible for surface water pollution and aid in pathogen contamination. Solid manures contribute to air pollution by releasing carbon dioxide and methane up to 18 and 37%, respectively (Holm-Nielsen et al. 2009).

The agro-wastes from commodities such as fruits and vegetables include unprocessed fruits and vegetables. In India, these wastes are mostly dumped on the

Fig. 13.3 Cultivation of rice plant and the fate of postharvest residues: (**a**) cultivation of rice crops on a field; (**b**) postharvest rice residues; (**c**) people engaged in the collection of postharvest rice residues; (**d**) environmental pollution by dumping of postharvest residues with other organic matter; (**e**) rice straws being used as a cattle feed; (**f**) separation of rice grains from husks. These husks are disposed of as they do not have any such significant usage

outskirts of cities. They are a major contributor to environmental pollution, which may be attributed to their highly putrescible nature.

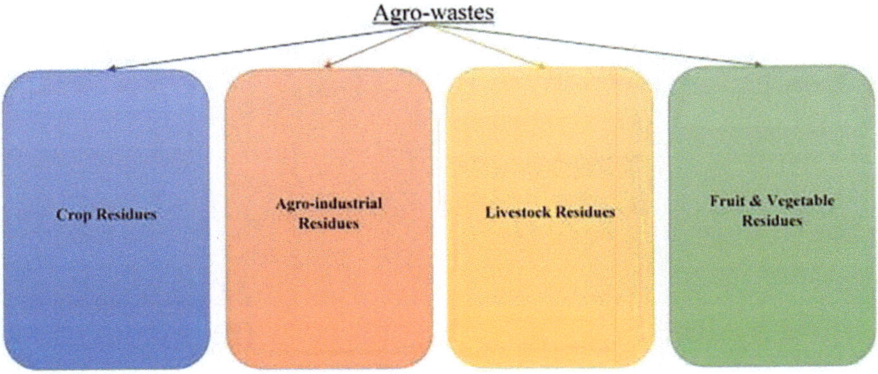

Fig. 13.4 Classification of agro-wastes based upon the source of production

13.4 Socioeconomic Problems Caused by Generated Agro-Wastes

The proper disposal of agro-waste is the cause of major environmental concern as tons of agro-waste, generated on a daily basis by the production and finishing of edible agro yields. Highly populous nations such as China and India have a wide acreage of lands that is used for paddy cultivation. The preservation, processing, and storage of edible cereals are difficult to handle in these countries. It is a source for the generation of nutrient-rich biodegradable wastes in tons. The edible cereal wastage is more in China (50 million tons/annum) as compared to its wastage in India (23 million tons/annum). African countries are also not spared from this problem with almost one-fourth of crop wastes being generated due to bad harvesting practices (Panahi et al. 2022). It is a problem that tends to compound with the increase in population and demand (Anwar Saeed et al. 2018).

These unaddressed problems are responsible for environmental degradation, economic loss, and public health issues. Transportation cost is a major hurdle for the safe disposal of agro-waste in rural areas. The economic condition of the marginal section of people acts as a hindrance to the disposal of agro-waste in an eco-friendly manner. To avoid the transportation cost, the open burning of agro-wastes on the crop fields by the concerned group of people is a common practice in rural areas. The resultant excessive release of greenhouse gases is responsible for environmental pollution, scaling up of mean earth temperature, and erratic monsoons (Sharma et al. 2021). Some of the major socioeconomic problems that cropped up in this regard are highlighted below.

13.4.1 Emerging Challenges to the Environment

Good environmental health is of prime concern for the safety of public health. The accumulation of agro-wastes may lead to the release of VOCs (volatile organic compounds) such as alcohol, aldehydes, and ketones in the environment. These wastes are usually generated during the production, processing, transportation, storage, or consumption of agricultural commodities. The secondary environmental pollution caused due to the accumulation of these wastes is one of the possible factors for ozone depletion in the stratosphere, pollution, greenhouse effects, global warming, acid rain, and climate change (Sharma et al. 2021).

13.4.2 Degradation in Soil Quality

Improper/conventional agro-waste management leads to the accumulation of agro-waste/partially degraded biowastes on crop fields. It can cause alteration of particle size, porosity, aeration, field capacity, temperature, pH, and levels of mineral nutrients in the soil. Subsequently, it changes the composition of soil microbial flora and ruins the soil quality (Sharma et al. 2021). Alterations in the soil quality may not allow the growth of the cultivars on the same piece of land, which had supported its growth prior to such changes.

13.4.3 The Declining Trend of Crop Yield

The stagnation of generated agro-wastes can lead to the loss of cultivable areas for cultivation purposes. The declining trend of crop yield may be due to the alteration in quality and quantity of cultivable lands caused by the waste load.

13.4.4 Immobilization of Resources

The accumulation and improper management of agro-wastes lead to the unavailability of these resources for mineral cycling in nature. After the prolonged continuation of this trend, it can lead to stagnation of minerals in reservoirs with a long residence time. The accumulation of disproportionate fractions of nutrients in the soil is one of the possible reasons behind undesirable soil quality and its adverse effects.

13.4.5 Additional Financial Burden

The normal cost involved with agricultural production processes such as irrigation, nutrient supplementation, processing of raw agro-products, and transportation is high. Farmers incur huge losses with the wastage of unused food products and the mid-failure of crops. The farmers have to take an additional financial burden for the safe disposal of agro-residues (Sharma et al. 2021). The in situ disposal of agro-wastes is a costly process and the marginal farmers have to bear the additional expenditure for it. The financial burden on farmers is still high at the time of ex situ remediation due to the involvement of transport costs. The additional expenditure can be curtailed by the production of biofuel using agro-wastes. The cogeneration of multiple value-added products by involving the intermediates of a reaction for one product as reactants for the generation of other products is beneficial in this context.

The declining trend of crop yield is a possible factor for the loss of earnings, faced by marginal rural farmers. The loss of fertility of the soil by the conventional open burning of agro-wastes on crop fields may require the purchase of fertilizers by the farmer to restore the soil fertility of cultivable lands. For it, the farmers incur more expenditure. The socioeconomic conditions of rural farmers may be ruined by these factors.

13.5 Agro-wastes as a Source of Biofuel

The agro-wastes represent the differences between the volume of produced biomass and that of edible biomass, generated during agricultural practices. Primary and secondary agro-wastes management is a real problem for farmers, during agronomical practices. The clearing of accumulated agro-wastes on agricultural fields, the in situ disposal of these wastes at the site of waste accumulation, and the shifting of these wastes for ex situ disposal are examples of agro-waste-based expensive, time-consuming, and labor-intensive environmental problems. The primary agro-wastes such as roots, stems, stalks, and leaves are produced during crop harvesting and the accumulation of those wastes occurs during postharvesting stages.

The agro-wastes have the energy potential to transform their stored energy pool into value-added energy sources such as bioethanol, biodiesel, biohydrogen, and biogas. The waste biomass from cereal crops such as rice and maize is a suitable feedstock for bioethanol production by hydrolysis, fermentation, and distillation. Similarly, the wastes from oilseed crops such as oil palm, soybean, peanut, and rapeseed are rich in fatty acids and glycerol. These wastes are suitable as feedstock for fatty acid methyl ester (FAME)/biodiesel production by transesterification (Schwaiger et al. 2011).

The agro-wastes are generated as field residues (roots, stems, and leaves), process residues (husk and bagasse), and kitchen residues (vegetable peels, leaves, and seeds) (Li et al. 2016). The generation of waste from unused food items creates a problem for waste management and the occurrence of environmental pollution across the world. More food waste is generated from unused root crops that account for almost half of its production per annum. Almost one-third of unused dairy, cereals, and meat have been added to the soil directly or indirectly, thereby increasing the volume of generated biowaste per annum (Jin et al. 2018).

After harvesting the crop, the agro-wastes are present as residues either on the crop fields or accumulated at the processing sites until transformation to other value-added products such as the spectrum of biofuels useful in different contexts. Without it, the crop productivity of the soil slowly drops down.

The molasses of sugarcane and purging nuts are huge raw sources of biofuel production. The residues (50–80%) are used for biofuel purposes in Africa (Sekoai and Yoro 2016; Oluseun et al. 2014). The production of bioethanol is possible using organic wastes such as fermentable sugars, cellulose, and hemicelluloses (Karaosmanoglu et al. 1999).

13.6 Mechanism of Generation of Biofuel from Agro-Wastes

Agro-wastes largely contribute to environmental pollution and need to be disposed of in a sustainable manner. Keeping in mind the current scenario of the global fuel crisis due to dwindling stocks of fossil fuels, the generation of biofuels from agro-wastes offers to be an excellent alternative. The current biofuel industry is mainly focused on the production of comparatively cleaner fuel sources that include bio-char (solid biofuel), bioethanol and biodiesel (liquid biofuels), biogas, and biohydrogen (gaseous biofuels).

Solid biofuels are composed of solid organic matter and include wood wastes, charcoal, animal excreta, and other biological biomass of nonfossil origin. These biofuels are mostly consumed for heating or operation of cooking and electrical appliances (Dahman et al. 2019).

Liquid biofuels are generally produced from natural plant and animal biomasses rich in lignocellulosic content and/or lipids (Alalwan et al. 2019). These types of biofuels are comparatively far less polluting than solid biofuels. Moreover, they provide high energy in comparison to both solid and liquid types of biofuels along with being easy to store and transport (Mahapatra et al. 2021). The liquid biofuels mainly used are bioethanol and biodiesel.

Gaseous biofuels have low density and include biogas and biohydrogen (No and No 2019). Gasification and pyrolysis are the two most commonly followed methods for the generation of these biofuels. These energy forms have the potential to meet the requirements for heating and electrical purposes. In comparison to gaseous biofuels, solid and liquid biofuels are mostly preferred for electricity generation. Table 13.1 provides information on the year-wise increase in global electricity generation (up to the year 2020) collectively from solid and liquid biofuels keeping 2010 as the base reference year.

Table 13.1 Annual increase (2011–2020) in global electricity generation collectively from solid and liquid biofuels in comparison to that of the reference year (2010)

Year	Increase in electricity generation (in GWh) collectively from solid and liquid biofuels
2010	1869 (BRYV)
2011	BRYV+38
2012	BRYV+197
2013	BRYV+172
2014	BRYV+469
2015	BRYV+550
2016	BRYV+453
2017	BRYV+1364
2018	BRYV+1349
2019	BRYV+1686
2020	BRYV+1686

Source: Mahapatra et al. (2021). BRYV stands for base reference year value

Table 13.2 Comparison of different thermochemical processes for the generation of biofuels

Process	Method	Major Products	Drawbacks	References
Pyrolysis	Application of heat to the feedstock under anaerobic conditions	Syngas	A huge amount of heat is required	Ruan et al. (2019)
Carbonization	Slow pyrolysis/ controlled combustion	Charcoal	Secondary pollution	Ruan et al. (2019)
Combustion	Burning biomass/ feedstock in the presence of air	Heat and electricity	Large-scale emission of pollutants such as CO_2, nitric acid, and particulate matter	Hassan et al. (2020)
Gasification	Converts biomass fuel into rich gaseous products	Syngas, fuel gas, methanol, and H_2	High demand for energy	WBA (2020)
Liquefaction	Application of low temperature and high pressure using hydrogen to organic biomass	Bio oils	Technical problems and limited utility	Hassan et al. (2020)

The global electricity generation from solid and liquid biofuels was reported as 1869 GWh in 2010. There has been a subsequent increase in electricity generation in the following years with a 90.2% spike in 2020 as compared to 2010.

Generation of biofuels from agro-wastes can be carried out via two different processes such as thermochemical and biochemical methods. The thermochemical processes involve techniques such as pyrolysis (WBA 2020), carbonization (Ruan et al. 2019), combustion (Hassan et al. 2020), gasification (WBA 2020; Patel et al. 2016), and liquefaction (Hassan et al. 2020). Table 13.2 provides a comparative account of thermochemical processes for biofuel generation. The thermochemical processes are fast but have major drawbacks in the form of high energy requirements and chances of secondary pollution. This is the reason for us to lay more emphasis and concentrate our discussion on the biochemical mechanisms of biofuel generation.

13.7 Biochemical Processes in the Generation of Biofuels

The biochemical process is a cleaner approach to generate biofuels. It involves several steps such as fermentation (for ethanol production), anaerobic digestion (for the generation of biogas), transesterification (for biodiesel production), and dark fermentation (for biohydrogen formation). However, prior to the initiation of the production process, there are several initial procedures that need to be carried out for processing the raw materials into usable forms. These include steps such as pretreatment and hydrolysis (Pattanaik et al. 2019). Figure 13.5 provides an overview of different biofuel generation techniques.

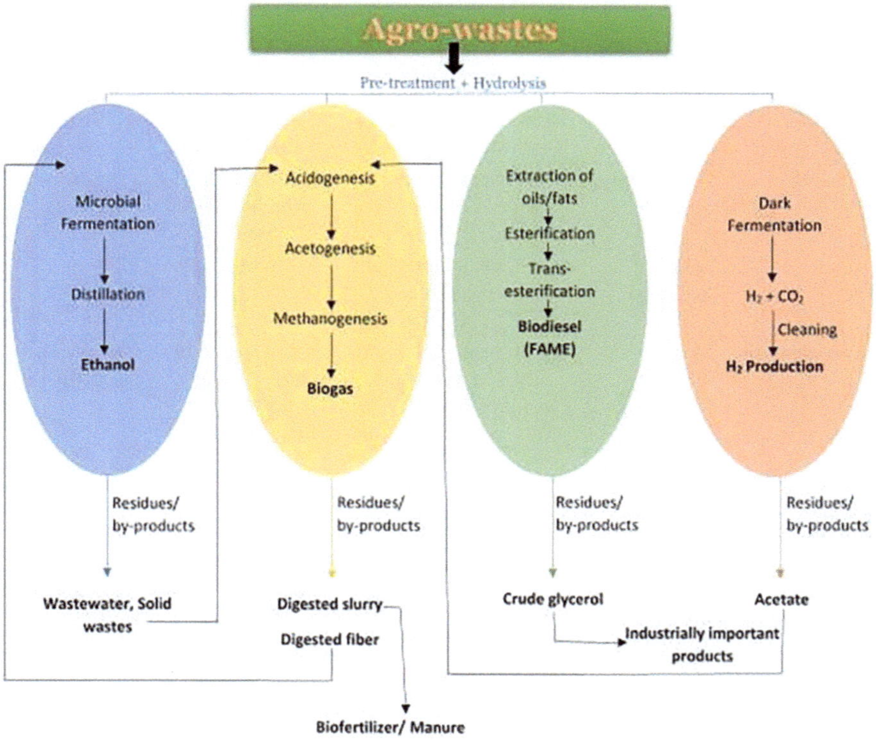

Fig. 13.5 Overview of techniques involved in the generation of biofuels from agro-wastes, residues/by-products formed, and their sustainable utilization

13.7.1 *Pretreatment of Agro-Wastes*

The lignocellulosic biomass of agro-wastes is composed of lignin as a protecting layer besides hemicelluloses and celluloses. The lignin is difficult to degrade components of those wastes and a pretreatment process is required for biomass delignification. The pretreatment process also helps in the decomposition of hemicelluloses and subsequent increase in porosity of the biomass. Table 13.3 provides an account of several pretreatment techniques along with their operating conditions and respective functions.

Table 13.3 Pretreatment techniques employed in the biofuel generation process

Pretreatment technique	Methods	Treatment conditions	Function	References
Physical	Milling (wet, dry, compression)	Grinding of crop biomass	Reduced size of biomass Decreased cellulose	Zabed et al. (2016)
			Crystallinity	
	Microwave irradiation	Use of electromagnetic radiation on crop residues	Weakened interaction between components of biomass	Ethaib et al. (2015)
	Ultrasound treatment	Exposure of biomass to ultrasound waves generated from a sonicator	Weakening of cellulose bond and delignification of biomass	Subhedar and Gogate (2016)
Chemical	Acid treatment	Treatment with acids such as HCl and H_2SO_4	Hydrolyzation of hemicellulose and cellulose	Nwosu-Obieogu et al. (2016)
	Alkali treatment	Treatment with alkalis such as NaOH and KOH	Dissolving lignin	Nwosu-Obieogu et al. (2016)
Physiochemical	Steam explosion	Treatment with steam at high temperatures and pressure	Hydrolyzation of hemicellulose Increased xylan yield	Alvira et al. (2010)
	Supercritical CO_2	Treatment in CO_2 medium at temperature >31.1 °C and pressure >73 bar	Removal of lignin	Alvira et al. (2010)
	Subcritical water	Treatment with hot water at 170–230 °C and 5 bar pressure	Hydrolyzation of hemicellulose	Pérez et al. (2008)
	Ammonia fiber explosion	Similar to the subcritical method with ammonia replacing water	Decomposition of hemicellulose, Removal of lignin	Teymouri et al. (2004)
Biological	Microbe assisted	Treatment with microbial enzymes	Delignification Decomposition of hemicellulose	Taha et al. (2016)

13.7.2 Hydrolysis

The pretreatment step is followed by the hydrolysis of biomass. The process of acid hydrolysis or enzymatic hydrolysis converts the pretreated lignocellulosic biomass, rich in celluloses and hemicelluloses, to their respective monomeric forms (Shi et al. 2023). However, enzymatic hydrolysis is more efficient as compared to the conventional acid hydrolysis process (Table 13.4).

Table 13.4 Comparison between the techniques of acid hydrolysis and enzymatic hydrolysis

Acid hydrolysis			
Dilute acid hydrolysis	Concentrated acid hydrolysis	Two-stage hydrolysis	Enzymatic hydrolysis
Employs Low acid concentrations (0.5–1.5% H_2SO_4), high temperature, high pressure, short duration	Employs high acid concentrations (41% HCl, 70–90% H_2SO_4), longer reaction time	1st stage operates at a lower temperature of <180 °C followed by 2nd stage at 230–240 °C	Employs commercial enzymes and/or microbes (cellulases, hemicellulases, ligninases)
Degradation of hemicellulose	Converts hemicellulose and cellulose to sugars	Degrades hemicellulose followed by conversion of cellulose to glucose	Cellulases convert cellulose to glucose Hemicellulases convert hemicellulose to pentose sugars, hexose sugars, and uronic acid Ligninases degrade lignin and provide better access to other cellulases and hemicellulases
Causes corrosion of equipment, environmentally hazardous, forms inhibitory by-products			Noncorrosive, nontoxic, comparatively less inhibitory products formed
Depends on biomass structure, crystallinity, rigidity, acid concentration, temperature, and applied pressure			Depends on sources and efficiency of the enzymes involved. It also depends upon biomass structure, crystallinity, particle size, and porosity

13.7.3 Generation of Bioethanol from Agro-wastes (Fermentation)

Sugar yields during the process of hydrolysis can be converted to bioethanol using the fermentation process (Das et al. 2019). Fermentation requires the presence of microorganisms in a bioreactor during batch/continuous/semicontinuous mode of operations. The microbes such as *Saccharomyces cerevisiae*, *Pichia stipites*, *Clostridium thermocellum*, and *Klebsiella aerogenes* are involved in the fermentation process. The fermentation process yields crude ethanol that can be further distilled to produce pure ethanol. A distillation tower is employed for the purification of ethanol. In it, the water is drained out using the bottom outlet and the ethanol through the top outlet. The evaporation of ethanol at 78 °C is a possible mechanism for the elimination of insignificant water present in it at the post-distillation stage (Das et al. 2020).

13.7.4 Generation of Biogas from Agro-wastes (Anaerobic Digestion)

The production of biogas following anaerobic digestion involves a series of steps. The initial step is hydrolysis and it involves the conversion of polymeric structures of macromolecules such as carbohydrates, proteins, and lipids into their respective monomeric constituents. This step is catalyzed by an enzyme called hydrolase, secreted by the anaerobes. It is followed by acidogenesis. The monomeric units formed during the previous step are converted to organic acids, alcohol, CO_2, and H_2 in the acidogenic phase. Besides, other products such as volatile fatty acids, H_2S, and NH_3 are also produced in this step. The H_2 and CO_2 produced during the previous phase undergo reduction to form acetic acid by homoacetogenic microbes in the acetogenic phase. The acetate is produced in this step from the earlier formed organic acids and alcohols. The generated acetate as a substrate leads to the production of methane during methanogenesis. The CO_2 and H_2O are produced during the production of methane in the methanogenic phase (Chandra et al. 2012). Hydrogen is one of the major unwanted by-products released, and inhibits acetate formation, during digestion under anaerobic conditions. The favorable conditions promote the production of CH4 with the consumption of hydrogen by the methanogenic bacteria. Employing further purification techniques can improve methane formation up to 90%. CH4 can also be converted into liquid methanol following a biorefinery approach (Naik et al. 2010). Moreover, the waste slurry containing the digested biomass generated during the process can be used as manure for increasing soil health.

13.7.5 Generation of Biodiesel from Agro-wastes (Transesterification)

Transesterification is the most widely accepted and followed method for biodiesel production, which may be attributed to its higher productivity and low cost of production. In this process, generally, seeds from oil-yielding plants such as *Pongamia* and *Jatropha* are used. Oils from the seeds are extracted using mechanical pressure, solvents, or suitable enzymes. The oils tend to be rich in lipids and fatty acids and undergo a transesterification process to generate the least toxic fatty acid methyl ester (FAME) compound or biodiesel, which is viscous and biodegradable in nature (Das et al. 2022a, b).

13.7.6 Generation of Biohydrogen from Agro-wastes (Dark Fermentation)

The suitable method for the production of hydrogen from organic wastes is dark fermentation (Ghimire et al. 2015). In the process, carbohydrate-rich biomasses are degraded by facultative/obligate anaerobes to produce molecular hydrogen (Villanueva-Galindo et al. 2023). The entire process is similar to the anaerobic digestion process excluding the methanogenic step. Electrons are generated upon oxidation of organic substances under anaerobic conditions. These electrons are accepted by protons to form hydrogen (Pattanaik et al. 2019).

13.8 Factors Involved in the Valorization of Agro-wastes to Biofuel

The optimized production of biofuel from agro-wastes is the key to sustainability (Sharma et al. 2022). The valorization of agro-wastes to biofuel is dependent upon certain factors, including the type of biomass, techniques used, by-product utilization, optimum temperature, pH, and the use of suitable enzymes and/or microbial species.

Biomass type plays a major role in providing higher biofuel yield. Agro-residues with high content of carbohydrates and lipids will definitely provide a higher yield of biofuels. Almost 33% of produced grains on a global scale remain unutilized or damaged during the processing of food grains. It leads to the addition of 1.3 billion tons of food wastes to the pool of agro-wastes (Singh et al. 2022). Pretreatment techniques used also decide the efficiency of the biofuel generation process. Physical or chemical pretreatment techniques have lesser yield as compared to biological techniques. This may be attributed to the secondary by-products produced during the process that acts as a limiting factor. Proper utilization of by-products minimizes the production loss and improves the biofuels yield. Maintaining optimum pH and temperature conditions in the bioreactor is the key to the successful generation of biofuels. Selection of suitable enzymes and/or microbes also ensures high production yield (Singhania et al. 2022).

13.9 Safety Measures Required During Valorization of Agro-Wastes

The radiant energy of solar light is the ultimate source of energy stored in organic materials of agro-wastes as chemical energy after transformation. The valorization of agro-wastes to biofuel looks promising. The biofuels such as bioethanol, biodiesel, biohydrogen, and biogas are the best and ultimate energy forms to replace the conventional energy requirements (Panpatte and Jhala 2019). To obtain sustainable development with public and environmental safety, the following measures have to be taken during the valorization of agro-wastes to biofuel.

13.9.1 Reduction in Greenhouse Gas Emission

The reduction in the emission of greenhouse gases to the environmental components is an essential feature and it needs to be monitored for protecting environmental components from global warming. Generally, the average energy contents of biofuels such as bioethanol and biodiesel are less than the respective conventional fuels. The energy contents of bioethanol and biodiesel are 66 and 88–95% of conventional petrol and diesel forms, respectively. Upgradation of the octane level of these fuels has the possibility to carry out complete combustion and minimization of pollution due to the release of fewer greenhouse gases. The application of the multienzyme system for waste treatment at the site of waste generation is preferable as compared to its offsite application. It may be due to the 40% less release of greenhouse gases during the application of multienzyme systems at the site of waste release (Singh et al. 2022). The carbon dioxide gas released during the process is less and biogenic (Soam et al. 2016). So, the transformation of agro-waste to bioenergy is treated as a green technology.

13.9.2 Consistency in Biofuel Production

The residues from crops and their products, with or without processing, generate voluminous agro-wastes. The agro-residues obtained from grasses such as the straw of rice, wheat, corn, and molasses of sugarcane are used as biofuel feedstock

(Escobar et al. 2009). Besides, unused edible foods such as grains, oilseeds, raw vegetables, and pulses add tons of biowastes to the soil resources and degrade their quality over the years. It is good to follow a circular bioeconomy approach by adopting waste disposal and energy generation at the same time (Mak et al. 2020). The bulk generation of agro-residues such as sugar, as well as molasses, have the requisite potential to make biofuels such as the production of ethanol in a consistent manner. The limitations in policies, availability of residues as feedstock, and expenses involved with the process are certain constraints for bioethanol production from biomass. The rectification in policies and long-term planning can improve the success of the biofuel program in India. It is required for maintaining an equilibrium between energy availability and the socioeconomic status of people in India (Gopinathan and Sudhakaran 2009).

13.9.3 Selection of Specific Consortia of Microbes

The individual steps of biofuel production from agro-waste are guided by a specific group of microbes. In a specific consortium, the microbes present should be compatible and capable of sharing a common bioreactor environment. Deviation from it can affect the efficiency of the productivity of biofuel. The enzyme complex produced from the microbial consortia is effective against specific substrates of agro-waste during catabolism (Singhania et al. 2022).

13.9.4 Maintenance of Bioreactor Parameters

The bioreactor parameters at pilot and field scale conditions required to be monitored include optimal levels of pH, aeration, temperature, growth media concentration, compartmentalization, and microbial population. Deviation from the optimal range can adversely affect the production process (Aryal et al. 2018).

13.9.5 Quality of Biofuel

The regular monitoring and optimization of factors controlling upstream and downstream processing is highly essential. It is required for upscaling the production of biofuel qualitatively and quantitatively.

13.10 Agro-wastes to Biofuel: Opportunities and Challenges

Generation and proper disposal of agro-wastes are major concerns. The principle of waste to energy was therefore developed to tackle not only the load of agro-wastes but also other solid wastes generated (Brunner and Rechberger 2015). The conversion of agro-wastes to biofuels is one of the most decorated waste-to-energy processes. With the utilization of agro-residues for energy production, the carbon footprint can be reduced on a global scale. The European Union (EU) currently utilizes 22% of renewable energy and achieved the target of reducing CO_2 emissions by 20%. It further targets achieving a 32% utilization of nonconventional energy and a 55% reduced emission of greenhouse gases by 2030 (Commission and Energy 2022).

The generation of biofuels from agro-wastes has several advantages. The process helps in reducing the accumulation of organic waste loads in the environment. This helps in preventing air and soil pollution. Farmers are also benefitted from the procurement of agro-waste by biofuel-producing companies, thus providing them another source of earning. Increased utilization of wastes means increased biofuel generation and use and thus a reduced dependence on fossil fuels. Reduced combustion of fossil fuels keeps a check on the release of toxic pollutants and gases in the atmosphere.

There are several challenges that crop up in the case of developing nations like India. These are uncertainty in regulations and policies, lack of financial support, and sustainability (Yap and Nixon 2015). The generation of biofuels from agro-wastes reduces the carbon footprint as compared to other conventional waste utilization methods such as a landfill. However, these waste-to-energy facilities still emit several toxic pollutants into the atmosphere. The solid residues generated from the production plants are still dumped at landfilling sites, thus causing soil pollution (Karmakar et al. 2023). The global emission standard is better as compared to its standard in India (Nixon et al. 2017). This requires proper framing and implementation of regulations to mitigate the hazards of the pollutants released from the production plants on the local public. India also lacks technologically advanced and indigenous infrastructural facilities and equipment. The current equipment for the generation of biofuels is mostly procured from American and European nations that do not have the capability of processing biomasses with high moisture content (40–50%) as mostly found among Indian agro-wastes. Unavailability of skilled technicians, operators, and a high cost of production are some of the other major challenges being faced in this regard.

13.11 Future Prospects

Biofuels are suitable energy forms to support the present energy-demanding situation. The transformation of stored biochemical energy of agro-wastes to heat or electrical energy of biofuel is promising but requires certain improvements in the process of its generation. Further study on this aspect is required to improve biofuel production efficiency.

13.12 Conclusion

The change in lifestyle, especially at the urban sites, is due to rapid change in socio-economic developmental patterns. It is closely linked with the overexploitation of traditional limited energy forms. The modern energy consumption sectors are depleting the reservoir of nonrenewable energy forms as well as responsible for upscaling in environmental pollution. It is widening the gap between the production and demand situation in the energy sector. These problems can be solved to a certain extent by the valorization of agro-waste into bioenergy following a circular bio-economy approach. This gap can be bridged by contributions from nonconventional, renewable energy resources. Biofuel is a better form of renewable energy resource obtained from biomass. The use of carbon-based agro-wastes is a good alternative energy source to replace the use of traditional fuels for developmental purposes. Biofuels such as bioethanol and biodiesel have the ability to run the vehicles in their available present technology. The biofuels from agro-waste consume those wastes and help in agro-waste management. It helps the rural sector in converting wastes to energy and improving the environmental parameters of the area accumulated with agro-wastes. The developed countries are working on the production and consumption of biofuels such as bioethanol to phase out the consumption of fossil fuels from different sectors. It is beneficial to see the integration of compatible multiple production units through cogeneration to obtain resource mobilization and waste minimization.

Acknowledgements It is an opportunity for the author(s) to acknowledge the inspiration and support provided by the Siksha 'O' Anusandhan (Deemed to be University), Bhubaneswar, Odisha, to carry out this work in its present form.

References

Adejumo IO, Adebiyi OA (2020) Agricultural solid wastes: causes, effects, and effective management. In: Strategies of sustainable solid waste management, vol 8. https://doi.org/10.5772/intechopen.93601

Adhikari S, Nam H, Chakraborty JP (2018) Conversion of solid wastes to fuels and chemicals through pyrolysis. In: Waste biorefinery, pp 239–263. https://doi.org/10.1016/B978-0-444-63992-9.00008-2

Ajayi VA, Lateef A (2023) Biotechnological valorization of agrowastes for circular bioeconomy: melon seed shell, groundnut shell and groundnut peel. Clean Circ Bioecon 20:100039. https://doi.org/10.1016/j.clcb.2023.100039

Alalwan HA, Alminshid AH, Aljaafari HA (2019) Promising evolution of biofuel generations. Subject Rev Renew Energy Focus 28:127–139. https://doi.org/10.1016/j.ref.2018.12.006

Alvira P, Tomás-Pejó E, Ballesteros M, Negro MJ (2010) Pretreatment technologies for an efficient bioethanol production process based on enzymatic hydrolysis: a review. Bioresour Technol 101(13):4851–4861. https://doi.org/10.1016/j.biortech.2009.11.093

Anwar Saeed M, Ma H, Yue S, Wang Q, Tu M (2018) Concise review on ethanol production from food waste: development and sustainability. Environ Sci Pollut Res 25(29):28851–28863

Aryal N, Kvist T, Ammam F, Pant D, Ottosen L (2018) An overview of microbial biogas enrichment. Bioresour Technol 264:359–369

Barr MR, Volpe R, Kandiyoti R (2021) Liquid biofuels from food crops in transportation-a balance sheet of outcomes. Chemi Eng Sci X 10:100090

Bracco S, Calicioglu O, Gomez SJM, Flammini A (2018) Assessing the contribution of bioeconomy to the total economy: a review of national frameworks. Sustainability 10(6):1698. https://doi.org/10.3390/su10061698

Brunner PH, Rechberger H (2015) Waste to energy–key element for sustainable waste management. Waste Manag 37:3–12. https://doi.org/10.1016/j.wasman.2014.02.003

Chandra R, Takeuchi H, Hasegawa T (2012) Methane production from lignocellulosic agricultural crop wastes: a review in context to second generation of biofuel production. Renew Sust Energ Rev 16(3):1462–1476. https://doi.org/10.1016/j.rser.2011.11.035

Dahman Y, Syed K, Begum S, Roy P, Mohtasebi B (2019) Biofuels: their characteristics and analysis. In: Verma D, Fortunati E, Jain S, Zhang X (eds) Biomass, biopolymer-based materials, and bioenergy. Woodhead Publishing, pp 277–325. https://doi.org/10.1016/B978-0-08-102426-3.00014-X

Das BP, Roy AT, Dash P (2008) Effect of seedling age and submergence on chlorophyll content of rice cultivars. Oryza 45(2):169–172

Das BP, Dash P, Roy AT (2009) Role of total sugar and starch content of rice seedlings at different ages in variable submergence tolerance. Oryza 46(4):304–309

Das PK, Das BP, Dash P (2019) Role of energy crops to meet the rural energy needs: an overview. In: Kumar RP, Bharathiraja B, Kataki R, Moholkar VS (eds) Biomass valorization to bioenergy. Springer, pp 11–30. https://doi.org/10.1007/978-981-15-0410-5_2

Das PK, Das BP, Dash P (2020) Potentials of postharvest rice crop residues as a source of biofuel. In: Kumar RP, Gnansounou E, Raman JK, Baskar G (eds) Refining biomass residues for sustainable energy and bioproducts. Academic press, pp 275–301. https://doi.org/10.1016/B978-0-12-818996-2.00013-2

Das PK, Das BP, Dash P (2021) Chromite mining pollution, environmental impact, toxicity and phytoremediation: a review. Environ Chem Lett 19(2):1369–1381. https://doi.org/10.1007/s10311-020-01102-w

Das BK, Das PK, Dash P (2022a) Evaluation of phenotypic responses of selected rice (Oryza sativa L.) cultivars to hexavalent chromium stress in soil. Nat Environ Pollut Technol 21(3):1341–1347. https://doi.org/10.46488/NEPT.2022.v21i03.041

Das PK, Das BP, Dash P, Gurunathan B (2022b) Production of biofuel from genetically modified microalgal biomass and its effects on environment and public health. In: Gurunathan B, Sahadevan R, Zakaria ZA (eds) Biofuels and bioenergy opportunities and challenges. Elsevier, pp 505–519. https://doi.org/10.1016/B978-0-323-85269-2.00004-6

Dey S, Sreenivasulu A, Veerendra GTN, Rao KV, Babu PA (2022) Renewable energy present status and future potentials in India: an overview. Innov Green Dev:100006. https://doi.org/10.1016/j.igd.2022.100006

Dixit S, Yadav VL (2020) Comparative study of polystyrene / chemically modified wheat straw composite for green packaging application. Polym Bull 77(3):1307e1326

Elliott D (2000) Renewable energy and sustainable development. J Future 32:261–274

Escobar JC, Lora E, Venturini OJ, Yanez EE, Castillo EF, Almazan O (2009) Biofuels: environment, technology, and food security. Renew Sustain Rev 13:1275–1287

Ethaib S, Omar R, Kamal SM, Biak DA (2015) Microwave-assisted pretreatment of lignocellulosic biomass: a review. J Eng Sci Technol 10:97–109

European Commission, Directorate-General for Energy (2022) EU energy in Tableures – statistical pocketbook. Publications Office of the European Union. https://doi.org/10.2833/334050

Food and Agriculture Organization of the United Nations (FAO) (2017a) Strategic work of FAO for sustainable food and agriculture. Available from: http://www.fao.org/3/a-i6488e.pdf. Accessed: 13 Aug 2020

Food and Agriculture Organization of the United Nations (FAO) (2017b) The future of food and agriculture: trends and challenges. Available from: http://www.fao.org/3/a-i6881e.pdf. Accessed 13 Aug 2020

Food and Agriculture Organization of the United Nations (FAO) & Organization for Economic Co-operation and Development (OECD) (2019) Background notes on sustainable, productive and resilient agro-food systems: Value chains, human capital, and the 2030 agenda. A Report to the G20 Agriculture Deputies. Available from: https://www.oecdilibrary.org/docserver/dca82200en.pdf?expires=1563959111&id=id &accname=guest&checksum=5BD0A7A51327DB165936B4AE57A0E5CE. Accessed 13 Aug 2020

Ghimire A, Frunzo L, Pirozzi F, Trably E, Escudie R, Lens PN, Esposito G (2015) A review on dark fermentative biohydrogen production from organic biomass: process parameters and use of by-products. Appl Energy 144:73–95. https://doi.org/10.1016/j.apenergy.2015.01.045

Gopinathan MC, Sudhakaran R (2009) Biofuels: opportunities and challenges in India. In Vitro Cell Dev Biol Plant 45(3):350–371

Hassan NS, Jalil AA, Hitam CNC, Vo DVN, Nabgan W (2020) Biofuels and renewable chemicals production by catalytic pyrolysis of cellulose: a review. Environ Chem Lett 18:1625–1648. https://doi.org/10.1007/s10311-020-01040-7

Hijosa-Valsero M, Paniagua-García AI, Díez-Antolínez R (2017) Biobutanol production from apple pomace: the importance of pretreatment methods on the fermentability of lignocellulosic agro-food wastes. Appl Microbiol Biotechnol 101:8041–8052. https://doi.org/10.1007/s00253-017-8522-z

Himri Y, Malik AS, Stambouli AB, Himri S, Draoui B (2009) Reviews and use of the Algerian renewable energy for sustainable development. Renew Sust Energ Rev 13:1584–1591

Holm-Nielsen JB, Al Seadi T, Oleskowicz-Popiel P (2009) The future of anaerobic digestion and biogas utilization. Bioresour Technol 100(22):5478–5484. https://doi.org/10.1016/j.biortech.2008.12.046

Ji A, Zhang S, Bhagia S (2020) 3D printing of biomass-derived composites: application and characterization approaches. RSC Adv 10:21698–21723

Jin Q, Yang L, Poe N, Huang H (2018) Integrated processing of plant-derived waste to produce value-added products based on the biorefinery concept. Trends Food Sci Technol 74:119–131

Jose R, Ramakrishna S (2021) Comprehensiveness in the research on sustainability. Mater Circ Econ 3:1–2

Karaosmanoglu F, Tuter M, Gollu E, Yanmaz S, Altintig E (1999) Fuel properties of cotton seed oil. Energy Sources 21:821–828

Karmakar A, Daftari T, Sivagami K, Chandan MR, Shaik AH, Kiran B, Chakraborty S (2023) A comprehensive insight into waste to energy conversion strategies in India and its associated air pollution hazard. Environ Technol Innov 29:103017. https://doi.org/10.1016/j.eti.2023.103017

Kumar S, Paritosh K, Pareek N, Chawade A, Vivekanand V (2018) De-construction of major Indian cereal crop residues through chemical pretreatment for improved biogas production: an overview. Renew Sust Energ Rev 90:160–170. https://doi.org/10.1016/j.rser.2018.03.049

Kumar JA, Sathish S, Prabu D, Renita AA, Saravanan A, Deivayanai VC, Anish M, Jayaprabakar J, Baigenzhenov O, Hosseini-Bandegharaei A (2023) Agricultural waste biomass for sustainable bioenergy production: feedstock, characterization and pre-treatment methodologies. Chemosphere 27:138680. https://doi.org/10.1016/j.chemosphere.2023.138680

Li Y, Jin Y, Li J, Chen Y, Gong Y, Li Y, Zhang J (2016) Current situation and development of kitchen waste treatment in China. Procedia Environ Sci 31:40–49

Mahapatra S, Kumar D, Singh B, Sachan PK (2021) Biofuels and their sources of production: a review on cleaner sustainable alternative against conventional fuel, in the framework of the food and energy nexus. Energy Nexus 4:100036. https://doi.org/10.1016/j.nexus.2021.100036

Mak TMW, Xiong X, Tsang DCW, Iris KM, Poon CS (2020) Sustainable food waste management towards circular bioeconomy: policy review limitations and opportunities. Bioresour Technol 297:122497

Naik SN, Goud VV, Rout PK, Dalai AK (2010) Production of first and second generation biofuels: a comprehensive review. Renew Sust Energ Rev 14(2):578–597. https://doi.org/10.1016/j.rser.2009.10.003

Nixon JD, Dey PK, Ghosh SK (2017) Energy recovery from waste in India: an evidence-based analysis. Sustain Energy Technol Assess 21:23–32. https://doi.org/10.1016/j.seta.2017.04.003

No SY, No SY (2019) Production of liquid biofuels from biomass. In: Application of liquid biofuels to internal combustion engines, pp 1–33. https://doi.org/10.1007/978-981-13-6737-3_1

Nwosu-Obieogu K, Chiemenem LI, Adekunle KF (2016) Utilization of agricultural waste for bioethanol production-a review. Int J Curr Res 8(19):1

Oluseun AI, Durojaiye OA, Ajibade AI, Idowu OS (2014) Status of exposure of bio-systems to restricted aluminium phosphide pesticide in Kano State, Nigeria. Int J Sci Knowl 2(7):306

Panahi HKS, Dehhaghi M, Guillemin GJ, Gupta VK, Lam SS, Aghbashlo M, Tabatabaei M (2022) Bioethanol production from food wastes rich in carbohydrates. Curr Opin Food Sci 43:71–81

Panpatte DG, Jhala YK (2019) Agricultural waste: a suitable source for biofuel production. In: Prospects of renewable bioprocessing in future energy systems. Springer, pp 337–355

Pappas D (2017) Energy and industrial growth in India: the next emissions superpower? Energy Procedia 105:3656–3662

Patel M, Zhang X, Kumar A (2016) Techno-economic and life cycle assessment on lignocellulosic biomass thermochemical conversion technologies: a review. Renew Sust Energ Rev 53:1486–1499. https://doi.org/10.1016/j.rser.2015.09.070

Pattanaik L, Pattnaik F, Saxena DK, Naik SN (2019) Biofuels from agricultural wastes. In: Dalena F (ed) Basile A. Elsevier, Second and third generation of feedstocks, pp 103–142. https://doi.org/10.1016/B978-0-12-815162-4.00005-7

Pérez JA, Ballesteros I, Ballesteros M, Sáez F, Negro MJ, Manzanares P (2008) Optimizing liquid hot water pretreatment conditions to enhance sugar recovery from wheat straw for fuel-ethanol production. Fuel 87(17–18):3640–3647. https://doi.org/10.1016/j.fuel.2008.06.009

Prasad M, Ranjan R, Ali A, Goyal D, Yadav A, Singh TB, Dantu PK (2020) Efficient transformation of agricultural waste in India. In: Naeem M, Ansari AA, Gill SS (eds) Contaminants in agriculture. Springer, pp 271–287

Ragaert K, Delva L, Van Geem K (2017) Mechanical and chemical recycling of solid plastic waste. Waste Manag 69:24–58

Ruan R, Zhang Y, Chen P, Liu S, Fan L, Zhou N, Ding K, Peng P, Addy M, Cheng Y, Anderson E (2019) Biofuels: introduction. In: Pandey A, Larroche C, Dussap C, Gnansounou E, Khanal SK, Ricke S (eds) Biofuels: alternative feedstocks and conversion processes for the production of liquid and gaseous biofuels. Academic Press, pp 3–43. https://doi.org/10.1016/B978-0-12-816856-1.00001-4

Ruggero B (2007) World geothermal generation in (2007) Geo-heat Centre Quarterly Bulletin (Klamath falls, Oregon: Oregon Institute of Technology). 28(3):8–19

Saini JK, Saini R, Tewari L (2015) Lignocellulosic agriculture wastes as biomass feedstocks for second-generation bioethanol production: concepts and recent developments. 3 Biotech 5:337–353. https://doi.org/10.1007/s13205-014-0246-5

Schwaiger H, Pena N, Mayer A, Bird DN (2011) Technologies to produce liquid biofuels for transportation - an overview. Working paper 72. CIFOR, Indonesia, p 13

Sekoai PT, Yoro KO (2016) Biofuel development initiatives in Sub-Saharan Africa: opportunities and challenges. Climate 4(2):33

Sharma S, Kundu A, Basu S, Shetti NP, Aminabhavi TM (2020) Sustainable environmental management and related biofuel technologies. J Environ Manag 273:111096

Sharma P, Gaur VK, Sirohi R, Varjani S, Kim SH, Wong JWC (2021) Sustainable processing of food waste for production of bio-based products for circular bioeconoomy. Bioresour Technol 325:124684

Sharma V, Tsai ML, Chen CW, Sun PP, Patel AK, Singhania RR, Nargotra P, Dong CD (2022) Deep eutectic solvents as promising pretreatment agents for sustainable lignocellulosic biorefineries: a review. Bioresour Technol 360:127631

Shi Q, Abdel-Hamid AM, Sun Z, Cheng Y, Tu T, Cann I, Yao B, Zhu W (2023) Carbohydrate-binding modules facilitate the enzymatic hydrolysis of lignocellulosic biomass: releasing reducing sugars and dissociative lignin available for producing biofuels and chemicals. Biotechnol Adv 14:108126

Singh A, Singhania RR, Soam S, Chen CW, Haldar D, Varjani S, Chang JS, Dong CD, Patel AK (2022) Production of bioethanol from food waste: status and perspectives. Bioresour Technol 360:127651. https://doi.org/10.1016/j.biortech.2022.127651

Singhania RR, Patel AK, Singh A, Haldar D, Soam S, Chen CW, Tsai ML, Dong CD (2022) Consolidated bioprocessing of lignocellulosic biomass: technological advances and challenges. Bioresour Technol 354:127153

Soam S, Kapoor M, Kumar R, Borjesson P, Gupta RP, Tuli DK (2016) Global warming potential and energy analysis of second generation ethanol production from rice straw in India. Appl Energy 184:353–364

Subhedar PB, Gogate PR (2016) Use of ultrasound for pretreatment of biomass and subsequent hydrolysis and fermentation. In: Mussatto SI (ed) Biomass fractionation technologies for a lignocellulosic feedstock based biorefinery. Elsevier, pp 127–149. https://doi.org/10.1016/B978-0-12-802323-5.00006-2

Sukiran MA, Abnisa F, Daud WMAW, Bakar NA, Loh SK (2017) A review of torrefaction of oil palm solid wastes for biofuel production. Energy Convers Manag 149:101–120. https://doi.org/10.1016/j.enconman.2017.07.011

Taha M, Foda M, Shahsavari E, Aburto-Medina A, Adetutu E, Ball A (2016) Commercial feasibility of lignocellulose biodegradation: possibilities and challenges. Curr Opin Biotechnol 38:190–197. https://doi.org/10.1016/j.copbio.2016.02.012

Teymouri F, Laureano-Pérez L, Alizadeh H, Dale BE (2004) Ammonia fiber explosion treatment of corn stover. In: Proceedings of the twenty-fifth symposium on biotechnology for fuels and chemicals held May 4–7, 2003. Humana Press, Breckenridge, CO, pp 951–963. https://doi.org/10.1007/978-1-59259-837-3_77

Villanueva-Galindo E, Vital-Jácome M, Moreno-Andrade I (2023) Dark fermentation for H_2 production from food waste and novel strategies for its enhancement. Int J Hydrog Energy 48(27):9957–9970. https://doi.org/10.1016/j.ijhydene.2022.11.339

Wang Q, Dong Z, Li R, Wang L (2022) Renewable energy and economic growth: new insight from country risks. Energy 238, Article 122018:1–10

World Bioenergy Association, global bioenergy statistics 2020 (WBA). http://www.worldbioenergy.org/uploads/201210%20WBA%20GBS%202020.pdf. Accessed 9 July 2021

Yap HY, Nixon JD (2015) A multi-criteria analysis of options for energy recovery from municipal solid waste in India and the UK. Waste Manag 46:265–277. https://doi.org/10.1016/j.wasman.2015.08.002

Zabed H, Sahu JN, Boyce AN, Faruq G (2016) Fuel ethanol production from lignocellulosic biomass: an overview on feedstocks and technological approaches. Renew Sust Energ Rev 66:751–774. https://doi.org/10.1016/j.rser.2016.08.038

Chapter 14
Food Process Industry Waste Biomass as a Promising Alternative for Green Energy Production

Anjali Sethumadhavan, Siddh Rajesh Shah, Mani Jayakumar, Gnanasundaram Nirmala, and Magesh Rangaraju

Abstract The most significant renewable energy source in the world is biomass, and as national energy strategy and policy place more emphasis on renewable energy sources and conservation, biomass will become even more significant. Biomass has the potential to deliver a reasonable and long-lasting supply of green energy in the future. Green bioenergy synthesis from food waste is growing in popularity in recent days, gaining more recognition as a renewable energy source that could help manage rising energy costs and perhaps provide income to industries involved in food processing. Food waste can potentially be exploited for the synthesis of green fuel. Green bioenergy produced from biomass derived from food waste offers incomparable socioeconomic, environmental, and strategic advantages and can be viewed as a safe and cleaner liquid and gaseous fuel substitute for conventional fossil fuels. Renewable and sustainable energy sources are gaining importance due to global warming and the world's energy future. At the bench scale, there has been a significant advancement in the creation of green energy (biohydrogen, biomethane, biodiesel, biohythane, and bioethanol) from food waste. Based on information from recent literature, the purpose of this chapter is to investigate the possibility of producing green energy from food waste using modern conversion technologies. Additionally, the state of biomass-based green energy production

A. Sethumadhavan · S. R. Shah · G. Nirmala
Mass Transfer Laboratory, School of Chemical Engineering, Vellore Institute of Technology, Vellore, Tamil Nadu, India

M. Jayakumar (✉)
Department of Chemical Engineering, Haramaya Institute of Technology, Haramaya University, Dire Dawa, Ethiopia
e-mail: drjayakumarmani@haramaya.edu.et

Department of Biotechnology, Faculty of Engineering, Karpagam Academy of Higher Education, Coimbatore, Tamilnadu, India

M. Rangaraju
Department of Chemical Engineering, Wachemo University, Hossana, Ethiopia

E. Cherian, B. Gurunathan (eds.), *Value Added Products From Food Waste*,
https://doi.org/10.1007/978-3-031-48143-7_14

methods using food waste is thoroughly discussed, which might make it simpler for the scientific and industrial communities to properly utilize food waste.

Keywords Green energy · Food waste · Biohydrogen · Biomethane · Biodiesel · Biohythane · Bioethanol

14.1 Introduction

Due to ongoing population growth, the worldwide need for energy has been rising exponentially (Shanmugam et al. 2023; Jayakumar et al. 2023). By 2100, approximately six times as much energy will be consumed worldwide (Kothari et al. 2010). The fossil fuel reserves have been reduced, and due to the emissions of greenhouse gases (GHGs), when fossil fuels are burned, it has a severe impact on the environment (Kapdan and Kargi 2006). There is a great need for an alternative source of energy that could potentially substitute fossil fuel. This is the need of the hour.

Additionally, as the world's population has increased, food production has increased. As per studies, by 2050, global food demand has been expected to rise by 59–98% (Mong et al. 2022). As a result of this rapid growth, the food processing industry is experiencing an increase in demand, resulting in large quantities of food waste. In the food processing sector, waste may be generated directly or indirectly, through the separation of desirable products from undesirable by-products. These wastes may be product-specific (Barik 2019).

All processing industries (processing, packaging, storage, and transportation) produce substantial amounts of waste due to their inherent efficiency in waste management. Owing to their high water content and high accumulation rates, they pose severe problems such as bacteria contamination and so on. Due to the poor development of effective waste management measures for these food processing wastes, the problem of extensive waste build-up valorization in the environment has intensified (Ravindran and Jaiswal 2016). Biodiversity and the environment are also adversely affected by them. Hence, the valorization of these food wastes into useful products has been considered a promising approach since they offer solutions to waste management and disposal problems and the need for an alternative renewable energy source (Barik 2019).

In the food processing sector, waste may be generated directly or indirectly through the separation of desirable products from undesirable by-products. These wastes may be product-specific and are synthesized in liquid, solid and, semisolid forms. These effluents are rich in large amounts of organic matter, suspended solids, carbohydrates, fats, oils, and several other inorganic materials (Barik 2019). Food processing wastes (FPW) are mainly characterized by substrate concentrations of almost 19–346 g/L COD and BOD. According to studies, they have a higher carbon–nitrogen ratio of 14:37, with an inherent 75–90% moisture content and 25.5–143 g/L carbohydrate content. Despite the fact that most food wastes are utilized in some way, not all of them have been fully utilized. All of these characteristics make FPW a promising and potential approach for substituting fossil fuels and resolving the waste disposal problem (Shanmugam et al. 2023; Yasin et al. 2013).

Several renewable fuels that may be produced from FPW include biohydrogen, biogas (methane), biohythane, biodiesel, ethanol, biochar, and so on. These biofuels have gained a lot of importance over the years due to their enhanced properties and large energy yield. Since their conventional production methods may contribute to the contamination of the environment, biological methods have been investigated in search of environment-friendly and sustainable production methods. In this review, different biofuels, their properties, their different production methods, and the limitations encountered will be summarized.

This chapter addresses the detailed information connected to the food industry waste lignocellulosic biomass feedstocks and different biofuels, their properties, numerous production methods such as direct biophotolysis, dark fermentation, microbial electrolysis cells, indirect biophotolysis, photofermentation, anaerobic digestion, transesterification, microwave-assisted reaction, ultrasound-assisted reaction, supercritical fluid extraction and saccharification fermentation. In addition, future prospects and challenges are also discussed. The goal of the current chapter is to advance future research on the use of industrial waste biomass for the production of bioenergy by assisting the scientific community and related industries in understanding the potential and constraints of food waste biomass for the production and application of green energy.

14.2 Identifying Waste Sources in Food Processing

The food processing industry can be primarily divided into six categories, which are the dairy industry, fermentation industry, beverage industry, fruit and vegetable industry, meat and poultry industry, and seafood industry. These can be divided into further subsectors such as grain processing, confectionery manufacture, aquaculture, and so on. Each of these industries has separate operations that lead to the production of waste. Some of these operations include blanching, washing, cooling, packaging, rendering, and scalding (Khedkar and Singh 2017). Due to its lignocellulosic nature and high cellulose and lignin content, food processing waste has generated a lot of interest in the renewable energy sector. This is considered an attractive quality of feed that can be used to prepare a large range of biofuels (Ravindran and Jaiswal 2016).

Different wastes are utilized in different manners to produce different biofuels based on their chemical composition and properties (Kavitha et al. 2020). The meat and poultry sectors produce waste with high protein and fat contents, while the dairy industry leans toward high soluble sugar content. The fruit and beverage industry, coupled with the fermentation industry, produces large amounts of carbohydrate and lignin-rich waste. The Agri-based industry plays a crucial role in providing waste with lignocellulosic waste (Meena et al. 2020). All of these industries contribute their waste as feedstock to different biofuel production plants. For example, you would need some feedstock rich in carbohydrates, while for biomethane production, some feedstock rich in lipids (fats) is necessary.

14.3 Biofuel

A fuel made from biomass is known as biofuel. Biologically derived from plants and animals, it is a natural substance. It mostly consists of forestry by-products, animal wastes, wood, aquatic plants, waste and leftovers, and agricultural crops and products. In its broadest sense, a biofuel is any sort of fuel that can be produced from biomass, including solid, gaseous, and liquid fuels (Benti et al. 2021). The major classes of biofuels that are most frequently utilized are biohythane, biohydrogen biodiesel, biomethane, and ethanol. A comparison of the carbon emissions and energy output of various biofuels has been provided in Table 14.1.

14.3.1 Biohydrogen

Hydrogen is viewed as a promising substitute for fossil fuels in the search for more environmentally friendly energy sources. It is said to be the "energy carrier" of the future (Kapdan and Kargi 2006). With limited or close to zero utilization of hydrocarbons, hydrogen is recognized as a clean, renewable, and environmentally friendly fuel. When burned, it produces more water than greenhouse gases, unlike fossil fuels (Kothari et al. 2010). Further, hydrogen fuel has a 2.75-fold higher energy yield than hydrocarbon fuels (Kapdan and Kargi 2006). Since hydrogen is considered a climate-neutral energy source, according to predictions, it will be the main energy source by the year 2100, moving toward a greener and more sustainable economy (Kothari et al. 2010). In order to use hydrogen as fuel, however, expensive and pollution-causing production methods are required (Kapdan and Kargi 2006).

Table 14.1 Comparison of the carbon emissions and energy output of various biofuels

Type of fuel	Specific carbon emission (kg C/kg fuel)	Energy per unit volume (J/m^3)	Energy per unit mass (J/kg)
LH	0.00	10.10	141.90
GH	0.00	0.013	141.90
ML	0.50	18.10	22.30
EL	0.50	23.60	29.90
BD	0.50	33.00	37.00
NG	0.46	0.04	50.00

LH liquid hydrogen, *GH* gaseous hydrogen, *ML* methanol, *EL* ethanol, *BD* biodiesel, *NG* natural gas

Conventionally, the production of hydrogen involves processes such as steam reforming of hydrocarbons, autothermal reforming, non-catalytic partial oxidation, and conventional gasification. A majority of these practices are energy-intensive techniques, with a temperature requirement of more than 800 °C (Kapdan and Kargi 2006). These biological methods are highly advantageous due to their low energy requirements, low investment costs, and efficient production of a clean energy source from waste materials (Kothari et al. 2010). To move toward a more environmentally friendly approach, the synthesize of hydrogen from the biological degradation of renewable organic substrates has been highly sought after. Due to their high sugar content, food processing wastes serve as one of the most promising alternatives for the efficient synthesis of biohydrogen. It is considered the most suitable and efficient substrate to generate hydrogen.

Hydrogen is most commonly produced through dark fermentation, microbial electrolysis, photofermentation, and direct biophotolysis, followed by indirect biophotolysis (Kothari et al. 2010).

14.3.1.1 Direct Biophotolysis

Biophotolysis is the direct production of hydrogen from water through photolysis. Cyanobacteria, microalgae, and other biological organisms that can perform photosynthesis directly split the water molecules in the presence of sunlight without any intermediate reaction. As a result, in both the photosystem I's reducing site and the photosystem II's oxidizing site, hydrogen and oxygen ions are generated, respectively. The by-product electrons that are produced are used by the hydrogenase enzyme to convert hydrogen ions to hydrogen gas (Singh and Das 2019):

$$2\mathrm{H_2O} + \mathrm{Light} \rightarrow 2\mathrm{H_2} + \mathrm{O_2} \tag{14.1}$$

14.3.1.2 Indirect Biophotolysis

Hydrogen is produced in two phases during indirect biophotolysis. In the first phase, the autotrophic organisms, in the presence of sunlight, produce carbohydrate biomass by absorbing CO_2 from the atmosphere and water from the soil, which is stored (Kamran 2021). In the second stage, this carbohydrate substrate undergoes fermentation to produce hydrogen. These two stages are separated for the efficient production of hydrogen since oxygen evolution may pose problems in hydrogen production (Singh and Das 2019):

$$12\mathrm{H_2O} + 6CO_2 + \mathrm{Light} \rightarrow \mathrm{C_6H_{12}O_6} + 6\mathrm{O_2} \tag{14.2}$$

$$\mathrm{C_6H_{12}O_6} + 12\mathrm{H_2O} + \mathrm{Light} \rightarrow 12\mathrm{H_2} + 6CO_2 \tag{14.3}$$

14.3.1.3 Photofermentation

In this process, organic matter present in the waste is transformed into hydrogen in the presence of sunlight by the combined action of various microorganisms. There are numerous organic compounds present in the food processing wastes such as volatile fatty acids, glucose, and sucrose. Photosynthetic bacteria, which is purple non-sulfur bacteria (PNS), break down these organic compounds under anaerobic conditions and produce H_2 and CO_2. In this process, in the absence of molecular nitrogen, the nitrogenase enzyme produces hydrogen (Singh and Das 2019).

$$2\mathrm{H}^+ + 2\mathrm{e}^- + 4ATP + \mathrm{Light} + \mathrm{Nitrogenase} \rightarrow \mathrm{H}_2 + 4ADP + 4Pi \quad (14.4)$$

14.3.1.4 Dark Fermentation

Dark fermentation is one of the simplest and most promising method to produce biohydrogen. In this process, organic matter undergoes anaerobic digestion in the absence of sunlight to produce hydrogen. It takes place at normal temperature and pressure (Singh and Das 2019). It converts the organic substrates present in the waste to biohydrogen, CO_2, and compounds such as butyrate and acetate. This method is highly popular due to its simple reactor design and high hydrogen production rates (Ding et al. 2016):

$$\mathrm{C_6H_{12}O_6} + 2\mathrm{H_2O} \rightarrow 2\mathrm{C_2H_4O_2} + 2CO_2 + 4\mathrm{H_2} \quad (14.5)$$

14.3.1.5 Microbial Electrolysis Cell (MEC)

The most recent and promising method of production of biohydrogen is through the microbial electrolysis cell (MEC). Microbial fuel cells (MFCs) directly convert organic matter into electricity using several microbes. Upon modifying the MFCs into MECs, hydrogen can be effectively produced. In the anodic chamber, the microbes present ferment the organic substrate through dark fermentation, and the cathodic chamber acts as an electron acceptor. Upon supplying a small amount of electricity, hydrogen may be produced by electrohydrogenesis. Even though they have immense potential in large-scale hydrogen production, there are several limitations such as high cost and great current density, which must be overcome (Singh and Das 2019).

All the above technologies have their own disadvantages and limitations. In order to overcome these shortcomings and facilitate the smooth and efficient production of hydrogen, several integrated systems are proposed. They include photofermentation plus dark fermentation, dark fermentation plus MEC, and so on. They overcome the limitations of each individual process and allow the continuous and green production of hydrogen.

14.3.2 Biomethane

Lately, biomass from food processing has been researched with various processes for valorization in the search for a viable alternative to fossil fuels. The upgradation of these products, ranging from hydrogen to biomethane (biogas) has been well sought out since it presents enormous advantages with respect to production, distribution, and usage. The most significant processes for biogas production from food processing industry waste are pyrolysis, gasification, and anaerobic digestion. Pyrolysis and gasification are thermochemical processes that work on the principle of thermal degradation of feedstocks to produce more valuable products (Appels et al. 2011). These require copious amounts of energy to maintain the high temperature by burning fossil fuels and hence cannot be considered a source of green energy.

14.3.2.1 Anaerobic Digestion

Anaerobic digestion (AD), a biochemical process, is one of the most promising waste utilization methods, low residual waste production, and its utilization as a renewable resource (Chozhavendhan et al. 2022). AD greatly uses lipid-rich waste, which is produced in abundance by the food processing industries, meat and poultry industries, edible oil industries, dairy industries, and olive oil mill. Residues from food crops such as leaves and vegetable waste and from energy crops (maze, wheat, and beet) provide a large potential for the AD (Appels et al. 2011) as illustrated in Fig. 14.1.

The process of AD consists of four distinctive steps for converting raw material into biogas.

Hydrolysis

Hydrolysis breaks down large organic polymers such as carbohydrates, lipids, and proteins into simpler compounds such as amino acids, fatty acids, and simple sugars.

Acidogenesis

This is the conversion of low molecular weight compounds into volatile fatty acids, carbonic acids, and alcohols. Some other by-products that are generated are H_2, CO_2, and NH_3.

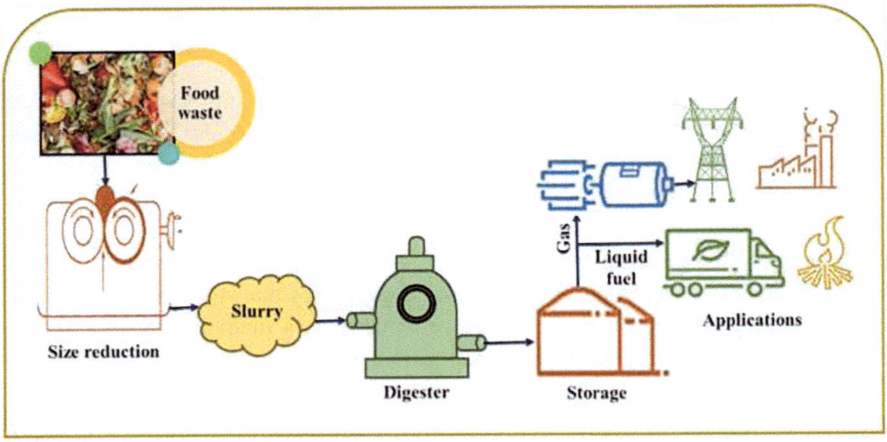

Fig. 14.1 Anaerobic digestion process

Acetogenesis

The primary function of this stage is to use acetogenic bacteria to produce acetic acid with CO_2 and H_2O.

Methanogenesis

With the action of methanogenic bacteria on the products of acetogenesis, methane is produced with CO_2 and traces of other gases. It can yield anywhere from 50% to 75% methane (Hall and Howe 2012).

The AD process configuration can be done in the following two ways:

- *Single-stage anaerobic digestion*
- The entire process takes place simultaneously in a single reactor, including hydrolysis, acidogenesis, acetogenesis, and methanogenesis. This is a relatively inexpensive process that does not have too many errors as well (Kiran et al. 2014).
- *Two-stage anaerobic digestion*
- The first stage in this process uses H_2-generating microbes for the synthesis of H_2, which is followed by the second stage, which is dedicated to methane production. This is a much more efficient method that results in greater yields of methane while also producing biohydrogen (Kiran et al. 2014).

AD is an extremely attractive method of biogas production since it is a well-established technology (Hall and Howe 2012). Moreover, the biomass feed can be extremely wet and still be useable without treatment unlike other methods (gasification and pyrolysis), which would greatly depend on the drying operation before the feed can be utilized. For gasification and pyrolysis, the energy efficiency drastically decreases if wet feed is used. The primary disadvantage of AD is that the process has very long retention times such as 20–30 days. This is largely because the action of anaerobic microorganisms is ineffective on certain types of biomasses. However, this can be reduced with pretreatment methods such as disintegration, which change the structure and substrate concentration of the biomass. Finally, anaerobic digestion offers some advantages over other available technologies (Appels et al. 2011).

14.3.3 Biodiesel

Biodiesel (fatty acid methyl esters; FAME) has greatly grown in popularity as one of the better alternatives to petroleum-derived fuels. Biodiesel production has greatly advanced over the years (Wang et al. 2014). The scientific community has become increasingly interested in biodiesel due to its similar physicochemical and fuel qualities to pure diesel fuel and the decrease in harmful gas emissions (except for NOx) (Ashine et al. 2023). Plant oils and nonedible animal fats are used as feedstocks for biofuel, as are algal oils, waste oils, and oleaginous microorganism oils (Amini et al. 2017). The feedstock must be rich in triglycerides, diglycerides, and monoglycerides since these are the molecules responsible for the production of FAME. Some organic materials could be directly used in diesel engines; however, they are highly viscous and hence not viable. A few methods that can be used to reduce viscosity are blending, emulsification, pyrolysis, and transesterification. Blending and emulsification are not popular since they can be used for a few selective feedstocks (Demirbas et al. 2009). Transesterification is therefore of greatest interest to the scientific community. It is also one of the most widely utilized practices for the production of biodiesel.

14.3.3.1 Transesterification (Alcoholysis)

This method aims to convert triglycerides to triacylglycerols. This can be achieved by using alcohol with a catalyst (Jayakumar et al. 2022a, b). The most common acceptors are methanol and ethanol because of their relatively low prices. Biodiesel is the result of the following three sequential reactions between triglycerides and methanol (Ferella et al. 2010):

$$\text{Triglyceride} + CH_3OH \rightarrow \text{Diglyceride} + \text{FAME} \tag{14.6}$$

$$\text{Diglyceride} + CH_3OH \rightarrow \text{Monoglyceride} + \text{FAME} \tag{14.7}$$

$$\text{Monoglyceride} + CH_3OH \rightarrow \text{Glycerol} + \text{FAME} \tag{14.8}$$

For the reaction to occur, the ratio of methanol to triglyceride must be at least 3:1. It is crucial to choose the right catalyst when designing an efficient process to maximize material value and minimize waste production. There is a vast array of catalysts to choose for transesterification: homogeneous, heterogeneous, and biocatalysts. It was found that a base catalyst was more suitable for transesterification compared to an acid catalyst (Amini et al. 2017).

14.3.3.2 Supercritical Fluids (SCF)

SCF is a green technology method in which transesterification takes place with alcohol at supercritical conditions (Patil et al. 2023). This poses several advantages such as better heat and mass transfer, better species mixing, and a greater rate of reaction. It was found that due to the ease of continuous production, it had good scalability. However, with these advantages come a number of disadvantages. To maintain supercritical temperature and pressure, a high amount of energy is required, which in turn means high operational costs. Moreover, maintaining such high temperatures and pressure also poses a safety issue (Amini et al. 2017).

14.3.3.3 Microwave-Assisted Reaction

In this method, microwaves are introduced in all steps of the process-feedstock preparation, extraction, and transesterification. Microwave irradiation applies heat directly to the reactant, which acts as a source of heating. This ensures better heat transfer than conventional methods of heating. This also leads to a fast reaction due to increased catalyst activity in the presence of microwaves. Overall, these advantages add up to a significant reduction in the operation cost, better efficiency, and increased safety (Amini et al. 2017).

14.3.3.4 Ultrasound-Assisted Reaction

It works on the same principle as microwave-assisted reactions. It enhances mass transfer, reduces reaction time, and hence the operation cost (Amini et al. 2017). The entire world has taken great interest in the production of biodiesel and has been trying to work on this technology for decades now. With more such technological advances, it is considered to be one of the most viable technologies.

14.3.4 Biohythane

Biohythane, a mixture of biohydrogen and methane and a more advanced biofuel, has gained considerable interest in recent years. This fuel consists of a maximum percentage of methane (60%), followed by 30% carbon dioxide and 10% hydrogen (Yeshanew et al. 2016). It displays remarkable advantages over several preexisting fuels, especially in vehicles and combustion engines (Yeshanew et al. 2016).

Biohythane is a better vehicular fuel, owing to the significant presence of H_2, which improves the performance of the fuel and stimulates the effective combustion of methane in the engine. As a clean, energy-efficient, and ecologically sustainable alternative to utilizing food processing waste, biohythane offers enhanced fuel and heat efficiency, as well as the ability to ignite engines with less input energy (Yeshanew et al. 2016). Additionally, the hydrogen present in the fuel also reduces the carbon content of the fuel, thereby reducing greenhouse emissions into the atmosphere (Bolzonella et al. 2018). Recently, hythane has also been commercialized as a vehicular fuel by countries such as India and the USA (Yeshanew et al. 2016).

In this current blend of hydrogen and methane, the molecular hydrogen present is mostly obtained from fossil fuels through environment-impacting methods such as steam reforming (Sarkar et al. 2021). Thus, an alternate and more sustainable method to produce hydrogen must be sought. The commonly used method for biomethane production is the two-stage anaerobic digestion (AD) process, which facilitates the simultaneous production of methane and biohydrogen (Yeshanew et al. 2016). In the first stage, biohydrogen is produced via dark fermentation (DF), followed by methane production by anaerobic digestion. This method is found to have a better performance than the single-stage AD due to higher energy recovery, which is related to higher methane production due to the enhanced methanogenic activities in the anaerobic bioreactors (Ghimire et al. 2021).

For this two-stage system, different reactor configurations have been implemented. Due to more biomass retention, better operational simplicity, and reduced energy consumption, the best configuration was an integrated system with a biohydrogen-producing CSTR (continuous stirred tank reactor) and a methane-producing AFBR (anaerobic fixed-bed reactor) arrangement, used for dark fermentation and methanogenic phases, respectively (Yeshanew et al. 2016). Yeshanew et al. (2016) tested the two-stage AD with a CSTR–AFBR configuration, and the reactors were found to operate smoothly for over 200 days. The CSTR depicted a promising hydrogen production rate of a maximum of 115 L of hydrogen and a methane production rate higher than 330 L/kg of food waste fed (Bolzonella et al. 2018). At the end of this dual-stage process, the final product is biohythane, which contains hydrogen, methane, and carbon dioxide in the ratio of 1:6:3.

14.3.5 Bioethanol

Another important product obtained from the valorization of food processing waste is alcohol. Alcohol production has become highly popular, with ethanol being the most widely utilized alcohol. However, several other biofuels such as propanol and 1-butanol are also becoming more prominent. However, there are several downfalls to utilizing propanol as a fuel, owing to its relatively higher viscosity, and butanol has also not been widely used as an efficient fuel blend. Methanol, produced by chemical synthesis, is used as a fuel to an even lower extent (Hegde et al. 2018).

However, ethanol has been gaining a lot of attention as a potential alternative fuel due to its production from renewable bio-based sources (Jayakumar et al. 2022a, b, 2023). Aside from reducing particulate emissions, its oxygenated nature could also reduce engine particulate emissions. Additionally, it is also used to synthesize ethylene, which is one of the key materials used in the production of polyethene and other plastics (Kiran et al. 2014). Thus, the synthesis of bioethanol from cheap and renewable feedstocks has become a matter of great interest.

Ethanol can be produced from different raw materials, with the most common being sugarcane molasses and corn (Hegde et al. 2018). Its production from food processing wastes has become a cheap and effective method, owing to the large proportion of carbohydrates present in the organic substrate.

14.3.5.1 Saccharification–Fermentation

Bioethanol is produced from the organic matter present in food processing wastes by a combined saccharification–fermentation process. They are directly converted using microbial strains such as *Saccharomyces cerevisiae*. Due to the complex nature of the conversion of lignocellulose in the waste to bioethanol, it undergoes the following three steps: (1) Delignification – the breaking of the bond between the carbohydrate and aromatic polymers, i.e., between the cellulose and lignin; (2) Saccharification and hydrolysis – the depolymerization of carbohydrate polymers into free sugars; and (iii) Fermentation – the conversion of released sugars such as pentose and hexose to bioethanol (main) (Demirbas et al. 2009) as presented in Fig. 14.2.

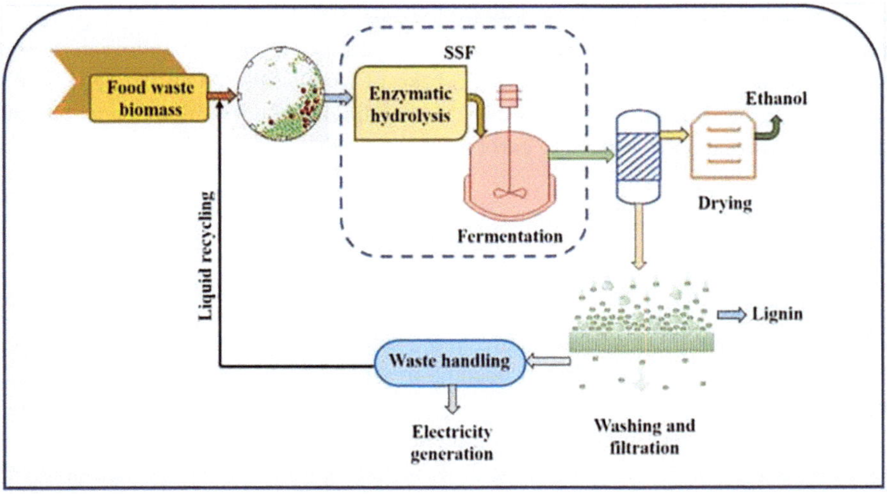

Fig. 14.2 Simultaneous saccharification fermentation process

Despite considerable research on this method, it is necessary to conduct more research on the integration of fermentation, enzymatic hydrolysis, and saccharification to advance the process and overcome any shortcomings for the sustainable generation of bioethanol from food industry waste.

14.4 Future Prospects and Challenges

14.4.1 Prospects

Using biomass waste from the food and feed processing sector could be a significant source of alternative renewable energy. Nonetheless, there is little doubt that green energy production from food waste has a bright future. More energy production, cheaper production costs, and greater efficiency can all arise from this. Additionally, this may pave the way for commercially effective green energy recovery. The potential of global food processing waste streams as sources of renewable green energy must be considered from an economic, ecological, and resource perspective. Biomass can be considered one of the best options, with immense potential to ensure future fuel and energy supplies. At present, it contributes to 10–15% of the world's energy usage, and various scenario studies predict the potential market share to rise as high as 50% by the year 2050. However, before reaching that, various obstacles such as environmental, technological, and social challenges must be overcome.

To attain sustainability, multiple factors play a role: electricity cost, efficiencies, GHGs, availability, and limitations. All these methods suffer from distinct disadvantages such as the low yield of the biofuel. Further extensive research and development studies are highly necessary to improve these novel methods for the efficient and sustainable production of biofuels and, by extension, green energy. Optimizing the digestion system also requires the development of mathematical models. It is widely accepted that biogas production can be improved by increasing conversion ratios, but pretreatment methods need to be improved. To achieve this, it is possible to pinpoint the specific changes in biomass structure that promote gas generation. Efforts need to be focused on more efficient bioreactors, optimizing the work conditions, effective microbe cultures, and so on. The world is heading on the right path by diverting its focus toward biofuels, which show remarkable promise for a greener future. Last but not least, it is advised that the stakeholders create marketing strategies to inform their clients of the advantages and benefits of using current technologies to recover green energy from food waste.

14.4.2 Challenges

It is necessary, however, to thoroughly understand the environmental impacts, energy policy, emission characterizations, and life cycle assessments (LCAs) associated with biofuels in order to make their use a legitimate energy source (Chung 2013). Currently, there are no comprehensive national policies that direct biomass energy activities in a well-thought-out manner. In some cases, when certain technologies need to be promoted, policies could be announced that do not adhere to renewable energy plans. Biomass is not governed by a defined framework (Benti et al. 2021).

It is important to develop institutions with long-term oriented plans and strong mandates. Interinstitutional coordination is lacking among the organizations, agencies, and stakeholders involved in the progress of lignocellulosic food waste biomass energy. It is due to this lack of coordination, collaboration, and delays that biomass energy production is limited. Investing in this area has been limited due to a lack of coordination in implementing policies. For small developers, or native producers, interested in undertaking biomass energy projects, including biogas, it may be challenging for the relevant states to prepare pre-feasibility reports. There are no established or properly functioning research centers for the development of renewable infrastructure. Additionally, renewable energy developers do not have access to customer service centers. Underdeveloped manufacturing industries and insufficient technology transfers lead to high foreign exchange expenditures. Particularly in rural areas, storing and converting biogas into electricity is impossible given a lack of equipment and facilities (Benti et al. 2021; Smith et al. 2014).

14.5 Summary

As a result of rapid industrialization's debt to energy consumption, currently, the world's primary energy source is fossil fuels. Due to global warming and the world's energy future, renewable and sustainable sources of energy are becoming increasingly important. Waste products from the food industry offer a lot of promise for making alternative energy. High levels of lipids, proteins, and carbohydrates can be converted into numerous forms of energy. Biohydrogen, biomethane, biodiesel, biohythane, and bioethanol have all been touted as potential energy sources for clean development in this area. Gaining access to a waste-to-well approach is made possible by the utilization of food waste biomass materials in green energy generation practices. Green energy has a significant potential to be developed and emerge as a clean, environmentally friendly energy carrier, despite the fact that it is expensive (due to low rates of green energy generation and a lengthy process). As a result, food waste biomass is a viable local source of sustainable green energy production.

References

Amini Z, Ilham Z, Ong HC, Mazaheri H, Chen W-H (2017) State of the art and prospective of lipase-catalyzed transesterification reaction for biodiesel production. Energy Convers Manag 141:339–353. https://doi.org/10.1016/j.enconman.2016.09.049

Appels L, Lauwers J, Degrève J, Helsen L, Lievens B, Willems K, Van Impe J, Dewil R (2011) Anaerobic digestion in global bio-energy production: potential and research challenges. Renew Sust Energ Rev 15(9):4295–4301. https://doi.org/10.1016/j.rser.2011.07.121

Ashine F, Kiflie Z, Prabhu SV, Tizazu BZ, Varadharajan V, Rajasimman M, Joo SW, Vasseghian Y, Jayakumar M (2023) Biodiesel production from Argemone mexicana oil using chicken egg-shell derived CaO catalyst. Fuel 332:126166. https://doi.org/10.1016/j.fuel.2022.126166

Barik D (2019) Energy extraction from toxic waste originating from food processing industries. In: Energy from toxic organic waste for heat and power generation, pp 17–42. https://doi.org/10.1016/b978-0-08-102528-4.00003-1

Benti NE, Gurmesa GS, Argaw T, Aneseyee AB, Gunta S, Kassahun GB, Aga GS, Asfaw AA (2021) The current status, challenges and prospects of using biomass energy in Ethiopia. Biotechnol Biofuels 14(1):1–24. https://doi.org/10.1186/s13068-021-02060-3

Bolzonella D, Battista F, Cavinato C, Gottardo M, Micolucci F, Lyberatos G, Pavan P (2018) Recent developments in biohythane production from household food wastes: a review. Bioresour Technol 257:311–319. https://doi.org/10.1016/j.biortech.2018.02.092

Chozhavendhan S, Karthigadevi G, Bharathiraja B, Kumar RP, Abo LD, Prabhu SV, Balachandar R, Jayakumar M (2022) Current and prognostic overview on the strategic exploitation of anaerobic digestion and digestate: a review. Environ Res:114526. https://doi.org/10.1016/j.envres.2022.114526

Chung JN (2013) Grand challenges in bioenergy and biofuel research: engineering and technology development, environmental impact, and sustainability. Front Energy Res 1:4. https://doi.org/10.3389/fenrg.2013.00004

Demirbas MF, Balat M, Balat H (2009) Potential contribution of biomass to the sustainable energy development. Energy Convers Manag 50(7):1746–1760. https://doi.org/10.1016/j.enconman.2009.03.013

Ding C, Yang K-L, He J (2016) Biological and fermentative production of hydrogen. In: Handbook of biofuels production, pp 303–333. https://doi.org/10.1016/b978-0-08-100455-5.00011-4

Ferella F, Mazziotti Di Celso G, De Michelis I, Stanisci V, Vegliò F (2010) Optimization of the transesterification reaction in biodiesel production. Fuel 89(1):36–42. https://doi.org/10.1016/j.fuel.2009.01.025

Ghimire A, Luongo V, Frunzo L, Lens PN, Pirozzi F, Esposito G (2021) Biohythane production from food waste in a two-stage process: assessing the energy recovery potential. Environ Technol 43(14):2190–2196. https://doi.org/10.1080/09593330.2020.1869319

Hall GM, Howe J (2012) Energy from waste and the food processing industry. Process Saf Environ Prot 90(3):203–212. https://doi.org/10.1016/j.psep.2011.09.005

Hegde S, Lodge JS, Trabold TA (2018) Characteristics of food processing wastes and their use in sustainable alcohol production. Renew Sust Energ Rev 81:510–523. https://doi.org/10.1016/j.rser.2017.07.012

Jayakumar M, Gebeyehu KB, Selvakumar KV, Parvathy S, Kim W, Karmegam N (2022a) Waste Ox bone based heterogeneous catalyst synthesis, characterization, utilization and reaction kinetics of biodiesel generation from Jatropha curcas oil. Chemosphere 288:132534. https://doi.org/10.1016/j.chemosphere.2021.132534

Jayakumar M, Vaithilingam SK, Karmegam N, Gebeyehu KB, Boobalan MS, Gurunathan B (2022b) Fermentation technology for ethanol production: current trends and challenges. Biofuels Bioenergy:105–131. https://doi.org/10.1016/B978-0-323-90040-9.00015-1

Jayakumar M, Gindaba GT, Gebeyehu KB, Periyasamy S, Jabesa A, Baskar G, John BI, Pugazhendhi A (2023) Bioethanol production from agricultural residues as lignocellulosic biomass feedstock's waste valorization approach: a comprehensive review. Sci Total Environ 879:163158. https://doi.org/10.1016/j.scitotenv.2023.163158

Kamran M (2021) Bioenergy. In: Renewable energy conversion systems, pp 243–264. https://doi.org/10.1016/b978-0-12-823538-6.00002-6

Kapdan IK, Kargi F (2006) Bio-hydrogen production from waste materials. Enzym Microb Technol 38(5):569–582. https://doi.org/10.1016/j.enzmictec.2005.09.015

Kavitha S, Yukesh Kannah R, Kumar G, Gunasekaran M, Rajesh Banu J (2020) Introduction: sources and characterization of food waste and food industry wastes. In: Food waste to valuable resources, pp 1–13. https://doi.org/10.1016/b978-0-12-818353-3.00001-8

Khedkar R, Singh K (2017) Food industry waste: a panacea or pollution hazard? In: Springer briefs in environmental science, pp 35–47. https://doi.org/10.1007/978-3-319-58415-7_3

Kothari R, Tyagi VV, Pathak A (2010) Waste-to-energy: a way from renewable energy sources to sustainable development. Renew Sust Energ Rev 14(9):3164–3170. https://doi.org/10.1016/j.rser.2010.05.005

Meena RA, Rajesh Banu J, Yukesh Kannah R, Yogalakshmi KN, Kumar G (2020) Biohythane production from food processing wastes – challenges and perspectives. Bioresour Technol 298:122449. https://doi.org/10.1016/j.biortech.2019.122449

Mong GR, Chong WW, Nor SA, Ng J-H, Chong CT, Idris R, Chiong MC, Wong S, Nyakuma BB (2022) Waste-to-bioenergy pathway for waste activated sludge from food processing industries: an experiment on the valorization potential under CO_2 and N_2 atmospheres through microwave-induced pyrolysis. Fuel 323:124380. https://doi.org/10.1016/j.fuel.2022.124380

Patil MA, Sarkate AP, Sakhale BK, Ranveer RC (2023) Supercritical fluid technology: recent trends in food processing. In: Advances in food process engineering: novel processing, preservation, and decontamination of foods, vol 25. Apple Academic Press

Ravindran R, Jaiswal AK (2016) Exploitation of food industry waste for high-value products. Trends Biotechnol 34(1):58–69. https://doi.org/10.1016/j.tibtech.2015.10.008

Sarkar O, Santhosh J, Dhar A, Venkata Mohan S (2021) Green Hythane production from food waste: integration of dark-fermentation and methanogenic process towards biogas up-gradation. Int J Hydrog Energy 46(36):18832–18843. https://doi.org/10.1016/j.ijhydene.2021.03.053

Shanmugam S, Mathimani T, Rajendran K, Sekar M, Rene ER, Chi NT, Ngo HH, Pugazhendhi A (2023) Perspective on the strategies and challenges in hydrogen production from food and food processing wastes. Fuel 338:127376. https://doi.org/10.1016/j.fuel.2022.127376

Singh V, Das D (2019) Potential of hydrogen production from biomass. In: Science and engineering of hydrogen-based energy technologies, pp 123–164. https://doi.org/10.1016/b978-0-12-814251-6.00003-4

Smith KR, Bruce N, Balakrishnan K, Adair-Rohani H, Balmes J, Chafe Z, Dherani M, Hosgood HD, Mehta S, Pope D, Rehfuess E (2014) Millions dead: how do we know and what does it mean? Methods used in the comparative risk assessment of household air pollution. Annu Rev Public Health 35:185–206. https://doi.org/10.1146/annurev-publhealth-032013-182356

Uçkun Kiran E, Trzcinski AP, Ng WJ, Liu Y (2014) Bioconversion of food waste to energy: a review. Fuel 134:389–399. https://doi.org/10.1016/j.fuel.2014.05.074

Wang R, Sun L, Xie X, Ma L, Liu Z, Liu X, Ji N, Xie G (2014) Biodiesel production from Stauntonia chinensis seed oil (waste from food processing): heterogeneous catalysis by modified calcite, biodiesel purification, and fuel properties. Ind Crop Prod 62:8–13. https://doi.org/10.1016/j.indcrop.2014.08.002

Yasin NH, Mumtaz T, Hassan MA, Abd Rahman NA (2013) Food waste and food processing waste for biohydrogen production: a review. J Environ Manag 130:375–385. https://doi.org/10.1016/j.jenvman.2013.09.009

Yeshanew MM, Frunzo L, Pirozzi F, Lens PNL, Esposito G (2016) Production of biohythane from food waste via an integrated system of continuously stirred tank and anaerobic fixed bed reactors. Bioresour Technol 220:312–322. https://doi.org/10.1016/j.biortech.2016.08.078

Index

E. Cherian, B. Gurunathan (eds.), *Value Added Products From Food Waste*, https://doi.org/10.1007/978-3-031-48143-7

MIX
Papier aus verantwortungsvollen Quellen
Paper from responsible sources
FSC® C105338
FSC
www.fsc.org

If you have any concerns about our products,
you can contact us on
ProductSafety@springernature.com

In case Publisher is established outside the EU,
the EU authorized representative is:
Springer Nature Customer Service Center GmbH
Europaplatz 3, 69115 Heidelberg, Germany

Printed by Libri Plureos GmbH
in Hamburg, Germany